Charles D.F Phillips

Materia Medica and Therapeutics-Vegetable Kingdom

Charles D.F Phillips

Materia Medica and Therapeutics-Vegetable Kingdom

ISBN/EAN: 9783743407558

Manufactured in Europe, USA, Canada, Australia, Japa

Cover: Foto ©berggeist007 / pixelio.de

Manufactured and distributed by brebook publishing software (www.brebook.com)

Charles D.F Phillips

Materia Medica and Therapeutics-Vegetable Kingdom

AND

THERAPEUTICS

VEGETABLE KINGDOM

BY

CHARLES D. F. PHILLIPS, M.D.; F.R.C.S.E.

LECTURER ON MATERIA MEDICA, WESTMINSTER HOSPITAL, LONDON

Edited and Adapted to the U. S. Pharmacopœia

BY

HENRY G. PIFFARD, A.M., M.D.

PROFESSOR OF DERMATOLOGY, UNIVERSITY OF THE CITY OF NEW YORK, SURGEON
TO CHARITY HOSPITAL, ETC.

NEW YORK
WILLIAM WOOD & COMPANY
27 GREAT JONES STREET
1879

COPYRIGHT, BY
WILLIAM WOOD & COMPANY,
1879.

Trow's
Printing and Bookbinding Co.
205–213 *East 12th St.*,
NEW YORK.

EDITOR'S PREFACE.

The design of the following pages will be most readily comprehended by the perusal of an extract from the author's original preface: "This work makes no pretension to bring forward a large mass of original research respecting abstract questions, although I believe it will be found to contain a considerable number of practical remarks on the use of drugs which are the genuine result of my own observations. It aims, however, at bringing together in a moderate compass a more extensive series of facts respecting the action of drugs, and especially a more enlarged view of what has been done in other countries, than will be found in the ordinary text-books."

The work of the Editor has been to condense the original so as to bring it within the limits of a single volume of the present series, and to render it more useful to the American practitioner, by adapting it to the U. S. Pharmacopœia. The reduction in size has been mainly effected by omitting the botanical descriptions that in the original edition preceded each of the drugs mentioned. A few unimportant articles of materia medica, *e. g.*, Cydonia, were omitted to make room for others that were deemed more useful. The additions, by the Editor, chiefly concern indigenous drugs, as Gelsemium, Hamamelis, Iris Versicolor, etc., as well as a few of foreign origin (Jaborandi, etc.). These have been briefly considered.

The sections on "Preparations and Dose" have been uniformly (with two or three exceptions) rewritten, in order to substitute (U. S.) officinal preparations for those of the British Pharmacopœia. After the dose, expressed in apothecary's weight or measure, will be found in brackets the corresponding dose according to the metric system.

The Editor has occasionally made additions to the notes, or to the text, concerning matters that have come within the sphere of his own experience, or are derived from other sources. Among the most important of these latter will be found extracts from the recent valuable researches of Prof. Rutherford and Dr. Vignal on the action of cholagogues. All additions to the original text are enclosed in brackets.

Information has been received from the Author, to the effect that the second portion of the work, devoted to the consideration of drugs of inorganic origin, is nearly ready for the press.

THE EDITOR.

NEW YORK, August, 1879.

MATERIA MEDICA

AND

THERAPEUTICS.

VEGETABLE KINGDOM.

RANUNCULACEÆ.

ACONITE (Aconitum Napellus).

ACTIVE INGREDIENTS.—The toxic and medicinal properties of Aconite are mainly, if not wholly, represented by the alkaloid Aconitia, $C^{30}H^{47}NO^7$, discovered by Pallas about the year 1770. [Besides Aconitia there is another body called Napelline, and in the Aconitum ferox, one differing somewhat from the others, to which the name of pseudo-aconitia has been given. For therapeutic purposes the most highly prized at present is Duquesnel's crystallized aconitia.]

PHYSIOLOGICAL ACTION.—The ancients, who were very imperfectly acquainted with mineral poisons, considered aconite to be the most deadly thing in nature, and their opinion is almost justified by modern experience. Every portion of the plant is more or less virulent; even the odor thrown out when the plant is in full bloom is said to operate injuriously upon susceptible constitutions. Sometimes it causes loss of sight for a day or two; in other instances it has been known to induce fainting-fits. The juice of the stem or leaves, accidentally introduced into a wound in the hand, affects the whole system; pains are produced in the hand itself, and in the arm; to these succeed cardialgia, a sense of suffocation, with great mental anxiety, and not uncommonly syncope.

If a leaf or a small scraping of the root be chewed, a sensation of numbness is quickly produced upon the lips and tongue, and this effect is still perceived after the lapse of many hours. A quantity sufficient to cause death, if received into the stomach, produces pungent heat in the palate and fauces, accompanied by a sensation of burning in the stomach itself. To these sensations are soon added a condition of painful numbness, which pervades the limbs, to the fingers and toes, and a general tremor of the whole body. Severe vomiting, attended by pain in the abdomen, quickly follows, and along with it an intermittent, weak, and irregular action of the heart. There is then an approach to suffocation, with great anxiety, restlessness, and vertigo; the limbs become cold and clammy, the pulse is more and more irregular, and death soon puts an end to the patient's sufferings. Neither convulsions, spasms, stupor, nor delirium can be reckoned upon as certain, though it is true that in several recorded cases one or more of these phenomena have been manifested, and it frequently happens that after full and poisonous doses the mind remains unclouded to the last. Pereira states that a dog under the influence of aconite will recognize and follow his master, but is insensible to any pain or feeling produced by running needles into the skin or paws.

Dr. Thompson states that the first results of swallowing aconite are acrid and burning sensations, accompanied by profuse salivation. He further tells us that if the extract be administered without the greatest caution, it operates first upon the stomach, then upon the nervous system, producing vomiting, hypercatharsis, vertigo, cold sweats, delirium, and convulsions which terminate in death. The same authority states that aconite placed upon the eyelids causes a flow of tears, but no sensation of heat; also that if the powdered leaves be sprinkled upon an ulcer, neither heat nor pain ensues. [It causes in the whole region of the trigeminus a peculiar pain, which at first is vague, but at length becomes constant. (Ott.)]

Hirtz, in a paper of great interest, considers that *quasi-electric twitchings beneath the skin* are a characteristic and necessary part of poisoning with a good specimen of aconite itself; and regards the fact of the alkaloid, aconitia, not producing this effect, to be strong evidence that that substance does not fully represent the plant in its powers.

The experimental researches of Achscharumow (Virchow's Archiv, 1866, p. 255) are the most complete that have been made upon the physiological effects of aconite. He experimented on both cold and warm blooded animals; and the general conclusions at which he arrived are the following. Fatal doses of the drug—

1. Produce death from asphyxia by cardiac paralysis.
2. They, in the first place, stimulate the medulla oblongata.
3. This stimulation affects the vagi, and is succeeded by paralysis of these nerves.
4. The cerebro-spinal motor nerves are paralyzed, and voluntary movements are abolished, the muscular substance remaining unaffected.
5. Finally, the reflex action of the spinal cord and the conductivity of the afferent nerves remain unaffected; while the conductivity of the sympathetic ones is stimulated.

These observations, however, are far from explaining the whole action of aconite; and, as bearing on its therapeutical use, the only points that they distinctly suggest are its power to diminish excessive action of the heart, and to relieve pain by partially and temporarily paralyzing the sensory nerves.

The general effects produced by taking aconite are remarkable for the great rapidity with which the symptoms follow each other. This may possibly be referable in part to a direct action of aconite through the medium of the nervous system, independent of the effects of actual absorption of the drug, which, however, itself takes place with great rapidity.

THERAPEUTIC ACTION.—The poison invested with these formidable characters has been reduced to so manageable a condition as to become a powerful and valuable therapeutic agent in some of the most troublesome and dangerous diseases to which the human frame is subject.

Baron Störck was the first to draw public attention to its value in therapeutics. In the year 1762 he published a little volume upon three or four of the principal vegetable poisons, aconite included, and described it as narcotic, diuretic, and diaphoretic. Himself a physician, the Baron administered it in intermittent fever, also in chronic rheumatism, gout, exostosis, paralysis, and scirrhus; and relates many instances of the success which attended its exhibition in these disorders. Being well acquainted with the potency of the drug, he recommends small doses at first, the increase, if necessary, to be very gradual. Störck's observations soon led to the employment of aconite in various other diseases; in many of which it was found useful. In consequence, however, of the terrible energy of the medicine, and the uncertainty of its operation, alarming symptoms have occasionally been produced, and hence upon the part of some practitioners there has arisen a certain distrust of it.

Amidst great doubts which still remain as to some possible actions of aconite in disease, there are certain main facts which can easily be verified by those who are willing to try the remedy perseveringly and with strict attention to the proper mode of administration. In the first place, by the employment of aconite, so much of the evil of inflammatory and febrile states as depends directly upon too rapid action of the heart and excessive temperature, may in a large number of cases be rapidly and effectually diminished or put an end to. Secondly, pain and spasm of various kinds not dependent on inflammation may be relieved either by the internal use of this drug, or more frequently by its external application.

In regard to its applicability to inflammatory and febrile affections, the conclusion seems justified that aconite is especially adapted to produce the kind of effects which were formerly aimed at by the now little used practice of bleeding (this is the opinion also of Dr. Ringer,[1] who has very extensively employed the drug in the same manner as is recommended in these pages), and that it is probably in the class of cases in which the latter remedy did good when its benefits were not outweighed by counterbalancing evils peculiar to itself that the employment of aconite proves valuable. A quick pulse with a certain strength of heart action as measured by the *resisting* quality of the pulse, and a certain attendant dry heat of the skin, seem the leading external characteristics of the typical condition which serves as our therapeutic indication. As regards the action of aconite in reducing temperature, we can but conjecturally suppose, but with much probability, that it effects this, firstly, by lowering the action of the heart, and secondly, in a minor degree, by restoring skin-transpiration.

Inflammatory Fevers.—In my experience I find aconite always indicated in the early stage of simple inflammatory fevers, where as yet little organic change has taken place; also in the early stage of pneumonia, and in most acute congestions. It should be given in all inflammations of serous membranes, before the exudation has passed the plastic stage, especially in pleurisy, pericarditis, &c. When administered soon after the first invasion of the disease, it quickly diminishes the action of the heart, calming and subduing it; and at the time moistens and often bathes the skin with profuse perspiration. Subsequently it allays the fever, and prevents the spread of any congestion which may already have taken place.

---

[1] Handbook of Therapeutics.

Aconite does not necessarily remove the exudations, but it checks and prevents the further development of the evil.

When aconite is given in the early stages of inflammatory fever, it not only abates the frequency of the heart's action, but reduces the temperature to its normal standard (which, indeed, is the powerful good to be gained by it), and, having accomplished this, it ceases to be of much further use.

Pneumonia.—I have now before me nine recorded cases of pneumonia in the first stage, several of them implicating the pleura more or less. The subjects of these were of ages ranging from twenty to sixty-two. The pulse varied from 110 to 140, and the temperature from 102° to 105$\frac{2}{3}$°. In every one of these nine cases the attack was ushered in by rigors followed by preternatural heat of the body; short, quick breathing; dry, hot skin; flushed face, headache, thirst, exhaustion, and short, dry cough, with more or less pain or uneasiness in the affected lung. The usual examination showed dulness over the congested part, with fine crepitation. Now, we are all aware that in some of the most favorable cases of pneumonia treated *without* medicine the fever subsides upon the third or fourth day, but that it far more frequently lasts from six to ten days, even though the ordinary remedies be administered. In these nine cases, however (which it is proper to remark are not selected, but have been taken consecutively as they occurred to me in practice), the fever in no instance lasted longer than six days, reckoning from the commencement of the rigors. In every one of these cases, moreover, in from three to six days after the temperature had fallen to 99°, or below, the lungs became almost natural. In none of them did the pulse fall to the normal standard without the temperature likewise falling to at least 99°; but in several instances the pulse remained high, and the temperature sank to below 99°, making it very probable that the aconite had done its required work. Eleven other cases of pneumonia, all of which, before I saw them, had passed into the second or consolidated stage, I also treated with aconite, but in none could I perceive that it had any effect in removing the consolidation. At the circumference of the consolidated lung there is generally a tendency to spreading of the congestion. Aconite will control and remove this tendency, but in comparison with certain other medicines it has no power over the actually consolidated portion.

Tonsillitis.—Aconite is valuable in tonsillitis, and in ordinary sore-throat. It relieves those irritable tickling throat-coughs so often met with in throat and lung affections; and is likewise useful to plethoric persons when suffering from asthma, accompanied by short, dry cough, an anxious look, a sense of great oppression, often amounting almost to suffocation, and a full strong pulse. I have often employed it also, with the best results, in croup, hæmoptysis, and epistaxis, and likewise in coryza, and in acute catarrh.

Rheumatic Fever.—In *rheumatic fever*, if aconite be used from the commencement, the heart is, in my experience, seldom affected, and the patient suffers much less from pain and swelling in the joints, while the duration of the fever is considerably lessened. It is rare indeed to meet with permanent organic disease as a result of rheumatic fever when the disorder is treated with aconite *from its commencement.*

The treatment of acute rheumatism with aconite was first prominently brought into notice by Dr. Lombard, of Geneva, whose very favorable statements were subsequently corroborated by Dr. Fleming, the inventor of the saturated tincture which bears his name. In the practice of the majority of English physicians,[1] however, the drug has not established a permanent reputation, either for efficacy or manageableness in rheumatic fever; and I cannot but suppose that this has been largely owing to the use of too high doses, which, by their nauseating and even dangerously depressing effects, rendered it difficult or impossible to carry out the treatment with that persistent regularity which is essential to a good result. It is highly important here to adopt the system of small doses, frequently repeated. No doubt also much undeserved discredit has been thrown upon aconite by the use of feeble and inefficient preparations (for example, the old extract of the London Pharmacopœia), as was remarked by Neligan, who had himself obtained excellent results from the use of the drug in rheumatic fever.

The extract of aconite, applied as a plaster to the joints in the form usually employed for belladonna, though less energetic than the last-named drug, is unquestionably very useful in rheumatic affections. But it will be more convenient to treat of these matters in connection with our general remarks on the use of aconite for relieving pain.

Erysipelas.—Erysipelas is a complaint which will frequently yield to aconite. I have now before me in my diary thirteen cases of erysipelas, or acute inflammation of the skin, some of them implicating the subcutaneous areolar tissue, and all attended by high fever. The temperature taken in all the thirteen cases varied from 102° to 105°; yet, although the only remedy employed was aconite, with an occasional dose of castor-oil, the whole number recovered in five days after being attacked. It is proper that I should add that these thirteen cases were all of a purely sthenic kind, as I do not believe that there is any virtue in aconite as a remedy for the low or asthenic description of erysipelas, equally common in our experience.

The mortality from erysipelas is in England very considerable, the deaths amounting to over 2,000 annually. If aconite were used *early* in this disease and persevered with, I believe, from my own observation, that the deaths would not exceed one-fifth of that high number; and that it only needs to be fairly tried, in the treatment of acute erysipelas, to be found worthy of very extensive adoption.

In confirmation of the favorable action of aconite in erysipelas, I may cite Dr. J. le Cœur (*Union Médicale*, 92, 1861), who employed the alcoholic tincture of the root. It is obvious, however, from the large doses that were given, that this cannot have been a very powerful preparation; but as it was, it seems that rather awkward and probably quite needless physiological phenomena were evoked.

Continued and Exanthematous Fevers.—In typhus and typhoid fever aconite is of little use, since it neither reduces the temperature nor diminishes the frequency of the pulse. I have tried it upon various occasions, and have always found it injurious rather than beneficial.

In scarlet fever and measles it is valuable, since it moistens the skin, and certainly helps the emergence and development of the eruption when due, though it seldom reduces the temperature before the eruption comes out. In four cases of measles recently attended by me, aconite reduced the pulsation in every instance to about 72 in the early part of the second day of the eruption, and at this point the pulse remained during the erup-

---

[1] See also Nothnagel (Arzneimittellehre), who says that neither its power to shorten the disease nor to avert heart complications has as yet been distinctly established.

tive stage, but the temperature stood at from 101¾° to 103° till decided defervescence set in.

Apoplexy.—In simple apoplexy, especially when occurring in stout and plethoric persons in whom the vessels of the brain are congested, but not ruptured, accompanied by a full, strong pulse, a hot, dry skin, a flushed and turgid face, and a tendency more or less to coma, aconite is decidedly the best remedy that can be employed. But in cases of apoplexy where there is a tendency to syncope, a pale face, a feeble and perhaps an intermittent pulse, and a cold and clammy skin, aconite should as decidedly be *avoided*, since the effect of this medicine upon the heart is certainly most depressing, and in some instances almost paralyzing.

Should a marked *reaction* take place, and we find it necessary to reduce the action of the heart, and to diminish the force of the circulation through the brain, then, however, we may wisely resort to aconite as a medicine pretty certain to render the most speedy and effective aid.

Palpitation.—Aconite is also of great use in those cases of palpitation of the heart which depend upon simple hypertrophy of the left ventricle. On the other hand, in hypertrophy of the left side of the heart, with diseased valves, admitting of regurgitation, aconite is dangerous.

Retention of Urine.—Aconite is of great service in cases of retention of urine, with spasmodic stricture, which have been caused by a chill (sub-inflammatory). It is also a most effective remedy in the febrile excitement sometimes produced by worms, used alternately with santonine.

Diarrhœa.—In non-tropical dysentery and dysenteric diarrhœa, when the patient suffers from high fever and pain in the abdomen of a griping and cutting character, these latter symptoms preceding a frequent inclination to stool, aconite will at once reduce the fever and remove the cutting pains. It is also of use in diarrhœa caused by a chill, especially in the young; and, on the other hand, we find it useful sometimes as a remedy for constipation in persons of plethoric habit, and when there is a dry skin and a feverish tendency.

Diseases of Women.—Turning our attention specially to the congestive and inflammatory disorders of *women*, we meet with cases in our every-day practice of sudden and abrupt suppression of the catamenia through a chill, though the flow has appeared at its usual time and in its normal character. In these cases there is no remedy which will act so promptly as aconite in removing the discomfort produced, and in quickly causing the flow to reappear, especially if the body be kept warm so as to favor any tendency there may be to perspiration. One drop every half-hour or every hour is generally quite sufficient to bring on the discharge in from four to eight hours after the first administration, and if given within the first few hours after the suspension has occurred.

Puerperal Fever.—Deaths during childbed are in the great majority of cases caused by puerperal fever. This is mostly accompanied by peritonitis, and usually comes on about the second to the fifth day after confinement, ushered in by severe rigors, followed by high fever. When we consider that more than 3,000 women die annually in childbed in England alone, any remedy deserves careful and patient trial which seems to give rational hope of reducing so sad a mortality. I must therefore lay stress upon the fact that I have recorded five cases of puerperal peritonitis of the usual type, and that all five of my patients recovered through the operation of aconite and an occasional dose of castor-oil. Two of the sufferers were attacked upon the third day after delivery, and three upon the second day. With two of them the pulse rose to 140, and the temperature to

105¼° and 105⅜°, while with the other three the pulse rose to about 120 to 135, and the temperature to 103¼° and a maximum of 104⅜°. I gave nothing but aconite and an occasional dose of castor-oil, the former in drop doses every one or two hours steadily through the day and night, and had the satisfaction of seeing my treatment perfectly successful. Repeated hot poultices or laudanum fomentations were employed throughout the whole of the period of the exhibition of the medicine. If we have recourse to aconite in puerperal fever immediately after the rigors set in, it will assert its beneficial power and declare its superiority over venesection, antimony, mercury, and any ordinary mode of treatment.

In puerperal mania, accompanied by high fever, restlessness, head symptoms, and scanty secretion of milk, I have frequently used aconite with speedy and marked success; and here again, in order to secure the full benefit of the medicine, it must be administered soon after the chill has occurred. I am fully persuaded that in puerperal convulsions aconite is one of the best agents we possess, although I have failed in one or two cases; and although chloroform was freely employed as a subsequent means of relief, I was obliged to resort to the lancet, followed by opium, before the desired effect was produced.

Surgical Fever.—The analogy between surgical and puerperal fever (the former of which is no less deadly) was pointed out by Sir James Simpson; and I am satisfied, from extensive experience, that were aconite given quickly and repeatedly in the early stage of surgical fever, during the chill or soon after, the mortality would be much reduced, and become relatively trifling.

As regards the influence of aconite in pyæmia, Dr. Ch. Isnard reports three interesting cases[1] of a traumatic character, and one of childbed fever, in which favorable results were obtained, the fever having quickly abated, and a critical profuse sweat appearing.

In all the latter class of cases, namely puerperal fever, mania, or convulsions, and also in surgical fever, one or two drops of tincture of aconite may be given every ten or fifteen minutes for the first hour, and afterwards every half-hour or hour, according to the severity of the symptoms. As the pulse falls, the interval between the doses should be prolonged; but if the fever should not abate, and the temperature should not fall within a reasonable time after commencing the medicine it would be almost useless to continue it, for this might easily bring about sudden prostration.

In conclusion, in all pyrexial diseases for which aconite is prescribed, I recommend that it be given from every half-hour to every one, two, or three hours, in single or two-drop doses of the Pharmacopœia tincture, according to the special necessities of the case. In very acute cases it may be given every fifteen minutes during the first hour. Very important is it to observe, also, that this medicine should never be given in any vehicle except *water*. [We have used aconite to advantage in gonorrhœa and epididymitis when the imflammation runs high, and in urethral fever. It is also of service as a preventive of chill after the passage of sounds, as first pointed out by Mr. Long nearly thirty years ago, a drop or two of the tincture being given after the passage of the instrument. When effective it is preferable to quinine usually given for this purpose.]

Pain and Spasm.—We have now to consider the therapeutic properties of aconite in painful and spasmodic affections which are uncon-

---

[1] Union Méd., 1861. pp. 132–134.

nected in their origin with inflammation or congestion. In neuralgia, the efficacy of aconite has been very variously estimated by different authors; but doubtless a good deal of this disagreement is to be traced to the very vague way in which the word "neuralgia" has been used. As Dr. Ringer correctly observes, the drug is of little service in those cases where the principal cause of the pain is the existence of some permanent peripheral irritation, like a decayed tooth. In the neuralgias which are more properly so called, especially those of the face or brow, aconite proves useful in a considerable number of cases. [The effect is probably due to its elective influence on the trigeminus, and Ott suggests that the relief is obtained through a benumbing influence on the peripheral ends of the nerve.]

In **Myalgia**, or so-called "muscular rheumatism," and in various forms of non-inflammatory aching pain in tendons, ligaments, and other fibrous structures, aconite may be of considerable service, though somewhat pushed out of the field by other remedies, and especially by the hypodermic injection of morphia and atropia. In tetanus we should as yet be chary of receiving with perfect faith the appearance of cure by any remedy whatever. Paget and others formerly recounted instances in which the pulse came rapidly down, and the spasms much subsided; but I believe their recent experience has not supported the pretensions of aconite in this direction. It was thought that this drug exerted a sedative influence upon the nerve-centres, and might thus control reflex action; but upon this point all that we accurately know of the physiological action suggests the contrary; for example, the above-quoted experiments of Achscharumow[1] seem distinctly to show that the reflex irritability of the cord remains intact even in fatal poisoning with aconite, while the centres of the voluntary movement are progressively paralyzed.

It is at any rate certain that aconite may relieve pain. Handfield Jones speaks of it as especially appropriate to the relief of superficial pains,[2] and of spasms which are not of too severe a type, both by internal and external administration. But it is the latter mode of employment which is by far the most frequently efficacious. Local pain of a nerve, or local spasm of a muscle, if not too severe and of too long standing, may often be relieved by liniments containing aconite, or by painting the simple tincture on the skin over the part, or by rubbing in a small piece of the ointment of the alkaloid aconitia. Dr. Ringer remarks, very correctly, that if aconite be likely to give relief to pain it will do so quickly. Dr. Anstie[3] places it among those remedies for neuralgia which act by locally interrupting the conductivity of a sensory nerve, and thus giving the nerve-centre time for rest and self-recovery; he does not think it can ever be more than a subordinate agent, except in mild cases. In connection with the external use of aconitia it is important to remark that the German specimens of that alkaloid, as admitted by German experimenters, are totally inoperative when applied to the skin.[4] The external application of aconitia ointment has been tried not only in the milder local spasmodic affections, but in those severe and intractable maladies which belong to the same group with torticollis; but so far as I know it has not proved more successful in the latter terrible maladies than other remedies.

---

[1] Virchow's Archiv, loc. cit.
[2] Med. Times and Gazette, vol. i., 1863.
[3] Neuralgia and Diseases that resemble it, 1871.
[4] Nothnagel, op. cit.; Achscharumow, loc. cit.

It is right to state that certain eminent authors give a much wider scope than is above assigned to the therapeutic action of aconitia in painful and spasmodic disorders. M. Gubler, in particular,[1] assigns it an important *rôle* in the treatment of a class of neuralgic affections which he calls *congestive*, and also in that of the neuralgias which he calls *acrodynic*, which have their seat in the extremities of the limbs, in which situation the Pacinian corpuscles are numerous. In the latter cases he declares that it has succeeded where large doses of morphia and of atropia had failed. He also believes it to be appropriate to the treatment of angina pectoris, spasmodic asthma, convulsive cough, &c. But indeed, in France, aconitia is freely employed internally, which is not the case in England. The *granules de Hottôt* (made from an aconitia which is said to be nearly as powerful as the best English aconitine of Morson) are the usual form in which it is administered and in this way doses of $\frac{1}{80}$ grain of the alkaloid are given, at first twice a day, and afterwards more frequently. In England the majority of physicians distrust this exceedingly powerful agent as an internal medicine. [Much of the commercial aconitia is an impure mixture of alkaloids, and is believed to be largely derived from the A. ferox. Its composition and energy therefore are so uncertain that the commencing dose should be quite small until the sample has been fairly tested.]

For practical purposes we may fairly say that all the benefits that are to be expected from aconite may be obtained from its internal use in the form of the tincture of the British Pharmacopœia (which is the tincture uniformly referred to above) and the external use of aconitia in the form of the pharmacopœial ointment, or of the tincture of aconite, simply painted on the skin, or diluted with soap liniment, etc., and applied by friction. An advantageous combination can sometimes be made by adding chloroform, or laudanum, or both, to such a liniment. It is scarcely necessary to add that in applying aconite, or still more aconitia, externally, we must avoid all abraded or wounded portions of the skin.

Besides the well-established efficacy of aconite in relieving congestions and inflammations on the one hand, and pains and spasms on the other, there are sundry other actions that have been attributed to it on more doubtful evidence. Among these one of the most important, if it were established, would be its alleged diuretic effects. Fouquier is the strongest advocate of this power of the drug, which however has been denied by several good authorities, and of which there cannot be said to be any certainty at present.

PREPARATIONS AND DOSE.—The officinal preparations (U. S.) are extractum aconiti, tinctura aconiti radicis, linimentum aconiti, emplastrum aconiti, and aconitia. The properties of the root are similar to those of the leaves, but stronger. On this point Hirtz, among others, gives strong testimony.[2] He administered the extract of the *leaf*, in large doses, in various cases of lung and bronchial affections; but no effect, either physiological or therapeutic, was produced unless as much as 15 or 16 grains were given at once. The extract of the *root*, on the contrary, in doses of only $\frac{1}{4}$ grain, produced prompt physiological effects; and smaller quantities than these were found to act best medicinally. Aconite leaves dried and powdered may be prescribed in doses of gr. $\frac{1}{2}$–gr. ij.

---

[1] Commentaires Thérapeutiques, Paris, 1868.
[2] Gaz. méd. de Strasbourg, 1861, No. 1.

In all acute pyrexial affections of adults, one minim of the tincture every half-hour or hour, up to a total of 5 to 10 minims, or until the pulse or temperature sensibly falls; after this, about 2 or 3 minims every four hours. In children under two years of age, not more than one-fourth or one-sixth of these doses should be given; and from two years old to ten, one-fourth to one-half of the adult dose, according to the susceptibility of the child to the influence of the medicine.

In non-febrile pain or spasm, 5 to 10 minim doses of the tincture at long intervals.

Linimentum aconiti is a very strong preparation for external use, and should not be applied where there is any sore or abrasion of the skin, nor should it be used near the mouth. It should be applied with a sponge or soft brush, but this should not be continued long after numbness is produced.

Unguentum aconitiæ (B. Ph.) should be used with caution. It contains 8 grains of aconitia to one ounce of prepared lard. This external application is highly beneficial in neuralgia, especially facial neuralgia, also in sciatica and muscular pains.

Aconitia is rarely prescribed internally, as the one-fiftieth part of a grain, if of good quality, produces poisonous symptoms of a dangerous character.

PULSATILLA (Pulsatilla Nigricans).

Active Ingredients.—The camphor-like crystalline body known as anemonine appears to represent all, or nearly all, the active properties of the *P. nigricans*. This substance, which is contained also in the fresh root of *Anemone pratensis, A. nemorosa, Ranunculus Flammula, R. sceleratus,* and *R. bulbosus,* has for its formula $C^{15}H^{14}O^{7}$ (Fehling), and is obtained by distilling the root with water, and concentrating the distillate till the anemonine and anemonic acid distil over. Anemonine is separated from the latter (inert) substance by alcohol, which takes up anemonine and leaves the other behind. Anemonine occurs in colorless, shining, orthorhombic prisms which easily crumble, and are tasteless and neutral in reaction; in a molten state it has a burning taste, and leaves the tongue numbed for days together. In cold water or alcohol it hardly dissolves at all; in both of them when boiling it dissolves more readily. Cold ether does not dissolve the slightest portion; boiling ether dissolves a little; chloroform and lavender-oil dissolve the whole.

Physiological Action.—The local irritant action of pulsatilla can be produced either by direct application to the skin, or by breathing the dust in pulverization. Bulliard relates the case of a man who applied the bruised root to the calf of his leg to relieve rheumatism, and in consequence got inflammation and gangrene of the whole limb. The inhalation of the dust has produced itching of the eyes, colic, vomiting, frequent diarrhœa, &c.

Anemonine has been shown, by the researches of J. Clarus[1] and others, to possess powerful toxic properties; 4½ grains affected rabbits, and 9 grains killed them in three or four hours. The special phenomena were diminution of the frequency and strength of the heart's pulsations (some-

---

[1] Journal für Pharmadynamik, i. 439; Husemann, *Die Pflanzenstoffe.*

times preceded by a period of excitement) and slackening of respiration; finally, diarrhœa and stertorous breathing, sinking of temperature, semi-paralysis of the hind, and then of the fore limbs; stupor; mydriasis and (immediately before death) myosis; but no convulsions, such as have been seen to be produced by *extract* of pulsatilla. The liver, spleen, kidneys, and abdominal canal were found quite healthy: there were more or less congestion and œdema of the lungs; and relaxation of the heart walls: the heart cavities and great vessels were full of dark clotted blood, while the blood everywhere else was fluid; there was also marked hyperæmia of the membranes of the brain and cord, especially in the neighborhood of the medulla oblongata. Applied to the conjunctiva of a rabbit, anemonine produced slight inflammation; applied to the human tongue, it only left a slight burning. Murray states that the vapor of melting anemonine causes pricking in the tongue, followed by an abiding numbness and white patches. Heyer saw the same vapor produce intense irritation of the eyes.

THERAPEUTIC ACTION.—Pulsatilla may be employed in most of those acute and sub-acute inflammations of the mucous membrane generally in which the discharge is of a mucous or muco-purulent character, such as the early stage of purulent ophthalmia in children, and even in adults; also in gonorrhæal ophthalmia. It may also be employed with advantage in inflammation of the external auditory canal, so often met with in children, where the lining membrane is red and swollen, and severe pain is experienced, while later on a thin acrid discharge appears, often tinged with blood, and soon becoming puriform. It may be used also in inflammation of the nasal passages, accompanied by profuse mucous or muco-purulent discharge, the smell of which is offensive.

Inflammations.—In cases of inflammation of the conjunctiva and of the auditory and nasal passages, I recommend a wash composed of from one drachm to two drachms of tincture of pulsatilla to four ounces of water. The strength of the lotion must be determined by the sensibility of the inflamed surface and by the age of the patient.

In ophthalmic cases the lids of the eyes should be carefully opened and the conjunctiva freely washed with the lotion from eight to ten times every twenty-four hours. Drop doses should be given to infants every three hours; and adults may take five to ten drops every four hours in an ounce of cold water.

In otitis and in coryza the lotion should be warmed and syringed into the ear or the nose four or five times a day. Internally the same dose may be taken as in ophthalmia, according to the age of the patient.

Dyspepsia.—Pulsatilla is a good medicine in many of those cases of dyspepsia, or of subacute gastritis, met with in phlegmatic temperaments, when we find some or all of the following symptoms present, namely: depression of the nervous system, with fear of death; loss of appetite; white and thickly coated tongue; little or no taste, or, if taste be present, a sensation in the palate of greasiness; sensation of mucus about the mouth and gums; nausea, with an inclination or wish to vomit; flatulency; heartburn; occasional pain and flatulent colic in the epigastrium; sick headache; dry cough; coldness and clamminess of the extremities, and often likewise of the entire surface of the body, generally accompanied by constipation or by diarrhœa. When the diarrhœa is attended by mucous discharges or by active piles, the pulsatilla quickly removes them. Five drops taken as a dose every four hours, in a tablespoonful of water, will soon give relief in this form of dyspepsia.

Affections of the Uterus.—Pulsatilla exerts a peculiar action upon the uterus. In functional amenorrhœa with absence of catamenia, or if the catamenia be scanty or delayed, or even in suppression induced by fright or chill, pulsatilla is often of the greatest value in establishing the flow at the proper time and in its natural quantity. It is also of much benefit in functional dysmenorrhœa, where the discharge is scanty, or even when profuse, but blackish and clotted. Even though at the first period this medicine should fail to restore the menstrual flow to its normal standard, by persevering in the use of it for two months or more the desired effect is almost certain to be produced.

Leucorrhœal discharges attended by pain in the loins, feeling of weariness, depression of spirits, loss of appetite, and derangement more or less extensive of the nervous system, are also quickly removed by a steady course of pulsatilla taken internally in five-drop doses, three times a day, and continued for a few weeks. A teaspoonful of the tincture should also be put into a pint of cold or tepid water, and be used as an enema for the vagina every day.

[In epididymitis we have had the happiest effects from pulsatilla given in very small doses, a few drops of the tincture in a glass of water, and a teaspoonful given every two or three hours, according to circumstances. The frequency of the dose should be diminished as soon as the pain begins to subside. The drug appears to influence the afflux of blood to the part, and not unfrequently the pain will be almost gone in twenty-four hours, after which the tumefaction gradually subsides. If the inflammation is very acute, with rise of temperature, it is better to give a few doses of aconite before or in alternation with the pulsatilla.]

Tapeworm.—In addition to my experience of the value of pulsatilla, I may state that an extract of the root employed internally has been found valuable in cases of tapeworm.

Cough, &c.—Clarus found anemonine, in half-grain and one-grain doses, very useful in irritative cough, asthma, and hooping-cough. Clarus and Schroff agree in the statement that even larger doses (two grains) produce no physiological symptoms in men.

PREPARATION AND DOSE.—There are no officinal (U. S.) preparations. The German Pharm. contains an extract, and the Codex (Fr. Ph.) an "alcoolature" from the fresh plant. The extract is an unreliable preparation, as most of its activity has been lost in the process of making. The most available preparation at present attainable in this country is the imported homœopathic tincture (equal parts of expressed juice and alcohol), or a tincture made from the freshly gathered native Pulsatilla Nuttalliana (Anemone Ludoviciana). The tinctures and fluid extracts from the dried (imported) plant are not trustworthy. The dose of a good tincture is from $\mathfrak{m}\ \tfrac{1}{10}$ to $\mathfrak{m}\ v.$, to be repeated according to circumstances.

HELLEBORE (HELLEBORUS NIGER).

ACTIVE INGREDIENTS.—The question which of the constituents of hellebore root are active, has attracted renewed interest of late years on account of the discovery of the glucoside *Helleborin*, by Bastick (1853), the confirmation of this by Marmé and Husemann, and the discovery by the latter observer (1864) of a second glucoside, *Helleborein*. The former of these is not found so largely in the *Helleborus niger* as in certain allied species. It is crystalline, in concentrically grouped shining white needles, odorless, and of neutral reaction; in the solid state it is tasteless, but the alcoholic solution is very acrid and burning to the mouth. It is insoluble in cold water, little soluble in ether and in fats, readily soluble in rectified spirits and in chloroform. Formula $C^{18}H^{14}O^6$. Helleborein ($C^{26}H^{44}O^{15}$) crystallizes out of a very concentrated alcoholic solution in collections of fine, nearly colorless transparent needles, which quickly turn chalk-white under the influence of the atmosphere. The evaporation of a watery solution leaves a grayish-yellow resinous powder, which has no distinct smell, but is strongly provocative of sneezing. Helleborein is very soluble in water, little so in alcohol, not at all in ether. Since the discovery of these two substances (which will be again considered under the heading of Physiological Action), it is doubtful whether the older conjectures, that the action of hellebore depends upon an acrid oil and a "volatile acid," can have been just. The active volatile ingredient may be helleborein, which is faintly or not at all acid in reaction, and is not affected by boiling in the alkalies or baryta; but its volatility is limited.

PHYSIOLOGICAL ACTION.—The *Helleborus niger* emits no particular odor until bruised. The fibres of the root, which are the part employed in medicine, have a disagreeable smell; their taste while fresh is penetrating, and, though neither bitter nor very warm, it leaves a long persisting impression upon the tongue. Grew, in his work upon Tastes, describes its effects as benumbing and as inducing a paralytic stupor—comparable to that produced by drinking or eating anything too hot. When the fibres have become dry, their acridity is much diminished, but the nauseousness and the bitterness are more marked.

Whichever may be the constituents that determine the energy of hellebore, there is no doubt of its being a violently acrid poison, for even the flowers when crushed and applied to the skin produce redness and vesication. This is remarkable, since in these external effects the root and the leaves do not participate. The effects produced by swallowing an overdose are vomiting, colic, purging, great thirst, sense of suffocation in the throat, and of burning in the œsophagus and the stomach: these are accompanied by a painful pricking and swelling in the tongue and fauces, and the secretion in the mouth of a considerable quantity of viscid mucus; and are eventually followed by convulsions, cold sweats, a feeble pulse, and death. Post-mortem examinations show, that when death is caused by hellebore the stomach and intestines are much inflamed, and that the rectum more particularly becomes affected; thus indicating a certain analogy between the action of hellebore and that of colchicum.

It is interesting to append to this description of the physiological action of hellebore, as such, the observations that have been made on the effects of the two glucosides. Helleborin is found to be a very energetic substance. Marmé found that 1¼ grain, administered subcutaneously, killed frogs; half the quantity, by the stomach, killed pigeons; ¾ grain to 1½ grain killed rabbits; less than ¼ grains were fatal to dogs. It had no effect upon entozoa. It was found to exercise no action upon the skin, and to exert a less intense action on mucous membranes than helleborein. Helleborein, administered by the stomach to mammalia, caused licking and chewing movements; teeth-grinding; a certain amount of salivation in cats and dogs; vomiting in dogs and birds; pains in the belly (apparently) in dogs and rabbits; efforts at evacuation in dogs. On dissection the mouth and œsophagus were healthy; in pigeons the crop was always extensively inflamed; in mammalia the stomach and intestines showed signs of irritation in varying degrees, from simple increase of secretion to inflammation of high grade with extravasation of blood. The remoter actions of the drug fell upon the nerve-centres, especially the brain, paralysis of which appeared to be the cause of death. In animals, a period of excitement and restlessness was followed by paresis of the hind limbs, with tremor and vacillation of the whole body; a further stage was marked by profound paralysis and anæsthesia: cats alone recovered comparatively soon from this state. Marmé and Husemann always detected marked congestion of the membranes of the brain and cord: in rabbits there was also diminished consistence of the cord, and extravasation of blood into the cranial cavity. Functions of other organs were affected in much the same way as by narcotic agents in general. Urinary secretion somewhat increased (in cats only); respiration slowed during narcosis; hypostasis and hyperæmia appeared in the lungs after death; the heart's action was not reduced except by the heaviest doses; especially in frogs and dogs the cessation of pulsation was very late. The pupils widely dilated in narcosis, but contracted *post-mortem*, under electric stimulus.

The experiments of Marmé with helleborein show interesting differences of effect from the above. The action is both local and remote; locally it does not affect the skin, but it is intensely irritative to the mucous membrane. Absorbed, it exerts a powerful influence on the heart and the intestinal canal; yet it does not seem so deadly to animals, in equal dose, as helleborin; perhaps because only a part is absorbed unchanged, the remainder being split up into products which are inactive (e. g., *helleboretin*) when taken into the blood. Very small doses administered by the stomach for a long period produce a cumulative effect, shown by loss of appetite, nausea, and vomiting, which disappear rapidly on suspending the drug; and occasionally this mode of administration has produced (as larger doses will) pain, increased secretion, and gastro-enteritis. These effects, however, are perhaps not so liable to result from helleborein obtained from the *black* hellebore as from that of other species, especially the *H. viridis*. The special action of helleborein on the heart resembles that of digitalis, but is quantitatively much weaker: small repeated doses slow the heart; larger doses hurry its action, and then usually arrest it suddenly; the action is through the vagus, and blood-pressure is heightened both in the slowing and in the hurrying grade. Respiration lasts longer than the heart's action; respiration is nevertheless affected. The glands (especially the salivary and the kidneys) are affected, and the uterus. Respiration is at first quickened, then made slow and difficult. Salivation is always produced, whatever way the drug is introduced into the system; diuresis is constant, and the kidneys are hyperæmic after death; in female animals the uterine mucous membrane is invariably congested. There is semi-paralytic weakness of the limbs, and (after very large doses) there are severe convulsions.

THERAPEUTIC ACTION.—Previously to the introduction of medicines derived from the mineral kingdom, the value attached to hellebore was immense. The early physicians commended it in the treatment of insanity, both maniacal and melancholic, and there is no doubt that great benefit has followed its employment in the last named. It has also been given with apparent success in dropsy, worms, and cutaneous disorders.

and as an emmenagogue; but this last-named use may be reasonably supposed to depend upon the great cathartic powers inherent in hellebore.

Stillé says that hellebore in small doses appears to stimulate the abdominal organs, augmenting the secretion of the liver and of the pancreas, quickening the peristaltic movements of the bowels, and increasing the catamenial discharge. Hæmorrhoidal discharges are likewise increased by its use.

I have often prescribed the tincture of hellebore in doses of from five to fifteen drops every two or three hours, in cold water, with complete success in dropsical effusions, and particularly in general anasarca following scarlet fever uncomplicated by organic disease.

From what has been above mentioned as to the physiological action of the glucosides helleborin and helleborein, it is possible that we may hereafter derive new and valuable therapeutic benefits from these substances. Helleborein, in particular, seems marked out, as Marmé observes,[1] for employment in heart complaints and dropsy; its solubility in water, and the impunity with which it can be subcutaneously injected, give it important advantages.

PREPARATIONS AND DOSE.—Tinctura Hellebori, ♏ v.- ʒ ss. (.30 -2.); Extractum Hellebori, gr. j.-v. (.06—.30).

PODOPHYLLUM PELTATUM (THE MAY APPLE).

ACTIVE INGREDIENTS.—The rhizome, which is the only officinal portion, contains about 3 to 3.5 per cent. of a peculiar resin, called *Podophyllin*. It also contains a certain proportion of the alkaloid Berbera (the main active ingredient of Calumba, under which head it will be discussed), and it is possible that the tonic properties of this latter may render it, in some cases, a valuable adjunct to the resin, so that a preparation of the root which should contain them both in known proportions is perhaps a desideratum. Podophyllin is a greenish yellow powder, with a bitter and acrid taste: it is entirely soluble in rectified spirit and ammonia, and nearly so in pure ether.

PHYSIOLOGICAL ACTION.—Given in poisonous doses (two grains and upwards), podophyllin attacks the gastro-intestinal mucous membrane, by whatever channel it may be introduced into the system. Percy[2] administered an alkaline solution subcutaneously to dogs, and (after a few hours) observed the animals to suffer from colic, tenesmus, and vomiting; and Anstie[3] injected an alcoholic solution into the peritoneum of dogs, cats, and rats with the uniform result of provoking vomiting, bloody stools, and death from exhaustion, with some appearances of a very peculiar respiratory paralysis. Ulceration of the duodenum was found in several, and inflammation in all cases; but the absorption had been so prompt and complete that no inflammation of the peritoneum was produced. This is the more remarkable since both the rhizome and the resin are undoubtedly powerful local irritants, and in some cases, when applied to mucous

---

[1] Zeitsch. f. rat. Med. (3), xxvi. 1; Husemann, op. cit.
[2] Amer. Med. Times, iv. 243.
[3] Med. Times and Gaz., 1863, p. 326.

membrane, even act as escharotics. Other effects of physiological doses that have been noted, are profuse sweats and salivation.

[The action on the liver of podophyllin and several other American chalagogues has been carefully studied by Rutherford and Vignal by means of experiments on dogs. They say: "1. Podophyllin, when injected into the duodenum of a fasting dog, increases the secretion of bile. It is inferred that the increased biliary flow in the preceding experiments was due to increased *secretion*, and not merely to expulsion, because the gall-bladder had been well emptied by compression, and the cystic duct had been clamped, moreover the increased flow was far too prolonged in some of the experiments to be attributable to spasm of the larger bile-ducts; therefore an increase in secretion must have been the case. 2. When the bile is prevented from entering the intestine, podophyllin acts less powerfully and less quickly than when bile is introduced. 3. Augmentation of the biliary secretion is most marked when the purgative effect is not severe; indeed, if the purgative be very decided, diminution and not augmentation of the biliary secretion may be the chief result. 4. Podophyllin purgation is apparently due to local action, for the irritation of the intestinal mucous membrane extends gradually from above downwards. 5. The bile secreted under the influence of podophyllin, although it may be in increased quantity, contains as much of the special biliary matter as bile secreted under normal conditions." Brit. Med. Jour., Oct. 30, 1875.]

THERAPEUTIC ACTION.—The rhizome of the podophyllum was employed by the aborigines of America as a vermifuge; and in modern times has been greatly esteemed by the profession as a hydragogue purgative, having a special reputation as a cholagogue, whence the name sometimes given to it of "vegetable calomel."[1] In this country the cholagogue action of podophyllum has been denied by Dr. Hughes Bennett, on the strength of a number of experiments upon dogs with doses of from 2 to 8 grains, which diminished the secretion of bile. Dr. Anstie also concluded (from finding that fatal doses injected produced no visible change in the liver) that no direct action on that organ is produced by the drug, and that the increase of bile, when it does occur, is an accidental result of the intestinal irritation. For my own part I am convinced that, whatever the *modus operandi*, podophyllin, in suitable doses, is often capable of correcting a deficient secretion of bile, especially in children and infants. When the motions have become white or clay-colored, podophyllin, in a few doses of $\frac{1}{20}$ to $\frac{1}{10}$ of a grain every six hours for a child, and of $\frac{1}{15}$ to $\frac{1}{12}$ of a grain for an adult, has frequently, in my experience, restored the natural character of the evacuations, at the same time regulating the bowels. So far are these doses from acting in a specially *irritant* manner, that they will often allay the vomiting and diarrhœa of gastro-enteric inflammation; and will also cut short the symptoms in the remittent fevers of children, with high temperature, headache, and delirium, dry, brown, and furred tongue, nausea or vomiting of bilious matter, pain or uneasiness in the stomach, sleeplessness, with a general sense of weariness, grinding of teeth in sleep, etc. The effect is much improved by the exhibition of occasional doses of aconite. As another example of the favorable action of podophyllin on the intestinal canal, I may mention that prolapsus of the rectum, in children, may sometimes be removed by similar small doses, administered night and morning.

---

[1] E. Schmidt: Bayer. aerzt. Intell.-Blatt., 1866, p. 13.

As a laxative in general, and especially for the purpose of removing habitual constipation, podophyllin should be given in moderate doses. A single dose of half a grain (or at most an entire grain), though somewhat slow in acting, will usually, after some hours, produce decided watery and bilious purging; and often this effect, instead of being followed by a constipative reaction, will be succeeded by increased and long-sustained habitual activity of the bowels. For habitual constipation, it is perhaps better, however, to give from $\frac{1}{12}$ to $\frac{1}{8}$ grain every night and morning for a little time. This treatment is especially useful when the constipation is accompanied by nervous and bilious headaches. Moreover, in some fevers. as typhus, though decided purgation is not to be thought of, it sometimes happens that at the commencement, along with constipation and biliary derangement, there is much congestive headache and delirium; or, at any period of the disorder, marked constipation, with sleeplessness, may constitute an evil requiring to be dealt with. I do not for a moment believe, as has been asserted by some, that podophyllin will shorten the course of specific fevers; but I am convinced that $\frac{1}{8}$ to $\frac{1}{4}$ grain doses every six hours will often speedily relieve such symptoms as the above, and at the same time much relieve the general distress, and frequently lower the temperature by one or two degrees. It is needless to say that the action must be stopped as soon as it has produced mild laxation; otherwise, in such a disease as fever, the debilitating results might be mischievous.

In dyspepsia, and in hepatic derangement characterized by loss of appetite, acid regurgitation, putrid taste in the mouth, flatulence, and a tendency either to constipation or to diarrhœa, $\frac{1}{10}$ grain of podophyllin every night and morning will often produce the best results. I have often found it particularly useful in chronic vomiting after meals, and in obstinate heartburn connected with liver derangement.

In a variety of liver diseases, both acute and chronic, podophyllin will again often be found to justify the high reputation which it has acquired in America and elsewhere.

PREPARATIONS AND DOSE.—Extractum podophylli, gr. v.–x. (.30–.65); resina podophylii, gr. $\frac{1}{6}$–$\frac{1}{2}$ (.01–.03). These are purgative doses. It is somewhat the fashion to give a large dose and tie it up with opium, belladonna or hyoscyamus, a plan which can hardly commend itself to science or good sense. If the larger dose is too active, it is simpler and better to diminish it than to complicate its action with an additional ingredient. The tendency of the present age is toward mono- rather than poly-pharmacy, and prescriptions with the orthodox "adjuvans" and "corrigens" are less frequently seen than formerly.

HYDRASTIS (Hydrastis Canadensis).

[ACTIVE INGREDIENTS.—Hydrastis contains three alkaloids: Hydrastia, a white crystalline substance; berberia, a yellow cristalline body (found also in several other plants); and xanthopuccinia,[1] also yellow and crystalline. From these, various salts have been made. In addition there is an unnamed resin. Commercially there exists an article called "hydrastin," which is a mixture of the above-mentioned ingredients.]

---

[1] Am. Jour. Pharm., 1875.

PHYSIOLOGICAL ACTION.—The toxic effects of the alkaloid hydrastin, though not severe, are interesting, from their resemblance in some respects to those of quinine. Large doses produce noises and a sensation of rushing in the ears; they do not cause any considerable disturbance of the alimentary canal, but merely a sense of warmth at the epigastrium.

THERAPEUTIC ACTION.—In Europe the hydrastis has been used for disorders of the stomach and liver, and has been highly recommended in cancer. It has also been tried in the form of an infusion, in dropsy, being considered an efficient diuretic. Dr. Barton regarded this medicine as a good alterative in disordered conditions of the mucous membrane.

For my own part, I have employed it with excellent results in ulcers of the legs, the rectum, and the uterus. I have had great success with it also in prolapsus ani and in hæmorrhoids, though not greater than in headaches depending upon a constipated state of the bowels. In simple constipation referable to a sluggish state of the liver, hydrastis is likewise very valuable.

Glandular swellings frequently yield to its action, but I have never perceived that any advantage resulted from the employment of hydrastis in *cancer*.[1]

When the general system is debilitated, this medicine operates in a remarkably efficacious manner, and in its action seems not unlike quinine.

The value of hydrastis is further proved in chronic coryza, when this depends upon a syphilitic taint, and where the Schneiderian membrane is of a deeper red than is natural to that part, and when its surface is more or less studded with minute ulcerated patches, with accompaniment of profuse mucous discharge, the discharge itself varying in color and consistence, from thin, clear, and starchy to thick and greenish or yellow. In these cases, five drops of the tincture of hydrastis should be taken three or four times a day in a wineglassful of water, supplemented with the use of a lotion composed of one drachm of the tincture to eight ounces of water, with which the nose may be either bathed or syringed three times a day. This treatment will quickly set up a healthy action in the mucous membrane, and remove the troublesome and tedious disease to which I am referring.

Similar treatment will be found successful in cases of ulceration of the septum or any portion of the nasal fossæ.

Ophthalmic Cases.—In muco-purulent inflammation of the conjunctivæ, implicating the Meibomian follicles, and causing adhesion of the lids in the morning, a lotion prepared from hydrastis is likewise very serviceable. In cases also of catarrhal ophthalmia implicating the mucous membrane of the nose and throat, I have used a douche spray of hydrastis with good effect.

Inflammations.—Catarrhal inflammation, commencing in the duodenum and spreading up the bile-duct, is often cured by a steady course of hydrastis, while the jaundice which so often attends this affection is often rapidly removed.

Inflammation of the gall-bladder and of the gall-duct, caused by gall-stones, is also reduced by hydrastis, and a larger space is thereby allowed for the exit of the calculi.

Gonorrhœa.—Gonorrhœa is cured in many instances by employing

[[1] Kidd (Laws of Therapeutics, Lond., 1878) has recently reported some facts that give hydrastis a renewed interest in this connection.]

a lotion of hydrastis, one or two drachms of the tincture being added to a pint of water, and a syringeful injected up the urethra at first every half hour for a period of seven or eight hours; and subsequently, or for the ensuing two or three days, once every six or eight hours. [We prefer the infusion of hydrastis (ʒj. of powdered root to ℥viij. boiling water.) When cold, filter or inject three or four times a day in ordinary cases. As the infusion contains none of the irritant resin, it is better borne than an alcoholic preparation. In subacute cases and in gleet "hydrastin" (gr. ½–j. to f ℥j. of water) may be used. The patient should be warned that the solution stains the linen.]

Ulcers.—The chronic indolent ulcer so commonly met with upon the lower part of the leg is amenable to treatment with hydrastis, the lotion being applied externally, and the tincture given in a suitable dose.

Quite recently I have treated two cases of ulcer, one upon the nose and the other upon the eyelid. Both were of the type of the true rodent ulcer, so well described by Sir James Paget,—"The base of a dingy reddish-yellow color, dry, glazed, and free from granulations, and the discharge but slight." Both were cured by the treatment adopted in the case of the chronic indolent ulcer.

Cracks, &c., of the Nipple.—During lactation we frequently have to deal with cracks, fissures, and abrasions of the nipples. In such cases I recommend the use of a compress dipped in a lotion prepared from hydrastis, the compress to be renewed every four hours. While this treatment of the nipples is being prosecuted, the infant should be admitted to the breast not oftener than once every four or six hours, and the nipple itself should be protected with a calf's teat, so as to serve the double purpose of guarding it from injury by the child, and of defending the mouth from any consequences that might arise from direct contact with the hydrastis.

Hæmorrhoids.—Internal hæmorrhoids, which cause great prostration of strength, and are generally accompanied by various dyspeptic symptoms—usually giving rise to considerable pain during defecation, frequent and periodic attacks of bleeding, with a discharge from the anus of a considerable amount of mucus or of muco-purulent matter—are cured, or at all events relieved to a very material extent, by the use of hydrastis. In these cases a weak infusion should be employed as an injection every night and morning, and during its use the patient should take five drops of the tincture in a wineglassful of water three or four times a day.

In cases of *external* piles, hydrastis is also of great value, the lotion being used three or four times a day; or the drug may be applied equally well in the form of ointment.

Prolapsus of the Rectum.—Simple prolapsus of the rectum, occurring in children, yields to hydrastis, which in these cases relieves the congestion and the swelling of the mucous membrane, and hinders its protrusion. Now and then the prolapsus even of adults is successfully overcome by the same treatment, the form of exhibition in either case being lotion or enema, as circumstances may dictate. [It has also been found useful as an injection in *fistula in ano*.]

Miscellaneous Cases.—Lastly I may remark that hydrastis is a capital agent in cases of erosion and ulceration of the *cervix uteri;* and, in addition, it must not be overlooked that in America the alkaloid hydrastia has been strongly recommended for intermittent fevers, for typhoid fever with copious sweats, also for excessive diarrhœa, and tendencies to septic poisoning: the dose is from two to nine grains. It has also been employed in sunstroke and in chronic dyspepsia. In the shape of an ointment (one part to six or seven) or in a lotion (one part to fifty), it has been extensively used in a great variety of ulcerative affections of mucous membranes, and apparently with very good results. It is very necessary not to confound the *alkaloid* hydrastia with the *resin* called hydrastin, obtained by a different process from the root: the latter is a pale, straw-colored powder with a pure, bitter taste; it has purgative qualities, and is recommended in the constipation of old persons in doses of three or four grains.

PREPARATIONS AND DOSE.—The fluid extract is officinal. Commercially there are tinctures and fluid extracts from the dried and a tincture from the fresh plant; also the alkaloids hydrastia, and berberia, and their salts, and the mixed product hydrastin. The dose of the powdered root as a tonic stomachic and anti-dyspeptic is from gr. v.–x. (.30–.65), two or three times daily. The tinctures and fluid extracts in proportional quantities. The usual dose of hydrastin is gr. ss.–ij. (.03–.12).

STAVESACRE (DELPHINIUM STAPHISAGRIA).

ACTIVE INGREDIENTS.—The *Delphinium Staphisagria* contains two alkaloidal bodies, delphinine and staphisagrine, upon which its active properties depend. Delphinine, which, according to Erdmann,[1] constitutes about $\frac{1}{10}$ per cent. by weight of the seeds, is an amorphous white substance, which, when its ethereal solution is evaporated, remains as a somewhat resinous mass. It has an acrid taste and a strong alkaline reaction, is little soluble in water, soluble in ten parts of cold rectified spirit (80 per cent. rect.), and very soluble in ether, chloroform, and benzol. Erdmann makes its formula $C^{24}H^{11}NO^2$. Staphisagrine, which is left undissolved when the alkaline or ammoniacal precipitate of the two bases is treated with ether, is a yellow body, melting at 200° C., with a very acrid taste, insoluble in water and ether, but freely soluble in rectified spirit. The formula is uncertain.

PHYSIOLOGICAL ACTION OF DELPHININE.—The poisonous properties of stavesacre, as exhibited in the effects of delphinine, have received just attention and investigation of late years. Roerig and Falk, L. van Praag, Dorn, Albirs, and Dardel have experimented upon a very considerable number of animals. The general result so far established seems to be, that, after certain introductory local phenomena, a general paralysis of all movements is gradually developed, the breathing becomes labored, the beating of the heart is slow and weak, and sensation is annihilated. Death is followed by very marked rigor mortis; and dissection always discovers passive venous hyperæmia in all the cavities of the body; occasionally, also, there is local inflammation in the intestines.[2]

THERAPEUTIC ACTION.—The decoction of stavesacre has been occasionally employed as an anthelmintic, and with tolerable success. The

---

[1] Arch. Pharm. (2), cxvii, 43; Husemann: Die Pflanzenstoffe in chem., physiol, pharmakol. u. toxicol. Hinsicht (Berlin, 1871).

[2] See the summary in Husemann. op. cit., pp. 235–242.

nausea of pregnant women has also been found to be subdued by it when other medicines were of no avail; and relief has been given in the persistent vomiting which accompanies sea-sickness. In cases of amenorrhœa of long standing, I have known it to be an efficient emmenagogue, even when pulsatilla and other remedies obtained from ranunculaceous plants have proved useless. Van Praag,[1] on the score of some careful observations, recommends delphinine in acute rheumatism. It appears to act also as a sedative to the heart and to the muscular and nervous systems. Van Praag gave it in doses of about $\frac{1}{16}$ to $\frac{1}{8}$ grain, or less, three times a day.

It is in the form of a tincture that the seeds of this plant deserve special consideration. In obstinate neuralgia, affecting the facial and superficial spinal nerves of the neck, the tincture has frequently effected a cure when all other remedies have failed. In rheumatism it has been strongly recommended, though, as far as my own personal experience goes, the results in such cases have been far from satisfactory. In periostitis, and shifting pains in the long bones, stavesacre is to be favorably regarded.

Like several other plants of this natural order, staphisagria is useful in certain affections of the eyes, and especially in ophthalmia tarsi. It checks the superabundant secretion of the conjunctiva, the Meibomian follicles, and the ciliary glands, and renders the discharge less puriform. Irritation and itching of the eyes are allayed by it, and in case of ulceration it helps to promote an excellent healing action.

External Use.—Stavesacre seeds reduced to powder are valuable in scabies, fungous ulcerations, and humid sores of analogous character, and, above all, as an instrument for the destruction of pediculi in the head, whence their popular name of "lice-bane." To be used efficiently for this purpose, after being pulverized, the material should be incorporated with ordinary hair-powder. When employed for external purposes, the alcoholic solution (one part in sixty) is to be resorted to. It may be dropped into the cavity of an aching tooth; for neuralgia it may be painted upon the skin; and in rheumatism an ointment or liniment of similar strength may be employed.

PREPARATIONS AND DOSE.—Tincture of the seeds, dose 5 to 20 minims; decoction of the seeds, dose 1 to 2 ounces, as an anthelmintic; ointment of the seeds; liniment of the seeds; the alkaloid delphinine, dose $\frac{1}{16}$ to $\frac{1}{8}$ grain. None of these are officinal in the U. S. Ph.

CIMICIFUGA (C. RACEMOSA, ACTÆA RACEMOSA).

ACTIVE INGREDIENTS.—No perfectly satisfactory analysis of cimicifuga has yet been made, but it seems probable that there is no alkaloidal principle contained in it; and, on the other hand, that there are a volatile oil and two resins. A resinoid extract can be obtained by precipitation of the alcoholic tincture with water.

PHYSIOLOGICAL ACTION.—This is not by any means well understood. Among the various accounts given by Chapman, Davis, Young, etc., there is nothing very clear, except that in large doses cimicifuga pro-

---

[1] Arch. Path. Anat., 1854; Husemann, op. cit.

duces vertigo, dimness of vision, and a depression of the pulse, all of which symptoms are somewhat persistent. Dr. Young speaks rather of vague, uneasy feelings of the limbs than of these more positive symptoms, and says that it does not cause purging, vomiting, or diuresis.

THERAPEUTIC ACTION.—Had only a portion of its reputed properties been proved to exist in cimicifuga, we should be justified in regarding it as a medicine of great energy and importance. That it is of service in the treatment of humoral asthma, catarrh, and similar affections, is indisputable; and, as stated by Stillé, the success of cimicifuga as a popular remedy for chorea led to its being employed in the regular professional treatment of that disease. The last-named author goes on to say that it is one of the most valuable remedies which can be employed in chorea, the cases to which it is most peculiarly adapted being those in which the nervous derangement is independent of any definite disease in other parts of the body. Stillé insists also upon the importance of the remedy being exhibited in doses sufficiently strong for its specific effects to be developed, such, for example, as vertigo and confusion of sight. This idea is fully corroborated by other practitioners, since, when the administration of cimicifuga is prescribed by them, we find *large doses* recommended, its effect upon the head being considered as the test of the extent to which the medicine should be given.

Many writers upon cimicifuga give as their experience that they found no increase caused by it in any of the secretions, and that it does not possess stimulating qualities. But these conclusions must not cause surprise, since, before one can rely with any certainty upon the accomplishment of wished-for results, we must satisfy ourselves that there is some kind of intelligible relation between the drug and the disease which we hope to efface by the use of it; and this may not have been done by the writers who seem inclined to deny the efficacy of cimicifuga. For instance, if we wish to observe the beneficial effects of this medicine in cardiac affections, it must be administered in those cases where the heart and the pulse beat intermittently. And if we desire to witness the so-called diuretic effects, it must be given in cases of irregular action of the heart accompanied by general anasarca, where the secretion of the kidneys is very scanty. Here I may remark that the action of cimicifuga upon the heart and kidneys forcibly reminds us of the power which, under similar circumstances, is exerted upon those organs by infusion of digitalis. That the action of cimicifuga upon the heart is strongly stimulating and tonic, there can be no doubt; though this may be partly attributable to the action being exerted through the medium of the nervous system, since we frequently observe this medicine to prove singularly efficacious when the pulse is quick, or even *too slow*, with frequent intermissions, and accompanied often by much dyspnœa, and cold and rather clammy perspirations. I have frequently seen it relieve most distressing dyspnœa, when the heart has been weak and its action irregular, and this even when digitalis itself had failed to be serviceable.

In general anasarca, attending this same condition of the heart, when the urine is scanty, the pulse slow and irregular, and the breathing much oppressed, so that the patient is for nights together quite unable to lie down, I have prescribed cimicifuga with results most favorable in every way, the action of the heart becoming strengthened and quickened, the irregularity of the pulse being subdued, and a copious flow of urine being induced, with consequent early disappearance of the anasarca. It also appears to possess a powerful affinity for the *muscular system*. Hence, we may daily observe examples of its curative efficacy in removing localized rheumatic affections, such as wry-neck, lumbago, pleurodynia, and intercostal and abdominal rheumatism. It is valuable also in alleviating the pains which arise from rheumatism of the uterus.

Rheumatic Fever, etc.—In rheumatic fever I have, upon several occasions, found this drug very serviceable. It has quickly relieved the pain, and subsequently reduced the frequency of the pulse, causing profuse perspiration, and often changing the quality of the perspiration, which under the influence of cimicifuga has become less acid. In colds of a rheumatic and catarrhal kind, when the patient suffers from aching, or from severe pain in the upper or the lower jaw of a neuralgic character, and often attended with coryza or sore-throat, it is again to be regarded as an efficient means of relief. In rheumatic neuralgias generally, and in many of those nervous headaches familiarly termed "sick," and which baffle all the accustomed methods of treatment, it is also to be strongly recommended. Ordinary rheumatic headaches give way to it readily, and so do the headaches which occur during or about the menstrual period. Its utility in dispelling the pains attendant upon rheumatism of the uterus I have already mentioned.

Disorders of Females and Childbirth.—I wish to lay stress upon the influence which this medicine possesses in cases of spinal irritation, aggravated during the menstrual period, or very soon afterward, and which depends in all likelihood upon some sympathetic or rheumatic affection of the uterus. Several such cases have come under my personal observation, and I have had sincere pleasure in watching the salutary effects of the medicine as an alleviant of the symptoms most trying to the patient. Upon the uterus the effects of cimicifuga are truly remarkable, as also in puerperal mania and in peritonitis, especially in the rheumatic form of the last-named disease. I have known it check ordinary menorrhagia when the discharge was of a passive character, coagulated, and dark, the action again reminding us of that of digitalis in controlling this very troublesome disorder. My opinions in regard to its employment, in the class of disorders to which I am referring, are wholly borne out by Dr. Hale, of America, who tells us, in his work upon "New Remedies," that in the estimation of the Eclectic school of medicine no agent stands higher among those adapted to the treatment of diseases peculiar to women than cimicifuga, and its concentrated principle, *cimicifugin*. By the Eclectic physicians of North America it is constantly resorted to in uterine complaints, including some of the most opposite character. Dr. King, one of the best of the American [Eclectic] authorities, describes it as being useful in amenorrhœa, dysmenorrhœa, leucorrhœa, etc., and, as a *partûs accelerator*, he asserts that it may wisely be substituted for ergot, bringing on the expulsive action of the uterus both speedily and energetically; and, lastly, he assures us that after labor it will be found effectual in allaying the general excitement of the nervous system, and in relieving the after-pains. These results have the advantage over those attained by the employment of ergot, in being unattended by the powerful and continuous contractions of the uterus which ergot occasions, and consequently there is less danger to the child. Ergot, moreover, lessens the susceptibility of the organ to subsequent doses, should they be needed; whereas with cimicifuga there is no such result—the action of the last-named medicine being to excite the uterus freely and normally, but neither imperiously nor in such a way as to interfere with renewed use. Like ergot, cimicifuga is again useful in checking the hæmorrhage which follows parturition, especially when tediously prolonged, and either profuse or otherwise.

As would be expected from what I have said, it has proved itself a medicine of singular value, not only in connection with childbirth, but in puerperal hypochondriasis. A remarkable instance of this is reported by the late Sir James Y. Simpson, Bart. A lady, he tells us, the mother of several children, was twice the subject of the most painful mental despondency, commencing about a month after her confinement. Upon one of these occasions her delivery had taken place in London, where she had the advice of several eminent physicians; but her complaint took a very long and tiresome course, seeming to defy all remedies, and terminated only very gradually and slowly. It was arranged, upon a subsequent occasion, that the confinement should take place in Edinburgh, in order that the patient should be placed in the care of Sir J. Y. Simpson. Her child was born, and some weeks subsequently she returned to England, apparently in perfect health. At this point begins the very interesting aspect of the case in question. In the course of another month the lady again went to Edinburgh, but now in the lowest possible state of depression, a perfect embodiment of mental misery and unhappiness. "I tried many plans," says the eminent authority from whom I derive this account, "to raise her out of this dark and gloomy state;—all failed. At last, fancying from some of her symptoms and complainings that there might be a rheumatic element in the affection, I ordered her fifty drops of the tincture of cimicifuga daily. After taking one dose, she refused to continue the medicine, the drug having a taste similar to that of laudanum, and all opiates having a tendency to make her worse. On being assured that there was no opium in the medicine, she recommenced it, without any expectation, however, of good resulting to her, having lost faith in every species of medicine. When I next saw her, some eight or ten days afterward, she was changed in a marvellous degree, but all for the better. On the third or fourth day after resuming the drug, so she informed me, the cloud of misery which had been darkening her existence suddenly began to dissolve and disappear, and in a day or two more she felt perfectly herself again, in gayety, energy, and spirits. So pleased was she with the effects of the drug, that nothing could induce her to cease from the use of it for a further period of six or eight weeks; and the last time she passed through Edinburgh she told me she had prescribed her new-found and priceless remedy to more than one melancholic subject, with success nearly as great as she had experienced in her own person."

Returning to my own experience of the utility of cimicifuga, I have to add that I have prescribed it with success in suppression of the menses, brought on by cold, and attended by rheumatic pains in the thorax, the lumbar region, and the limbs, and especially when the subject has been one of a nervous habit. In this respect cimicifuga resembles pulsatilla, which itself, as we have seen, is an excellent remedy in the treatment of suppression of the lochia or of the catamenia.

Spermatorrhœa, etc.—I have prescribed cimicifuga for patients suffering from spermatorrhœa and nocturnal emissions, and who experienced the very usual accompaniments of melancholic hypochondriasis, and in most cases the effect of the medicine was highly beneficial. I am persuaded that it is an excellent tonic for the nervous system, and that its operation is good in delirium tremens, for which I have several times prescribed it. It has been used for these disorders in America, and with results of the most satisfactory nature. One recommendation, when compared with digitalis, is that it does not interfere with the digestive system in the way that foxglove does, but, on the contrary, rather increases and strengthens the appetite. The tremors and vertigo it certainly removes in a short space of time. While under the influence of cimicifuga, the patient often sleeps calmly, and awakes refreshed.

Hysterical Chorea.—In conclusion, I may mention that I have seen the effects of this medicine excellently illustrated in cases of long-standing hysterical chorea. In several instances the disease yielded to it rapidly. Such cases, as a rule, are attended by menstrual irregularity, and the patient often suffers from intercostal, or mammary, or uterine pains, of a rheumatic type. I recently attended a young lady of eighteen, who suffered from chorea ensuing upon an attack of rheumatic fever, with cardiac complication, which rapidly gave way before tincture of cimicifuga —a case the more important to be cited at the present moment, since the form of chorea with which I then had to deal is very often obstinate, and yields only with much difficulty to medical treatment, baffling, in many instances, all known agencies. Sympathetic pains and neuralgias, which arise from the so-called "irritable uterus," no matter what their precise character, are quickly relieved by tincture of cimicifuga—again proving this medicine, though hitherto so little regarded, to be one of the most valuable that we have at command.

PREPARATIONS AND DOSE.—The only officinal preparation is the extractum cimicifugæ fluidum, 3 ss.- 3 j. (2.–4.). Dr Phillips recommends a tincture (four ounces of dried root to pint of spirit), dose ten to fifty minims. Commercially there is a tincture (one part fresh root to two parts of alcohol) and also a mixture of the active principles obtained from the dried root, and sold under the names of cimicifugin or macrotin, the usual dose of which is gr. j.–ij. (.06–.12).

SOLANACEÆ.

BELLADONNA (ATROPA BELLADONNA).

ACTIVE INGREDIENTS.—It is now pretty certain that the poisonous action of belladonna leaves and root is dependent entirely upon the presence of the alkaloid atropia ($C^{17}H^{23}NO^3$), which forms silky, glistening crystals, in bundles of rods and needles (four-sided prisms), heavier than water, odorless, and with a very unpleasant bitter taste that lingers on the tongue. It has an alkaline reaction, and requires 300 parts of cold water, 58 parts of boiling water, 3 parts of chloroform, 40 of benzole, or 30 of ether, to dissolve it. Rectified spirit dissolves it completely, and amylic alcohol almost with equal facility. Its solution rotates the plane of polarization feebly to the left. Atropia is supposed to be identical with daturia, the active principle of stramonium, and probably of other species of datura; only, according to Schroff, daturia is much the more energetic of the two alkaloids. [There is a second alkaloid belladonnine, discovered by Hübschmann in 1858, about which very little is known.]

Something more, however, must be said respecting the active properties of the juice of the ripe fruit. There are reasons for thinking that this juice acts upon the nervous centres as a pure narcotic, and is devoid of the acrid properties of the preparations which are made from the leaves and the root.

In its effects upon the organs of deglutition the poison of belladonna berries seems to claim for itself a certain analogy with that of hydrophobia. All the phenomena are curious, and demand that further experiments should be made with the fruit (as distinct from the leaves and root), with the view to determine the therapeutic value of the former. The fruit, as might perhaps be expected, seems to promise a soothing narcotic medicine, the action of which would be unaccompanied (in general) by results of an irritative nature.

PHYSIOLOGICAL ACTION.—Belladonna is both locally and generally an "acro-narcotic," *i. e.*, it is capable of producing both narcotic and irritant effects. Scarcely a subject in the whole range of drug-action has been more elaborately discussed than the poisonous effects of belladonna. It will be well first to describe the more immediately obvious phenomena, as to many of which there cannot be dispute.

Like all the mydriatic solanaceæ, *i. e.*, those which characteristically dilate the pupil, belladonna produces a number of effects which group themselves around this well-known symptom, and seem to be inseparably connected with it. Mydriasis may indeed be produced alone by acting locally upon the eye with atropia, etc.; but when it arises as an effect of the internal use of belladonna or atropia, it marks the commencement of a very definite group of poisonous phenomena: dryness and heat of the fauces and pharynx, headache, flushing of face, dimness of vision, or actual amblyopia, and (if the poisoning be carried far enough) delirium, which is often of a noisy character, with bursts of laughter, but, at any rate, always busy, and usually attended by spectral hallucinations.

It is not to be supposed that all these symptoms necessarily occur when dilatation of the pupil is produced by internal doses of belladonna; but mydriasis, caused in that way, is always a sign that the other symptoms named are at hand, and would at once follow any considerable increase of the dose.

Among the phenomena of this stage of belladonna-poisoning is an appearance about which there has been considerable dispute, viz., scarlatina-like redness of the skin. That a more or less diffused erythematous blush upon the skin has frequently been noticed, is admitted by nearly all observers; but great difference of opinion has existed as to the proportion of cases in which it occurs, and also as to its persistence, and the extent to which it may spread. Dr. John Harley,[1] for example, says that "generally it is nothing more than a mere temporary blush; but in rare cases, and in persons who are liable to vascular irritation of the skin, the redness remains, and its disappearance is attended by slight roughness and desquamation." He himself mentions two cases, in one of which the patient was "scarlet from head to foot;" and in the other, after the fourth dose of the drug, there was a "scarlatinous tint of skin." Meuriot says that there is a contrast between the cutaneous and the mucous phenomena; the mucous membranes, *ten minutes* after a subcutaneous injection of atropia, are red, injected, and dry, while the skin is *pale* and dry, like parchment; sometimes, however, the skin, *soon afterward*, becomes covered with erythematous redness, and is thus uniform with the mucous

---

[1] Old Vegetable Neurotics, p. 223, London, 1867.

membranes. He says that the redness is most frequently seen upon the face and trunk, and does not spread to the limbs.

Many cases are reported in the English journals, and many more in the American ones. Dr. Sinclair's case, in particular, in the *Boston Medical Journal*, is very curious, the rash described by him being remarkably like scarlatina, and appearing to correspond with three instances in which I have myself seen a bright and livid eruption ensue upon the continued use of the tincture. A similar result followed in a case of the accidental poisoning of a child, three years and eight months old, which came before Mr. Holthouse. The dose so unfortunately administered was about half a grain of pure atropia. The child soon became strangely irritable and excited, though unconscious; the face was maniacally distorted; the pupils were widely dilated and immovable; the lids of the eyes were open, and not affected by passing the fingers before them; the pulse was 170, and somewhat feeble; the skin was pungently hot and dry, and "covered," says the historian of the case, "with a scarlatina-like rash." To these examples I may add the testimony of Dr. Garrod, who says that "belladonna produces redness of the skin."

For my own part, I am convinced that the scarlatina-like rash of belladonna-poisoning is far from being an accidental or rare occurrence. I believe it often occurs, and may occur in any case if the poisonous action of the drug be carried far enough. It is no doubt due to capillary congestion, and this congestion may equally invade the skin and the mucous membranes. The dryness of mucous membranes, and the arrest of secretion from various glands, are portions of the same effect. It is chiefly in actually fatal cases that the full development of this symptom is observed. The dilatation of the vessels which causes it is by some authors[1] considered to be a direct paralytic effect of the agent; by others[2] it is supposed to be paralysis following on a primary contraction.

The further symptoms in fatal cases of belladonna-poisoning are swelling of the face, protrusion of the eyeballs, great inflation of the conjunctiva, coma, and convulsions; besides which narcotic phenomena, there are usually signs of irritation in the alimentary canal, pain in the belly, nausea, vomiting, and occasionally diarrhœa, which justify the designation of the drug as an "acro-narcotic." As regards the effects of belladonna on the pulse, there is a rather remarkable conflict of evidence. While many authors (especially Schroff,[3] who made some 1,200 experiments) state that the pulse is at first diminished in frequency, and afterward (in the case of large doses) accelerated, Meuriot found constantly that in from eight to ten minutes after a subcutaneous injection of atropia there was an acceleration of the pulse, which lasted for one or two days, after which there followed a retardation. My own experience corresponds rather with the latter statement; but it is probable there are great differences in the affection of the pulse, according to the dose and the circumstances under which it is taken.

A summary of the more exact researches which have been made to determine the special physiological actions of belladonna and atropia on particular portions of the nervous centres, might lead us approximately to the following conclusions:

1. Large doses paralyze the peripheral ends of the motor nerves in

---

[1] Von Bezold and Bloebaum: Würzburg. physiolog. Untersuch. i., p. 1, 1867
[2] Meuriot, op. cit.
[3] Lehrbuch der Pharmacologie.

striped muscles and the peripheral ends of the sensory nerves in the skin; the muscular irritability proper remaining intact.

2. Atropia depresses and in large doses paralyzes the peripheral ends of the cardiac branches of the vagus and the motor cardiac ganglia, and also depresses the irritability of the heart walls; it has no influence upon the *cardiac depressor* nerve.

3. Upon the vaso-motor nerves atropia acts in such a manner as to produce ultimate vascular dilatation; but whether this is direct, or secondary to contraction, is still in dispute.

4. Atropia paralyzes the terminals of the vagus in the lungs, but only temporarily, and increases the excitability of the inspiratory centres.

5. Atropia, in very small doses, diminishes excitability; in larger doses it produces paralysis in the ganglionic apparatus of the intestinal canal, the bladder, the uterus, the ureters, and possibly in the unstriped muscular fibres themselves.

6. Atropia paralyzes the restraint influence of the splanchnics on the motor fibres which perform intestinal peristalsis,[1] while all the other muscular fibres of the intestines are unaffected.

7. It paralyzes the restraint nerves from the chorda tympani to the sub-maxillary gland.[2]

Besides the above, which are phenomena of general belladonna-poisoning after absorption into the blood, there are some local influences which require mention. The local influence of atropia, applied directly to the eye, upon the pupil, has been noticed. The dilatation of the pupil is strictly one-sided, and usually there are no phenomena of irritation. But occasionally (as was first pointed out by George Lawson[3]) the local application causes irritative symptoms, which are apparently due to peculiarities in the subject of the medication; these are redness and chemosis of the conjunctiva, erysipelatoid swelling of the lids, lachrymation etc. More decided irritation has been noticed by Meuriot, after the application of collyria containing atropia, in the skin of the face, eczema, boils, etc. A more remote action (after the application of atropia to the eye), which seems to be conveyed through the nerves of the part to which the substance has been applied, is instanced by dryness and redness of the throat in man, and by salivation in cats. As regards the local action of belladonna upon the sensory nerves, we have the statement of Fleming that the inunction of atropia upon the unbroken skin produces no benumbing effect. But Erlenmeyer showed that the subcutaneous injection of atropia produces marked and rapid lowering of the cutaneous sensibility in the neighborhood of the puncture.

[The quantity of atropia required to dilate the pupil has been estimated by several observers. Dr. H. C. Wood states that a drop or two of a solution containing $\frac{1}{30}$ grain to an ounce of water is sufficient in many cases. One drop of such a solution would contain about $\frac{1}{10000}$ of a grain; but as the entire drop is probably not absorbed, the actual amount of effective atropia is indefinitely less. Dr. D. B. St. J. Roosa (manuscript communication to editor) states that he has seen dilatation result from $\frac{1}{10000}$ of a grain, and Dr. Ely from $\frac{1}{40000}$. Trousseau and Pidoux refer to an instance in which a dog's pupil was dilated for eighteen hours by the

---

[1] [Other authors state the contrary. Vide Ott: Action of Medicines, p. 141.]
[2] The above summary, which is only approximate, is taken from Husemann (Pflanzenstoffe), and is based on the experiments of a large number of authors.
[3] Ophth. Hosp. Reports, ii., 2, 119.

$\frac{1}{128000}$ of a grain. Lastly, Dr. E. G. Loring, of this city, states (manuscript communication) that he has dilated his own pupil for twelve hours with the $\frac{1}{180000}$ of a grain.]

Post-mortem examinations of the bodies of persons who have been poisoned by belladonna show that putrefaction commences very soon after death. The smell is peculiar and intolerable, and the skin is covered with livid spots, while blood escapes from the mouth, the nose, and the eyes. Should the whole substance of the berries have been swallowed (as usually happens in such cases of poisoning), they are found to be very imperfectly digested, in consequence of the poison inducing extreme torpor of the stomach. The heart and lungs are livid; the latter are usually gorged with venous blood, and studded with black spots; and the blood itself is in an abnormal state, seeming to be dissolved.

In cases of the accidental swallowing of belladonna fruit, the first step should be to excite vomiting by means of small doses of sulphate of copper, or sulphate of zinc, repeated as often as may be needful. But care must be taken that these remedial agents do not themselves accumulate in the stomach, the torpidity of which is one of the special difficulties to contend with in belladonna cases. Sometimes the torpidity is so profound that to obtain an emetic action from the employment of any drug whatever is next to impossible, and the stomach-pump may not act well in emptying the stomach. Should the stupor induced by the poison be very alarming, it is necessary to relieve the blood-vessels of the head by opening one of the jugular veins, and by cold affusions. By adopting these measures, the stomach is likewise enabled to recover in some degree from its torpor, and the action of the emetic is considerably assisted. Stimulants may also be applied to the eyes and nose, and sinapisms to the feet; and friction may be used over the region of the heart. When, by the judicious administration of emetics, the stomach has been emptied, vinegar or some other vegetable acid may be given to the patient, and be followed up with diluents and saline purgatives.

The above remarks apply chiefly to poisoning with the fruit of belladonna, which is the most common accidental form, at any rate in country-places. But where the poison taken is an energetic preparation of the leaves or root, and we have consequently to deal with atropia in such a form as must be very rapidly absorbed into the blood, it is necessary to think of possible further antidotal treatment. Under these circumstances we shall have to turn either to morphia, or to physostigma, the supposed antidotal powers of which toward atropia will be discussed when I come to speak of these two agents respectively.

THERAPEUTIC ACTION.—It is a maxim in therapeutics that we may expect to find our most potent remedies among the deadliest poisons; and belladonna, which has been shown to exert such active and all-pervading poisonous influences upon the organism, affords a strong illustration of the truth of this saying. Its great energy makes it a somewhat uncontrollable remedy, even in the hands of those who have long familiarized themselves with its employment; hence by many practitioners it is seldom resorted to until other remedies have been vainly tried. The variety and certainty of its remedial effects are, however, so well established, that its intractableness *per se* should not be allowed to weigh as an objection to its use; rather ought we to strive more diligently to discover the laws of its therapeutic action, in order that it may be employed with more promptitude and confidence. That in special cases, or in particular temperaments, belladonna may seem not to justify its ancient repute, is quite possible. But, on the other hand, there are plenty of instances in which it can be pointed to as efficacious when other drugs fulfil no good purpose. In estimating the value of a medicine we are not to judge from negatives, which are far more likely to mislead as to its virtues than positive results, when the latter are obtained under due precautions against fallacy. It is scarcely necessary to remark that hardly a medicine can be named which is invariably efficacious, and in a uniform manner, even when the constitutions of the patients, and the symptoms, seem identical; and if belladonna does not at all times and under all circumstances accomplish what is justly expected from it, such failure is certainly not a defect peculiar to this remedy, or without a parallel.

In Pain.—One of the most characteristic effects of belladonna is its power to relieve certain kinds of pain. I say "certain kinds" (a necessary qualification), for belladonna by no means ranks with opium in quality as a universal anodyne.

It is interesting to note that the ascertained physiological action of the drug might prepare us for the therapeutic facts which we meet with in practice. Belladonna (according to Lemaire, Meuriot, and Botkin) in poisonous doses paralyzes the peripheral ends of the sensory nerves. Now it is a fact that the pains which we relieve with the greatest certainty by the use of belladonna are those which depend entirely or chiefly upon peripheral causes. Thus the pain of inflamed parts, especially gouty and rheumatic inflammation, can often be more speedily and effectually soothed by this remedy than by any other. And among neuralgic pains those are by far the most frequently and effectually relieved the main source of which is some peripheral disturbance. Belladonna is much more serviceable, for example, in the various painful affections which are produced by an irritated state of the pelvic organs (especially in females) than in neuralgia of the face.

At present there is a tendency to limit the employment of belladonna in cases of true neuralgias to the use of sulphate of atropia in hypodermic injection. A solution should be employed which contains one grain in two drachms of water; each minim thus contains $\frac{1}{120}$ grain; and it is advisable to begin with no larger dose than this; for though the great majority of patients will require, and bear quite well, a large dose equal to twice or three times this or more, yet we never know that the actual subject we are treating may not belong to that exceptional class with whom very minute doses cause unpleasant symptoms of atropism: *e. g.*, dry throat, disturbed vision, uncomfortable heat of body, headache, and even delirium; and it is very desirable to avoid the production of symptoms so alarming to the patient.

The local use of belladonna in the form of lotion, liniment, ointment, plaster, etc., has been found of eminent service in the relief of the dreadful pains of *cancer*. I have myself witnessed great benefit in cases of scirrhus, and also in cancerous and other painful ulcerations, from the use of a lotion composed of two drachms of extract of belladonna to a pint of water, the application being made two or three times a day; and if at the same time the tincture be given internally in five to ten drop doses, at intervals of a few hours during the day, the sufferings of the patient are still further alleviated. The pain of scirrhus of the pylorus, a very distressing malady, has been greatly relieved by the application of belladonna plaster to the epigastrium; and both in sciatica and lumbago local application of the plaster has done much good.

In Spasms.—The action of belladonna in relieving spasms, while very decidedly established as regards certain classes of affections, is more in dispute with regard to others. There is no doubt, for instance, that the local application of belladonna will calm spasms of particular muscles; and in this manner it comes to be of the greatest service in particular cases; as, for example, in the painful spasms of the sphincter which so aggravate the misery of fissure, or irritable ulcer of the rectum. Moreover, in spasmodic stricture of the urethra I have usefully assisted the effect of belladonna given internally by applying the fresh extract to the penis. The internal use of belladonna has a wider and more beneficial antispasmodic action; though there are very great differences in its relative success in particular cases.

The pelvic organs are an especially favored site of the antispasmodic action of belladonna, as in cases of spasms of the bladder, in which it is most useful. When there is simply much vesical irritability, with frequent micturition, without organic change, the tincture in five to twenty drop doses every three or four hours will give gradual but sure relief. If organic disease be present, the question of belladonna is more complicated, and a variety of individual circumstances must decide whether it shall be used. The power of belladonna to relieve general convulsions is much disputed, but upon this point I hold a strong affirmative opinion.

In General Convulsions.—Belladonna is useful in epilepsy, and especially when the attacks are brought on by fright. Trousseau long ago recommended it, while prescribing the treatment to be adopted for this disorder generally, though he does not specify the particular form of the disease in which the belladonna would prove most useful. "During the first month the patient takes a pill composed of extract of belladonna and of the powdered leaves of belladonna, one-fifth of a grain of each every day, if the attacks occur chiefly in the day-time; but in the evening, if they be chiefly nocturnal. One pill is added to the dose every month, and, whatever be the dose, it is always taken at the same period of the day. By this means the patient may reach the dose of five to twenty pills, or even more."

In enumerating the circumstances under which belladonna is specially useful, it is important not to overlook *puerperal* convulsions. The value of its action in many cases of the occurrence of such convulsions I can testify; but long before my own time it had been pointed out by M. Chaussier, and by various other distinguished practitioners. M. Chaussier, so far back as 1811, was accustomed to apply belladonna ointment to the uterus. This first and experimental employment of the drug for the purpose under consideration was followed, however, by uncontrollable uterine hæmorrhage, with other perilous symptoms, and the propriety of resorting to it was warmly and very reasonably disputed. Of late years quite a different method has been adopted, and to this there appears to be no possible objection, while the advantages are obvious. I refer to the internal administration of the tincture and the hypodermal injection of atropia, and it is of these that I desire to speak in terms of approval, their utility having been verified in my own experience.

In connection with this subject, I may remark that as a topical remedy belladonna is unquestionably of great service in those long-protracted and wearisome labors which arise from rigidity of the os and cervix uteri. In these cases, however, it is now seldom employed; other remedial agents usually claiming the preference. Dr. Conquest recommended that in such labors as those referred to, about half a drachm of the extract should be rubbed into the neck and mouth of the womb; the result of the application being, he assures us, that unproductive uterine action is suspended, and that there is an immediate relaxation of the parts, so that upon the recurrence of expulsive pains the *os* yields, and passage is allowed to the head of the child.

In all hyperæmic conditions of the brain and of the spinal cord, belladonna is one of the best medicines we can have recourse to. Among many admirable results which ensue on its employment, I may specify its beneficial effects in certain congestive kinds of convulsions, especially in fits produced by the irritation of teething. This kind of congestive convulsion may arise in various constitutions; for example, it may occur in apparently robust and healthy children; and, at the other extreme of the scale, it may frequently be met with in scrofulous and feeble children, with precocious development of cerebral activity; but, in fact, it may occur in all shades of temperament. In all these varied circumstances—granting a congestive origin—belladonna is of the highest possible utility; and its employment cannot be too highly recommended.

So with convulsions referable, proximately, to whooping-cough, and which often depend, like the former, upon a congested condition of the brain. In these, just as when induced by the excitement arising from dentition, belladonna rarely fails to give relief. On the other hand, I may add that when convulsions arise from irritation of the intestinal canal, the use of belladonna is relatively small.

Nocturnal Enuresis.—The very unpleasant nocturnal enuresis, to which children are so liable, is very usefully treated with this drug. I must acknowledge that I formerly experienced much disappointment in the effects of belladonna in this malady; but the failures were probably due to my using an insufficient dose—only five minims three times a day. It is only lately that I have been led to employ larger doses; the effect of these has given me confidence in the efficacy of the treatment.

Dr. Sydney Ringer states that, "in the incontinence of urine in children, belladonna acts more speedily and efficiently than any other medicine;" but that in order to be successful, it must be administered in full doses, namely, ten to twenty drops of the tincture thrice a day. "Small quantities," he continues, "often fail, while large ones immediately succeed." This positive assurance seems to explain my own ill-success when the quantity employed was only five minims.

Obstruction and Constipation.—Under this heading may be included the very remarkable action of belladonna in removing constipation or even positive obstruction of the bowels, which depends upon want of tone (mingled sometimes with partial irritative spasm) of the muscular coats of the intestine. The practice was introduced by Trousseau, who gave doses of $\frac{1}{4}$ to $\frac{3}{4}$ grain, repeated every few hours; and since that time it has been recommended by many authors. Fanciful attempts have been made to explain this action by the supposition that the belladonna suppresses the mucous secretions, and allows the fæces to come into direct contact with the mucous membrane, and thus to stimulate the bowel to act. There can be little doubt, however, that the drug acts directly upon the bowel as a stimulant and a co-ordinator of muscular action; and thus (secondarily) as a controller of any irregular (spasmodic) action that may be present.

Inflammation.—Besides its power to relieve functional affections, belladonna exerts a powerful influence over certain actual tissue diseases. Many forms of inflammation are amenable to its power, and among these may be mentioned certain diffuse inflammatory affections of the skin.

In cases of erysipelas, when the skin has been of a deep red hue, with accompanying heat, pain, and swelling, and when the constitutional disturbance has been considerable, I have administered this medicine with results that were perfectly satisfactory. Erysipelas exhibits itself under a diversity of forms, and is not amenable to any uniform mode of treatment. It becomes important, accordingly, that we should carefully note which of the forms give opportunity for the wise employment of belladonna—since I by no means wish it to be supposed that I am recommending belladonna for erysipelas without reservation. The forms in which belladonna may be employed with advantage are such as are marked by superficial inflammation—inflammation, that is to say, which does not much affect the subcutaneous areolar tissue, and in which the surface is free from vesicles. When these are the conditions, belladonna will rapidly quell the disorder. So when erysipelas attacks the brain. The power which this drug exerts in abating the delirium, and in promoting the quick subsidence of the disease, is manifested so energetically, and in so short a space of time, as to be quite astonishing. In erysipelas of a phlegmonous character, before suppuration or sloughing takes place, belladonna is likewise of considerable service; here, however, the operation of the drug cannot be relied upon with the absolute certainty that pertains to its use in the superficial and non-vesicular kind, and therefore I do not wish to say more than that in such cases we may at all events place faith in it. For erysipelas, five drops of the tincture should be administered in a little cold water every hour, to the extent of five or six doses; afterward every two or three hours, as may be deemed necessary, and at the same time the extract may be painted over the surface twice a day, if needful. [Mr. Liston called attention to this use of belladonna forty years ago (*Lancet*, April 16, 1836). He employed it alone, or in conjunction with aconite.]

Let me now speak of belladonna in connection with sore-throat. In all those forms of inflammatory sore-throat which have a prelude of more or less fever, with pain, redness, and swelling of the tonsils, and which are attended by difficulty of deglutition, belladonna operates with surprising efficacy. Five drops of the tincture, taken every one or three hours in half an ounce of cold water, will quickly remove the inflammation, and the other symptoms will soon disappear. In all ordinary cases of inflammation of the throat, whether accompanied or not by ulceration, this treatment will likewise prove successful, though altogether useless in sore-throats of the aphthous, diphtheric, and syphilitic kinds.

Catarrh of the bladder is another complaint which is very successfully treated with belladonna, the good effects being remarkable and permanent. The cases have been recent ones, and, where the medicine has shown itself to be most efficient, they have been such as have resulted from a chill, and have been attended by more or less pain of an undefined character in the hypogastric and perineal regions, and by frequent and painful micturition. Notwithstanding the severity of the symptoms ordinarily attendant upon this malady, belladonna overcomes them. The dose should be ten drops every one to three hours, in a little cold water, according to the condition of the patient; but as soon as relief has been obtained, the quantity should be reduced, and the intervals between the doses be extended. The perinæum may be smeared, night and morning, with a small quantity of the extract, the external effect well supplementing the internal one.

Belladonna will also relieve the violent contractions of the muscular coat of the bladder which so frequently accompany diseases of this organ. When the bladder is simply in a state of great irritability, and micturition is both frequent and painful, but where no organic change has taken place, belladonna, if given in five to twenty drop doses, every three or four hours, will soon bring relief.

In the early stages of encephalitis, and during the whole period of excitement, belladonna again shows its powers to advantage; the severe headache is relieved, the suffusion of the eyes is diminished, the delirium abates, and the nausea and vomiting are often rapidly subdued. Light and sound at the same time become less distressing to the patient, and the general result of the employment of the medicine is that the attack does not lead, as otherwise it probably would, to a state of great prostration, accompanied by unmistakable typhous symptoms. Usually, also, there is a complete absence of certain symptoms which ordinarily characterize the later stage of the disease, and some of which are occasionally manifested in the early one; namely, muttering delirium, strabismus, tremors, twitchings of the muscles, and incontinence of urine. In spinal meningitis, as a matter of course (the membranes being continuous with those of the brain), corresponding results may be confidently anticipated.

In cases of inflammation of the spinal cord and its membranes, when brought on by external violence, I am again disposed to rely strongly upon the efficacy of belladonna. Quite recently I have seen two patients suffering from this disorder, and the agony resulting from the violence of the pain in the lumbar region, whence it extended to the hips, the thighs, and round the abdomen, was simply indescribable. The pain was aggravated by pressure upon the lumbar vertebræ, and the pulse was quick and full: temperature 104°. Opium was administered in various ways, but the relief given by it was only temporary, and not until belladonna was tried had either patient any permanent alleviation from his sufferings. I ordered the tincture to be taken in five-minim doses in an ounce of water every half-hour or hour until relief should be experienced, and that, while the medicine was in course of operation internally, a belladonna plaster should be applied to the injured portion of the spine. The result of this treatment was that the pains were in a short time entirely subdued, and that the continuance of the medicine was rendered needless.

Pneumonia, Nephritis, and Iritis.—In the early stage of pneumonia, and in acute nephritis, belladonna has frequently proved useful. I have likewise successfully employed it in iritis. Taken internally in iritis, in five-minim doses, every three hours, and used as a lotion three or four times a day, in the proportion of one drachm of the extract to four ounces of water, belladonna not only dilates the pupil and breaks down the adhesion, but subdues the inflammation, helping in a two-fold manner to correct the mischief.

Belladonna is not merely useful in ordinary inflammations, but has a remarkable influence over ulcerative processes of various kinds. The extract, locally applied, will heal irritable ulcers of the rectum and fissures of the anus; or, if it fail to accomplish this end completely, it will, at all events, greatly mitigate the pain and trouble which arise from them. In fissure, the purpose may be accomplished in an easy and very satisfactory manner by passing a small portion up the back part of the rectum, the operation being repeated night and morning. For the removal of the ulcer the tincture should be administered internally, and the extract be smeared in small quantity upon the affected surface.

Belladonna is also an excellent agent for the mitigation, and even for the complete removal, of the acute and burning pain which so constantly follows defecation, when there is ulceration of the nature just described; it likewise prevents the recurrence of the distressing spasms of the sphincter ani to which the patient is subject. If it fails, when used alone, to heal the fissure, or to remove the ulcer, the extract of belladonna should be combined with mercurial ointment in equal proportions. The mercurial ointment, employed alone, does not exert the same salutary effects, nor is relief obtained so readily as when combined with belladonna. Both these last observations will apply also to chronic ulcerations of the rectum, arising from secondary syphilis; that is to say, while for such ulcerations mercurial ointment, employed alone, is of only limited utility, the mixture with it of an equal quantity of belladonna generally gives excellent results.

In recent induration and inflammation of the breasts, very remarkable effects are produced by belladonna, in consequence of its action in arresting secretion.

Among the most practically useful of these is its influence in arresting the suppurative inflammation which is so common in women who have been obliged suddenly to give up suckling. The local application of the ointment, or the fresh extract, together with the internal use of five to ten minims of the tincture, three or four times a day, very quickly relieves the distention and congestion, if the treatment be adopted sufficiently early. Another example of useful arrest of secretion is occasionally seen in the arrest of salivation by the use of belladonna. This has even been known to take place in mercurial salivation, and more decidedly in that which is occasionally produced by iodide of potassium.

In Fevers.—Not only in inflammations, but in certain kinds of fevers, belladonna is of great service.

Typhus fever gives occasion for the useful employment of belladonna. In the early stages of this formidable disorder, with its accompanying delirium, sleeplessness, painful sensitiveness to light and sound, and liability to convulsions, belladonna will reduce the severity of the symptoms. No medicine that I am acquainted with can be so thoroughly relied upon for averting the disastrous consequences which usually follow these head complications. One of the very special phenomena attending the exhibition of belladonna in typhus is seen in the altered condition of the tongue, a point to which John Harley (loc. cit., p. 251) directs attention. From being red and glazed, or dry, brown, and cracked, sometimes destitute of the least trace of moisture, the tongue of the patient suffering from typhus becomes, under the influence of belladonna, moist around the edges, if no farther, and, under favorable circumstances, absolutely wet in every portion. My own experience many years ago perfectly prepared me for Dr. Harley's statement.

Finally, I may allude briefly to several other disorders in which I have proved the efficacy of belladonna, but respecting which there is no need for extended remark.

One of these is **delirium tremens** accompanied by congestion of the brain.

Another is **gout of the stomach**, for which complaint I have administered the tincture in five-minim doses every one or two hours.

Asthma.—Dr. Hyde Salter has made some very important remarks upon the therapeutic action of belladonna in asthma. His observations have been verified by Dr. Sydney Ringer and others, and are well worth careful perusal.[1]

Belladonna has been much employed in whooping-cough, and with tolerable success, but the useful action of many other drugs is manifested more readily.

Contrary to the opinion expressed by Dr. Sidney Ringer, I believe that in the febrile state of the disease the value of belladonna is obvious; and that when whooping-cough is complicated with dentition, the action of this medicine is doubly valuable, the inflammation of the gums being removed. Belladonna relieves the congestive irritability of the respiratory passages; also the determination of blood to the head—both such common accompaniments of whooping-cough.

That belladonna is exceedingly useful when the severity of the convulsive attacks is abating, is generally acknowledged. I believe, for my own part, that it is equally useful during the previous or febrile stage.

The dose should be from three to ten drops of the tincture, administered every one to three hours, both quantity and interval being regulated by the susceptibility of the patient to the action of the drug.

As a Prophylactic against Scarlet Fever.—As a prophylactic against scarlet fever, I may add that, after many trials, I consider belladonna is often efficacious, provided the doses be sufficiently large to cause soreness or dryness of the throat, and provided also that the action be kept up during the time of exposure to the infection. The readiest and quickest process for bringing the individual under its prophylactic influence I believe to be the hypodermal injection of the sulphate of atropine. With children, of course, this is somewhat objectionable, on account of its giving some pain.

[The mooted question as to the prophylactic power of belladonna in scarlatina has given rise to much discussion and also to some very careful and thorough observations. When Hahnemann first asserted this power,[2] but five years had elapsed since the publication of his peculiar ideas concerning the remedial action of drugs. Nevertheless so strong was the feeling against him that his statements in this connection were almost unheeded, and little effort was made either to confirm or refute this pretended discovery, so important if true. Hufeland was one of the first to examine the question experimentally, and as a result gave his adhesion to the affirmative view. Since then much evidence has been collected, on the subject, and the preponderance is certainly in the same direction.

Cazin[3] in 1823 attended twenty-five indigent families, huddled together in contracted quarters at Calais. He admininistered the tincture of belladonna to sixteen children, of whom but one contracted scarlet fever, and this one but lightly, although they were in intimate association with other persons suffering from the disease.

Stievenart,[4] in an epidemic near Valenciennes (1840–'41), which had already claimed ninety-six victims, administered belladonna to four hundred persons, all of whom, without exception, escaped; while many others

---

[1] See also the very interesting paper of M. Sée (Practitioner, July, 1869), in which belladonna is strongly recommended in asthma, and its action explained by its physiological relation with the vagus nerve.

[2] Heilung u. Verhütung des Scharlachfiebers, Nürnberg, 1801.

[3] Les plantes médicinales, 4th Ed., Paris. 1876, p. 165.

[4] Bouchardat : Annuaire, 1845.

in the same locality, and subjected to the same influences, who did not take the drug, contracted the disease. Stievenart employed the tincture of belladonna in doses of two drops daily in children from one to three years of age, three drops to children of from two to six years, and after that age one drop additional for each year. Stillé,[1] after stating the evidence, concludes as follows: " On a review of the whole subject, we feel bound to express the conviction that the virtues of belladonna, as a protection against scarlatina, are so far proven that it becomes the duty of practitioners to invoke their aid whenever the disease breaks out in a locality where there are persons liable to the contagion, particularly in boarding-schools, orphans' asylums, and similar institutions, and among the families of the poor; whenever, in a word, it is difficult to place the healthy at a distance from the sick. The dose which it is proper to prescribe under such circumstances may be thus stated: dissolve from one to three grains of fresh and well-preserved extract of belladonna in an ounce of cinnamon-water, adding a few drops of alcohol to prevent fermentation. This solution may be given two or three times a day, a drop for each year of the child's age, to be so administered for two weeks or longer if the danger should continue."]

Antagonism between Belladonna and other Physiological Agents.—I have reserved to the last the very important subject of the antagonism between belladonna and other physiological agents. Three drugs, at least, are supposed, on more or less convincing evidence, to be antagonized by atropia, viz., opium, physostigma, and hydrocyanic acid.

The antagonism between belladonna and opium first began to be spoken of some twelve years ago; and from that time to the present there has been a steadily accumulating influx of affirmative evidence, intermingled with occasional attacks tending more or less to throw discredit upon the fact asserted. At present the state of the case is such that to me doubt seems no longer possible; and this appears to be the prevailing opinion. No doubt the manner in which the first suggestion of the antagonism was raised was such as might well have misled inquirers; for it was merely the fact that opium contracts and that belladonna dilates the pupils, which gave rise to the idea. It might well have happened that this was the only point of difference between the action of two drugs which, on the whole, were synergic ; and, in fact, this position has been taken up by some objectors. But there is now too large a body of cases recorded in which life has been saved by the use of belladonna under circumstances of severe opium-poisoning (coupled with the fact that belladonna-poisoning has been relieved by opium), to allow me to doubt that there is a general antagonism between the main features of the action of the two drugs. This does not prevent my believing that there may be some, perhaps many, individual points in which the action of the drugs is far from actually opposite, or even that they may occasionally be used in conjunction with advantage. The general view, which is taken up by Dr. John Harley, the principal opponent of the antagonism-theory, is, however, in my opinion, unjustifiable. According to this author, the recorded cases show that belladonna has *no influence whatever* in accelerating recovery from the poisonous effects of opium. He says that somnolency, stupor, narcotism, and coma are both intensified and prolonged by the concurrent action of belladonna; that belladonna is powerless to relieve the depression of respiration, which is the chief danger from opium; and that the results of

---

[1] Therapeutics and Materia Medica, 4th Ed., 1874, vol. i., p. 924.

the combined action of belladonna and opium are the same, whether given in medicinal or in toxic doses: in short, that belladonna will never antagonize opium by its poisonous action, but the reverse; though it is possible that the use of *very small doses* may aid to support the failing action of the heart. I am at a loss to know how these conclusions can possibly have been drawn by Harley from the recorded facts. The cases which he cites, and others which have been adduced, seem to point distinctly enough to a conclusion at least as strong as that expressed by Professor Fraser,[1] that "such knowledge as is already possessed renders it probable that a general antagonism does really exist, to the extent, at any rate, of the primary lethal action of opium or morphia being preventible by the physiological action of belladonna, hyoscyamus, or stramonium." Certainly, in the present state of things, no one would be warranted, I think, in omitting the free use of belladonna in any severe case of opium-poisoning. And even down to smaller (therapeutic) doses there can be no doubt that the general effect of opium on the *blood-vessels*, at any rate, is opposite to that of belladonna (*i. e.*, in doses of the two medicines which could fairly be compared with each other). Very positive testimony to the antagonism of belladonna and opium is given by Dr. Johnston in his Shanghai Hospital Reports, March, 1872.

Much more decidedly made out is the antagonism between belladonna and Calabar-bean, which, by the splendid researches of Dr. Fraser, has now been worked out in a manner that leaves little or nothing to be desired. The general nature of the facts elicited is fairly stated by Fraser[2] in the following words:—"In presence of the many obvious proofs to the contrary contained in this paper, I have considered it superfluous to enter into any discussion of the possibility of this counteraction being the result of some chemical reaction between atropia and physostigma, or of an increased rapidity in the elimination of the one substance produced by the action of the other. The conditions of the experiments, and the symptoms that were observed, render it certain that atropia prevents the fatal effect of a lethal dose of physostigma, by so influencing the functions of certain structures as to prevent such modifications from being produced in them by physostigma as would result in death. The one substance counteracts the action of the other; and the result is a physiological antagonism so remarkable and decided that the fatal effect, even of three and a half times the minimum lethal dose of physostigma, may be prevented by atropia." It may be mentioned also that Professor Bartholow, of Cincinnati (though this was unknown to Fraser), had, in 1869, already investigated the opposition of atropia and physostigma; but he had only been able to find an antagonism between their actions as regards the organic nervous system, and did not reckon atropia an available antidote for physostigma-poisoning. But the elaborate researches of Fraser, especially his more recent ones, leave not the slightest doubt of the practical applicability of atropia for this purpose.

Even more interesting than either of the above-named antidotal actions of atropia, is the power of antagonizing hydrocyanic acid, which was first assigned to it by W. Preyer[3] in 1868. Preyer discovered that prussic acid kills by embarrassing the heart and respiration; and that this ef-

---

[1] On the Antagonism between the Actions of Physostigma and Atropia, by Thomas R. Fraser, M.D., F.R.S.E. (Proceedings of Royal Society of Edinburgh, 1872).
[2] Fraser, op. cit.
[3] Die Blausäure physiologisch untersucht. See Review in Practitioner, vol. i., 1868.

fect is produced by an intense irritation of the inhibitory cardiac, and the pulmonary branches of the vagus. Atropia possesses the power of paralyzing these, and the result of this paralysis is to re-excite the action of the heart after it has been stopped by hydrocyanic acid. It is needless to say that this can only take place when the administration of the atropia is not too long delayed; so that the promptitude it calls for must often hinder the practical usefulness of this remedy. And it is right to mention that the experiments of Preyer as yet want full confirmation. The dose of atropia which he recommended in hydrocyanic acid poisoning is $\frac{1}{15}$ grain, which should be subcutaneously injected; and as, in so desperate an emergency, every auxiliary ought to be employed, it should be remembered that artificial respiration has itself been found by Preyer to save animal life; and that (as Dr. Ringer remarks) the only thing necessary is to keep the patient alive during a short period, after which the poison will have become eliminated, and danger will cease. It is therefore right, when we are called to a case of Prussic-acid poisoning, at once to set artificial respiration going, without waiting for the action of the atropia, which must occupy some time, be it ever so little.

I cannot conclude this brief notice of the singular antagonistic virtues of atropia without remarking that this modern addition to the other numerous excellences of belladonna which are available in medicine, raises this plant to an almost unexampled importance among remedies. With the single exception of opium, there is probably no vegetable medicine so important in existence; and already there seems a probability that many purposes to which opium has been applied will hereafter be found more efficiently discharged by belladonna.

For internal use atropia is dangerous. At first the dose should not exceed $\frac{1}{100}$ of a grain; but subsequently it may be increased, with caution, to $\frac{1}{20}$ grain, and in special [poison] cases even to $\frac{1}{10}$ grain.

PREPARATIONS AND DOSE.—Tinct. Belladonnæ, ℥v.—xxx. (.30 —2.); Ext. Belladon., gr. $\frac{1}{10}$—$\frac{1}{2}$ (.006—.03); Ext. Bell. Alcoholicum *idem*; Ext. Bell. Fl., ℥ i.—v. (.06—.30); Ungt. Bell.; Suppos. Bell.; Emplast. Bell.; Atropia rarely used; Atropiæ sulphas, gr. $\frac{1}{120}$—$\frac{1}{60}$ (.0005—.001).

HYOSCYAMUS (Hyoscyamus niger).

ACTIVE INGREDIENTS.—The principal activity of hyoscyamus resides in the alkaloid Hyoscyamia, which, after various erroneous processes, was obtained in unmistakable form by Geiger and Hesse (1833). Hyoscyamia, when pure, crystallizes slowly in stellate or tuft-like groups of silky odorless needles, which are occasionally transparent. It has a very sharp and unpleasant taste; is almost insoluble in cold water, more easily soluble in hot; and soluble in spirit, ether, chloroform, amylic alcohol, benzole, and dilute acids. The pure alkaloid is permanent in the air; the impure (which is amorphous) is hygroscopic, and much more soluble in water than the former; it has also a tobacco-like odor.

Besides the alkaloid, an empyreumatic oil has been obtained, by Morris, from the destructive distillation of the plant; it is identical in character with a similar oil obtained from digitalis, and is a powerful narcotic.

Hyoscyamia must not be confounded with an impure resinoid body, called "Hyoscyamin," which contains the true alkaloid, but from its uncertain strength is an objectionable preparation.

PHYSIOLOGICAL ACTION.—The offensive odor of the aërial parts of hyoscyamus is such as to deter, or at all events to disincline, any one from swallowing even a small portion, unless designedly. The roots, however, being sweetish, have sometimes been eaten by mistake for those of some esculent vegetable, in the same way as those of certain poisonous Umbelliferæ; and the results have then been vertigo, burning sensations in the lips, the tongue, and the throat, and severe pains in the iliac region, and in all the joints. The intellectual faculties soon afterward become perverted and the eyesight dim, the sufferer giving himself up to mad and ridiculous actions. Orfila and Christison give accounts of such cases; in some of them the patients recovered, while in others the symptoms increased and ended in death. Dissection of the bodies of those who have died from eating hyoscyamus shows considerable inflammation of the stomach, the surface of which is covered with gangrenous spots, while the brain exhibits appearances betokening great vascular excitement.

[Lawson (Report of West Riding Lunatic Asylum, 1876) says: " With regard to the physiological actions of the alkaloid, it was determined that the effect on man of a moderate dose was the production of a mental condition which partook of all the leading symptoms of simple mania, but which in addition was characterized by extreme physical helplessness, intermittent drowsiness, hypermetropia, and dryness of the mouth."]

A curious account is given in the "Philosophical Transactions," vol. xl. p. 446, of the effects of henbane upon a party of persons who ate the roots: some became speechless and convulsed, while others howled deliriously; in all of them there was protrusion of the eyes, with contraction of the mouth, and delirium.

To some of these unfortunates a certain measure of relief was afforded by the use of emetics; but their sight was impaired for many days, and every object that was looked at seemed to be scarlet. In the rare instances in which portions of the herbage of hyoscyamus have been mixed with green vegetable food and inadvertently swallowed, those who have eaten the leaves have suffered from vertigo, gripes, and purging, and (as in the cases where the root was eaten) they have made frightful grimaces.

Another effect produced by taking an overdose of hyoscyamus is remarkable dilatation of the pupil.

THERAPEUTIC ACTION.—Regarded as a medicine, henbane is one of the most valuable narcotics we possess. The principal uses to which it is applied are those in which it can be substituted for opium, which drug disagrees with the stomach in many instances, or is contra-indicated by peculiar symptoms. Hyoscyamus appears, moreover, to be tolerably free from the constipating effects of opium, especially when administered in large doses. In the maladies to which children are subject, hyoscyamus is decidedly preferable to opium. It is valuable during dentition and in convulsions, the properties being such as alleviate pain and subdue irritation. In moderate doses it is a powerful sedative, diminishing excessive irritability; it induces sleep, and relieves chronic and anomalous pains of the abdominal viscera. It is valuable, also, in irritable conditions of the kidneys, the ureters, and the bladder, and in hysteria, gout, rheumatism, chordee, palpitation of the heart, and " painters' colic."

In its operation, as would be anticipated from its near botanical affinities, hyoscyamus has a very close resemblance to belladonna and stramonium. All three of these drugs are mydriatics (*i. e.*, they dilate the pupil); they also produce dryness of the mouth, throat, and air-passages;

they also cause fulness, vertigo, hallucinations, and delirium; and act similarly upon the pulse, each of them reducing its force and frequency when administered in moderate doses, and quickening it when given in larger ones.

Not that all three medicines are correspondingly energetic in the same directions; hyoscyamus is said to dilate the pupil more powerfully than either belladonna or stramonium, but in all other respects it is the feeblest of this famous trio. Unlike the other two, hyoscyamus is not known ever to produce involuntary evacuations from the bladder or the rectum; nor, while belladonna and stramonium (when exhibited in large doses) are apt to produce a decided eruption upon the skin—belladonna, one of a scarlatinoid character, and stramonium an erysipelatoid one—is hyoscyamus likely to induce any eruption at all; and when an eruption does follow its employment, the color is less livid. All three medicines appear to have very much the same general action, the operation varying only in power and extent, as, for example, in cases of cerebral hyperæmia, the severer forms of which are removed by belladonna, while hyoscyamus proves its value when there is little or no congestion, but much excitement. So in the case of delirium: the forms of this disorder for which hyoscyamus is adapted are the milder and less inflammatory ones; whereas the severer cases are better dealt with by belladonna or stramonium. Hyoscyamus is specially useful again in those cases of delirium with hallucinations which are accompanied by little or no cerebral congestion, but where there is great excitability of the nervous system, and where there is reason to fear that the operation of opium would prove injurious. [Lawson (Practitioner, July, 1876) found hyoscyamia useful in the treatment of recurrent acute and subacute mania, monomania of suspicion, and the excitement of senile dementia, also in epileptic excitement. He gave it in doses of gr. $\frac{3}{4}$ to $1\frac{1}{2}$, of Merck's amorphous hyoscyamin.]

Febricula.—Hyoscyamus is very useful in cases of febricula. The usual symptoms of this malady are dry, hot skin, hard and full pulse, headache, restlessness and irritability accompanied by sleeplessness, slight delirium, thirst, dry fevered tongue, constipation, and scanty and high-colored urine. The head symptoms are speedily relieved by hyoscyamus, and the tongue becomes moist under its influence; and if the exhibition of the medicine be continued, the constipation is removed and a daily laxative action takes its place. (Belladonna in many instances produces similar effects.)

Monomania.—In hypochondriacal monomania, when the patient suffers from such mental symptoms as syphiliphobia, when really he has no reason to think himself the subject of any venereal taint, hyoscyamus will relieve the distressing despondency, and in many instances remove the hallucination.

Sunstroke.—I have used hyoscyamus with decided success in cases of mild sunstroke or heat-stroke, where the patient has suffered more or less from faintness, vertigo, headache, a sense of tightness across the forehead and chest, a quick and full pulse, sleeplessness, and much nervous irritability. In the convulsions of children, when brought about by undue exposure to the heat of the sun, I believe hyoscyamus to be a more valuable remedy even than belladonna.

Sub-acute Meningitis.—Hyoscyamus is valuable again in sub acute meningitis, and in the recent and less acute delirium which accompanies typhus.

Delirium Tremens.—In some forms of delirium tremens it plays its part admirably.

Coughs.—Spasmodic throat-coughs of a tickling and irritating character, such as occur principally during the night, debarring the patient from sleep for many hours together, are quickly mitigated by hyoscyamus. It is very useful also in night-coughs which are incessant for a long period and which actually trouble the patient during slumber.

Palpitation.—Nervous palpitations of the heart of a spasmodic character; violent palpitations also, which depend upon an excited condition of the brain, are frequently removed by hyoscyamus.

Breasts painfully distended with milk are quickly relieved by a plaster of hyoscyamus, like that of belladonna, being laid upon them.

Cancer and Hæmorrhoids.—So, in the form of cataplasm, the bruised leaves of the plant are often advantageously used, alike for the purpose last referred to and for cancers and scrofulous ulcers, as well as for hæmorrhoids and other painful complaints.

Constipation.—When the bowels are confined and irritable, and when it becomes necessary to promote their secretion by the use of colocynth or some other drastic, hyoscyamus is usually prescribed in addition; since it prevents the griping action of the drastic, and yet does not diminish the general effect, or at least not to any important extent. Hyoscyamus, when used for any complaint whatever in which opium is ordinarily employed, is superior to that drug, in so far as it possesses little tendency to confine the bowels. Hyoscyamus in some cases produces unpleasant symptoms, and the sleep which supervenes upon the exhibition of it is sometimes uncertain, labored, and unrefreshing; hence, it is generally resorted to as a *secondary* medicine, rather than as one to which we may confidently apply at first for its anodyne and hypnotic effects.

Toothache.—The good effects of smoking the seeds after the manner of tobacco, in the treatment of toothache, resemble those of smoking stramonium seeds, and are well known, both in professional and in domestic medicine. The remedy is one, however, which must be resorted to with caution; since the smoking of these seeds has been followed, in certain cases, by convulsions and temporary insanity. I may here remark that the properties of the seeds of henbane appear to differ from those of the root and of the leaves, the narcotic qualities of the two last-named being supplemented by others of an irritant character. Persons who have swallowed them have more frequently suffered from convulsions, and have experienced greater heat and dryness of the throat and more burning of the stomach; their thirst and delirium have also been intensified.

This circumstance may perhaps account for the untoward effects which have occasionally followed the exhibition of the *extract*, the seeds being sometimes mingled with the leaves through want of care in the collecting; though this can only be the case when the collecting of the leaves has been too long delayed—*i. e.*, until the flowering of the plant has long ceased.

PREPARATIONS AND DOSE.—Ext. Hyoscyami, gr. $\frac{1}{6}$—v. (.01—.30); Ext. Hyoscyam. Fl., ♏v.— ℨ ss. (.30—2.); Tinct. Hyoscyam., ℨ ss.—ℨ ij. (2.—8.) Ten drops of liq. potassæ destroys the action of ℨ j. of the tincture of hyoscyamus, a blunder we have more than once seen committed in copaiva mixtures. The alkaloid hyoscyamia may be given in doses of gr. $\frac{1}{6}$—j. (.01—.06).

STRAMONIUM (Datura Stramonium).

Active Ingredients.—Every portion of this plant, while fresh, has a heavy and disagreeable odor. In America the scent is said to be sometimes so powerful as to cause sickness in those who inadvertently expose themselves to it. The taste is bitter; and if the plant be chewed, the saliva acquires a green tinge. The active principle of stramonium is the alkaloid daturia, which, as already mentioned, is identical in formula, and in nearly all physical and chemical characters, with atropia, but appears to be more powerful than the latter.

Physiological Action.—The first mention of stramonium by a modern author occurs in the celebrated *Historia Stirpium* of Fuchsius, published in 1542. But the employment of it in regular medicine is of much more recent date. From time immemorial, however, it appears to have been used in some of the Asiatic islands as a soporific; and it would seem to have been long known that, if administered too copiously, the extract induces nausea, drowsiness, loss of sense, a sort of intoxication, and delirium. If not checked, these symptoms are followed by loss of memory, convulsions, a sense of suffocation, paralysis of the limbs, cold sweats, excessive thirst, dilatation of the pupil, and tremblings, which presently terminate in death.

Therapeutic Action.—The therapeutic action of stramonium, when employed as a remedial agent in disease, is apparently much the same as that of belladonna, a similarity at first sight very remarkable. Surprise, however, diminishes when it is remembered that the same principle exists in these two medicines, and that both belong to the same natural order of plants, as does also hyoscyamus. That all three should be resorted to in very much the same class of diseases follows almost as a matter of course.

Baron Störck, by whom aconite was introduced into modern practice, was the first also to point out the efficacy of stramonium. He commended it as well adapted for affections of the brain and nervous system; also for use in mania and epilepsy. Dr. Davey and many others have since declared its value in mania, by reason of its power in allaying irritation and inducing tranquil sleep.

The action, however, is so powerful that in England there is little disposition to employ it; and as with many other valuable drugs, the merits of stramonium, which are unquestionably high, do not receive general recognition.

Puerperal Mania.—Personally I can speak favorably of stramonium in cases of puerperal mania, when the delirium is of a wild and furious but intermittent character, attended by great restlessness, with sluggishness and scantiness in the secretion of milk, and when there is a tendency to suicide, or a disposition to destroy the child. Stramonium, under such circumstances, will often allay the cerebral excitement, and soothe the nervous system. Under its influence the flow of milk will soon be renewed, sleep will be restored, and a general calm will pervade the whole system. It must never be forgotten in these cases to watch the lochia and other secretions, nor to use poultices, hot and soothing fomentations, enemas or gentle laxatives, should occasion require; and, above all, we must not neglect to sustain the patient's sinking powers by means of nutritious and stimulating diet. From a quarter to half a grain of the extract of stramonium, or ten to twenty minims of the tincture, should be administered every three or four hours until relief is obtained.

Nymphomania.—Given in smaller doses, stramonium is very useful in nymphomania when unconnected with disease of the sexual organs, and attended by considerable mental and bodily depression.

Epilepsy.—The beneficial effects of stramonium in epilepsy are very doubtful. I am not aware of any case that has ever been palliated by its employment.

Tic Douloureux and Sciatica.—In tic douloureux, when exhibited in large doses and steadily administered for some time, stramonium often affords very decided relief. In such cases, and also in sciatica, doses of a quarter to half a grain should be given every three or four hours; but if the slightest symptoms of narcotism appear the medicine must immediately be discontinued. If relief be not obtained after the administration of four or five doses, to continue longer would be useless, and might even prove injurious; this is the case not only in tic douloureux and sciatica, but in any other disorders for which stramonium is eligible.[1]

Spasmodic Asthma.—In cases of spasmodic asthma stramonium is an old, often tried, and a frequently successful remedy. I have seen it produce speedy and permanent relief in several severe cases of this complaint when administered in doses of a quarter of a grain to half a grain at intervals of three or four hours. In America stramonium is a popular remedy for asthma, those cases in which it is found specially serviceable being the purely spasmodic ones. In these it acts, without doubt, by virtue of its sedative and antispasmodic properties.

Bigelow, in his "American Medical Botany," vol. i., p. 23, describes the excellent effects produced by smoking the seeds of stramonium in the same way as tobacco, when an efficient palliative is required alike for asthma and for certain other affections of the lungs. The practice was suggested by the employment, for this purpose, in India, of the seeds of another species of the genus, the *Datura ferox*. Bigelow admits that with plethoric and intemperate people this method of employing stramonium entirely fails. It fails also in cases where there is effusion of serum in the pleural cavity, or where there is other serious organic change implicating the lungs. But when the patient has no such lesions there seems to be good reason for believing that the reputed advantage may be derived.

Dr. Marcet publishes results of his experience with stramonium, which seem to prove this drug superior to any other narcotic; he states, that when employed internally, he found it paramount in many painful diseases. He says of it likewise, that although it sometimes excites disagreeable nervous sensations, alarming to the patient, these are by no means constant or serious. The effects upon the bowels he found to be relaxing rather than astringent.

Tumors.—The fresh leaves of stramonium made into a cataplasm, and applied externally, have been found successful in cases of inflammatory tumor; also for "discussing" indurated milk in the breasts of nurses.

Hæmorrhoids.—An ointment prepared from the fresh leaves is recommended to alleviate the pain of hæmorrhoids, and for other affections

---

[1] Gubler (Commentaires thérapeutiques) speaks highly of stramonium as a remedy for "congestive neuralgias."

of the rectum. [It is also useful in irritable ulcers, eczema, and some other pruriginous affections.]

Rabies.—In India stramonium is said to be successfully administered for rabies, and then given in large doses, so as to produce continuous intoxication.

Unfortunately for the disposition to allow it a fair trial, and to realize all the advantages which it offers, stramonium, like some other narcotic medicines, is very uncertain in its operation. In some cases the effect is all that could be desired; in others the operation is indifferent; or there may even be total and absolute failure.

As regards what is really one of the commonest and most useful applications of stramonium—its use by smoking in spasmodic asthma—it seems especially worth while to remember one point, viz., that the desired effects can be produced with much greater promptitude and certainty by the use of the seeds than by that of the leaves. Dr. Hyde Salter suggests that it would be a good plan to soak the leaves in an infusion of the seeds, and then dry them for use. At any rate, if the leaves alone be used, it is very necessary to employ them in a fresh state, as they soon become nearly inert.

In conclusion, let me remind English readers that though stramonium is much gone out of fashion, it was the deliberate opinion of Trousseau and Pidoux (Matière Méd., etc.) that it can do all that belladonna can do, *and is more powerful.*

PREPARATIONS AND DOSE.—Ext. Stramonii Foliorum, gr. ½—j. (.02—.06); Ext. Stramon. Semin., gr. ¼—½ (.01—.03); Tinct. Stramon., ℥v.—xxx. (.30—2.); Ungt. Stramonii.

DULCAMARA (Solanum Dulcamara).

ACTIVE INGREDIENTS.—The powers of dulcamara probably depend entirely upon the presence of the true alkaloid solanine, although the glucoside dulcamarine, which is also contained in the plant (and which Pelletier imagined to represent solanine combined with sugar), much more nearly represents the compound taste—sweet and bitter—of the wood. Solanine is bitter and rather burning in taste, weakly alkaline in reaction, and crystallizes out of its alcoholic solution in microscropic, white, four-sided right prisms, with a mother-of-pearl sheen upon them. It melts at 235° C., and coagulates amorphously as it cools. It is soluble in 8,000 parts of boiling water, 4,000 of ether, and in 500 of cold or 125 of boiling alcohol of .839 sp. gr. The amorphous form is said to be much more soluble in alcohol.

PHYSIOLOGICAL ACTION.—There can be little doubt that the only reason why dulcamara is not usually reckoned a poison is the smallness of the proportion in which solanine exists in the wood; for the alkaloid in question possesses unquestionable toxic properties, although there is considerable dispute as to their exact range.[1] Of its power, as a narcotic, to produce tremors, convulsions, hurried respiration and death, there seems

---

[1] It should be observed that some of the experiments were made with solanine from the potato and other species of Solanum, and it has been suggested that there are different varieties of the alkaloid; but there seems no evidence to support this idea.

no doubt; but it would appear likely that, to produce the severest results of this kind with certainty, it may be necessary to *inject subcutaneously, or directly into a vein.* The experiments of Fraas and Martin,[1] pursued in this way, are very striking. A dog died in seven minutes after injection of four and a half grains, the symptoms being sudden rapidity and convulsive embarrassment of respiration, general convulsions, tetanic spasms, and strong dilatation of the pupil, besides others to be mentioned presently. Clarus[2] also observed convulsions and tetanic cramps. Leydorf[3] administered it to pigeons (tying the œsophagus to prevent vomiting, which otherwise occurred), and observed hurried breathing, tremors, slight convulsions, and powerful dilatation of the pupil.

Whether solanine also acts as an irritant upon the alimentary canal is more disputed, but there is plenty of evidence to show that it will produce vomiting; and the presence of distinct inflammatory mischief (enteritis) was found by Malik and Spatzier (after very large doses of impure solanine), and also by Fraas and Martin.

As to its physiological action on the vascular system, there are cases recorded by several observers in which solanine has produced congestion of the vessels of the cranial meninges; and the vessels of the kidneys and liver have been found engorged simply,[4] or these again have been found actually inflamed.[5] The vomiting, on the other hand, which was suffered by Clarus in his experiments upon himself, seemed connected with the nervous system, and not at all with the stomach. As to the power of solanine to dilate the pupil, the statements are discrepant. It seems clear, from the experiments of L. van Praag,[6] that even moderate doses will sometimes produce this result. Clarus, on the other hand, applied solanine locally to the eye and got slight *myosis*, with severe irritative effects upon the conjunctiva; while Schroff, in cases where he gave doses (to human patients) sufficient to produce decidedly unpleasant toxic symptoms, observed no change whatever in the pupil, and Fronmüller[7] found it more frequently absent than present. One or two authors speak of salivation as a toxic effect of solanine, and several speak of a heat and dryness in the throat, more or less accurately resembling that produced by the more powerful Solanaceæ. Vertigo has very often been observed among the poisonous effects, and sometimes there has been produced an erythematous rash, which is even said to have proved fatal; though here it may well be doubtful if the dulcamara were really the cause of death.

The most distinct form of dulcamara-poisoning has ensued upon eating the berries, to which children are often attracted. There are well-established instances in which this has been followed by severe pain in the bowels, with great heat in the throat and chest. Pressure upon the abdomen cannot in such cases be borne, and there is much nausea, thirst, and prostration of strength. At the same time the pulsation at the wrist becomes hurried, and the breathing quick and painful. Sometimes these symptoms are exchanged for violent vomiting and purging, with profuse secretion of saliva. Sulphate of zinc relieves the stomach of its contents,

---

[1] Virchow's Archiv, iii. 225, 1854.
[2] Journ. für Pharmacodynamik, i. 2.
[3] Studien üb. d. Einfluss des Solanins auf Th. u. M., Mar., 1863.
[4] Clarus, op. cit.
[5] Malik and Spatzier, in Husemann, op. cit.
[6] Journ. f. Pharmacod., i. 2.
[7] Deutsche Klin., 1865, 40.

and the subsequent use of the warm bath and of purgatives brings about recovery after the lapse of a few days.

The above facts show that in dulcamara we have at all events an agent of considerable potency. Its most powerful action is said to be manifested in persons of fair complexion, light hair, and mild disposition, and especially in those who are subject to catarrhal affections in cold and damp weather; but this is doubtful.

THERAPEUTIC ACTION.—The medical properties of dulcamara are generally supposed to be very feeble; but they are unquestionably of sufficient energy to render the plant valuable in the hands of the physician. Linnæus, Carrère,[1] and others highly commended it, a century ago, in herpetic diseases, scabies, etc. It has also been employed with advantage in chronic rheumatism, gout, incipient phthisis, humoral asthma, jaundice, and other complaints. Bergius recommends it in rheumatism; Murray states that all the ordinary secretions are promoted by the use of it; and Sir Alexander Crichton, in a letter published in Dr. Willan's celebrated work upon skin diseases, stated that, out of twenty-three cases of lepra [psoriasis] which were treated with dulcamara, only two failed.

Psoriasis and pityriasis appear likewise to be amenable to the influence of dulcamara. Dr. Neligan recommends the infusion, copiously administered, as an excellent vehicle for the preparations of iodine and arsenic employed in obstinate cutaneous affections. For pustular affections, and for vesicular and scaly ones, and in all diseases which arise simultaneously with the suppression or the disappearance of cutaneous eruptions, dulcamara is again a very useful remedy.

Personally, I have employed it in cases of humid asthma (especially when the disorder appeared after the repercussion of nettle-rash or some other eruption), and always with unequivocal success; also in the diarrhœa of children, when brought on by chill in damp weather, or when caused by dentition, using the *infusion*, which I found to quickly check the symptoms.

Dulcamara has long been used as a diaphoretic in rheumatic and venereal affections, and is strongly to be recommended likewise in nasal, pulmonary, and vesical catarrhs, attended by general dryness of the skin. Dr. G. B. Wood, in his "Dispensatory," recommends dulcamara as a fit medicine for subduing the sexual appetite in maniacs and others in whom this desire is morbidly active.

Bigelow and Bateman confirm the utility of this medicine, the latter declaring it to be one of the most effectual remedies for lepra [psoriasis], in all the varieties of that disorder which are at present known to medicine. He prescribes a decoction of the twigs and leaves. As in England, however, lepra is a disease which generally originates in a want of tone or vigor in the whole system, it is probable that a general mode of treatment would be more efficacious than a specific one. [Clarus (*Arzneimittellehre*) found it useful in many eruptions, but *not* in scaly ones.]

In delicate constitutions, and in hysterical women, it is proper to observe, the exhibition of dulcamara has sometimes been followed by syncope and slight palpitation of the heart. If under such circumstances the dose be diminished, the objectionable results no longer ensue.

The circumstance of dulcamara being a plant indigenous to our own

---

[1] [Traité des proprietés, usages, et effets de la douce-amère. Paris, 1787.]

country ought to be an argument for giving it every species of fair trial. [The plant grown in America is probably equal to the European in activity.]

Different samples of the herb do certainly indicate various degrees of strength; but as it is easy to judge of the goodness of any particular sample by the degree of narcotism (usually slight) which may follow upon its use, all that is needed is to regulate the dose according to its quality.

By some practitioners dulcamara is considered a valuable auxiliary to mercury.

While such are the positive claims of dulcamara, it is necessary to notice the opinion arrived at by Dr. Garrod, from his own experiments, that the action is almost *nil.* He states that he has given as much as sixty fluid ounces of the infusion in the course of a single day, and that no symptoms, unpleasant or otherwise, have resulted. [" The American plant, however, when gathered in full vigor, does not set easily on the stomach in large doses. I have known vomiting produced by a few grains of the powdered leaves, and by a small cup of the decoction. The strength of the plant seems to vary in some degree with the time of gathering and mode of preserving." Bigelow, *op. cit.*, i. 174.]

PREPARATIONS AND DOSE.—The best preparation (Phillips) of dulcamara appears to be that obtained by infusing an ounce of the fresh stalks, chopped small, in ten fluid ounces of boiling distilled water. This is much more trustworthy than the *decoction;* since, by the operation of boiling a considerable amount of the active principle becomes dissipated. Infusum Dulcamaræ, dose 3 ss.—2 fl. ounces, or more. The officinal are Decoctum Dulcamaræ, ℥ i.—ij. (30.—60.); Extract. Dulcam., gr. x.—xx. (.65—1.30); Ext. Dulcamaræ Fluid., 3 ss.—ij. (2.—8.).

Farquharson[1] dismisses this important drug with the brief comment: " Is never used." Girtanner,[2] however, ninety years before, formed a higher estimate of its value, but insisted on its being used in very small doses (*Sehr geringer Dosis*).

CAPSICUM (CAPSICUM ANNUUM).

ACTIVE INGREDIENT.—The physiological properties of capsicum are represented by *capsicine*, a yellow, thick semi-solid, which melts with heat, and is soluble with difficulty in water, but easily in ether, turpentine, and rectified spirit.

PHYSIOLOGICAL ACTION.—The local irritant action of cayenne pepper is well known; it excites a burning sensation in the mouth and tongue, and, if swallowed, in the fauces and throat. Should this action be carried to intensity severe inflammatory swelling of the mucous membrane is produced, with copious out-pour of saliva. In the stomach small quantities excite a gentle and not unpleasant feeling of warmth; very large doses cause severe gastric and intestinal inflammation; the prolonged use of large (though not acutely poisonous) doses diminishes appetite and digestive capacity: these evil effects, however, are not so readily produced

---

[1] Guide to Therapeutics, London, 1877.
[2] Abh. u. d. ven. Krankh, Göttingen, 1788.

in persons who reside in the tropical countries to which the capsicum is indigenous. Applied to the skin, especially in concentrated solution, capsicum is a powerful rubefacient, and will even blister if applied continuously.

As regards the more remote physiological actions of capsicum nothing can be said to be known with certainty. It does not appear, in any dose, to exercise narcotic power, but large doses would seem to increase the perspiration and the flow of saliva, whether by simply producing vaso-motor paralysis, or in some other way, is not known. The sweating of the brow and the salivation, which are immediately produced by a very large dose of cayenne, are probably due to reflex paralysis of the vaso-motor fibres which run with the branches of the fifth nerve.

Finally, it may be mentioned that capsicum is eliminated, in part, in the urine.

THERAPEUTIC ACTION.—The most important medicinal action, probably, of capsicum, is one which has only been observed of late years. It had been suspected, indeed, that this drug was a general nervous stimulant, but no proof existed of its acting in this way upon the nervous centres, and especially on the brain, until the remedy came to be used for delirium tremens. The remarkable success which has attended the use of cayenne pepper in this disease seems to have made strangely little impression upon British practice. Kinnear, Lawson, and Lyons have administered the pepper in doses of from 20 to 80 grains, single or repeated doses; the last-named authority believing that the effect is produced in a reflex manner, the nerve ends of the vagi in the stomach being the point of original impression.[1] He considers that capsicum has many advantages over digitalis, especially in those recurrent and similar cases which are so often attended by fatty heart. Kinnear and Lawson treated not less than seventy or eighty cases successfully with capsicum alone.

Other examples of the general stimulant action of capsicum, more commonly known, but not so decisive in their character, are its employment in atonic gout, in paralysis, in dropsy, in tympanitis, and in the debilitated stages of fever. Much smaller doses ($\frac{1}{4}$ grain to 2 grains every four hours) than are needed in delirium tremens are here available. For scrofulous constitutions capsicum may be usefully combined with iron. In intermittents it is said to be a useful adjunct to cinchona.

In scrofulous and fistular ulcerations a weak infusion of capsicum becomes a useful stimulant.

In the coma of fever, when rubefacient cataplasms are required for the feet, the same medicine furnishes a desirable ingredient.

Capsicum enters likewise into the composition of various kinds of throat-lozenges, and of preparations sold under the name of "cardiac tinctures."

Capsicum is a useful stimulant, again, in dyspepsia; and in cases of flatulency arising from vegetable diet it is an excellent carminative.

In *cynanche tonsillaris*, or common inflammatory sore-throat, which manifests itself with fever, pains and swelling of the tonsils, a thickly-furred tongue, a profuse discharge of viscid saliva, and the very disagreeable symptoms of deglutition accomplished with much difficulty, and a return through the nostrils of the fluids attempted to be swallowed, a gargle of capsicum rarely fails to give relief, provided it be used in the stage of the disorder which precedes suppuration.

---

[1] Med. Press and Circular, April 18, 1866.

In cases likewise of *cynanche maligna*, so commonly met with in conjunction with scarlatina, capsicum is highly beneficial.

In the form of cayenne pepper it is useful in ordinary sore throat; also in relaxed sore throat, and in relaxed uvula. Care should be taken when this gargle is employed, as it occasionally induces violent inflammation, which is by no means easily subdued.

Spasmodic, irritating coughs, which arise from elongated uvula and relaxed throat, are again amenable to capsicum; as are likewise atonic dyspepsia, accompanied by heartburn and diarrhœa. In the former cases the gargle should be prepared with honey.

I have employed capsicum in tincture, and have also used it externally for habitual pain in the loins, attended by sluggish circulation in the renal vessels, and where there was a slight but persistent trace of albumen in the urine. The symptoms in these cases had baffled all previous treatment, yet were speedily removed by a steady course of capsicum, the dose being five minims taken thrice a day, and a capsicum cataplasm being applied for two or three nights in the week. [Capsicum decidedly promotes the action of quinine in malaria, and, together with a little opium, may be combined in the following proportions: quinine, ten grains, capsicum, five grains, opium one-half to one grain, these to be mixed and divided into such doses as may be required.]

PREPARATIONS AND DOSE.—Capsicum, gr. i.—v. (.06—.30); Infusum Capsici, ℨij.—iv. (8.—15.); Oleo-resina Capsici, gr. ss.—j. (.03—.06); Tinct. Capsici, ♏v.—xxx. (.30—2.)

TOBACCO (NICOTIANA TABACUM).

ACTIVE INGREDIENTS.—To consider the whole subject of the active elements of tobacco, as used in its different forms, would be a task of much difficulty,—so many vexed questions have lately arisen. Until recently it was the custom to speak of tobacco as possessing two active ingredients,—nicotine, the alkaloid, which is present in the leaf, and an empyreumatic oil, which is generated only in combustion. The former was supposed (and the experiments of Brodie seemed to confirm this idea) to act chiefly as a paralyzer of the heart, while the oil exerted a general narcotic influence, without specially engaging the heart. According to these notions nicotine was the sole agent in the case of tobacco administered in watery solution (tobacco chewing, tobacco enemata), while tobacco smoke owed its power to the presence both of nicotine and of the empyreumatic oil. Snuff (as far as it is tobacco only) would owe its power to nicotine alone. More recent researches have rendered it at least very probable that tobacco smoke owes very little of its potency to nicotine, and very much to the combustion products, of which pyridine seems to be the most powerful. At any rate, it seems that to these substances the most formidable toxic effects may be ascribed.

To go into all these refinements is unnecessary. Pyridine and the other combustion-products of tobacco seem only to differ from nicotine in their action, so far as that they are milder, and less rapid in their toxic operation. It will be sufficient accordingly, for the purpose of this work, to regard nicotine as the active principle of tobacco. It is a powerful base, and completely neutralizes acids. It exists as a colorless, transparent, mobile fluid, possessing a well-marked tobacco odor (which is strong-

.y developed when it is heated), and a sharp and burning taste long retained by the tongue. The specific gravity curiously [?] diminishes at successively higher temperatures; the vapor is exceedingly heavy; it turns the plane of polarization strongly to the left; it is very soluble in water (which it also absorbs freely from the atmosphere), in ether, in rectified spirit, in turpentine, and in fatty oils. Nicotine which has been kept some time is apt to turn a brownish-yellow color. [The whole question of the tobacco derivatives is in the utmost confusion, no two chemists agreeing in the matter.]

PHYSIOLOGICAL ACTION.—The most familiar of the physiological effects of tobacco are those which are experienced by young smokers, who rarely fail to poison themselves, to a greater or less extent, in their first trials. Nausea, giddiness, vomiting, or a feeling of deadly sickness, with cold sweatings, and exceedingly feeble pulse, are the ordinary results of first attempts to smoke tobacco which is even moderately strong. Of these effects it is difficult to say how much is due to nicotine, and how much to the empyreumatic products. It has even been stated by recent experimenters[1] that nicotine is not present at all in tobacco smoke; and at any rate it is certain that the empyreumatic products are powerfully narcotic. In the few cases of fatal and nearly fatal results that have ensued from smoking,[2] the symptoms have probably been due to the inhaling of the smoke into the lungs, whereby the empyreumatic products must have been largely deposited upon the extensive surface of the bronchial ramifications, and thus have been absorbed in considerable quantities. In the infusions and decoctions of tobacco leaves, which have been the ordinary mode of medicinal and criminal administrations of the drug, and the usual source of accidental poisoning, the active principle is undoubtedly nicotine.

The soluble parts of tobacco are easily absorbed, not only from the alimentary canal but also from the skin. The latter fact is proved by the case related by Dr. Blanchard.[3] This author narrates the case of a man and his wife, who, to cure themselves of the itch, vigorously rubbed into the skin a decoction of tobacco. In a short time they were both affected; the man (who had scarcely wiped himself) more seriously than the woman. The latter was chiefly troubled with headache and vomiting, violent pains in the stomach, constant micturition, jaundice, muscular tremors, and painful cramps, besides constant vomiting and frequent stools, resembling those of cholera. Both patients recovered under the use of stimulants.

Fatal results have not unfrequently followed the administration of the decoction of tobacco in enema. Mr. Ede[4] reports the case of a girl who in half an hour after the administration complained of faintness, and of feeling sick, and in another half-hour became quite collapsed, with cold sweats; she vomited, was slightly convulsed, and died in an hour and a half after the first reception of the poison into her system. On post-mortem examination the heart was found quite flaccid; neither stomach nor intestines presented any trace of inflammation. A case of gastric administration was that of a lunatic,[5] who swallowed an ounce or an ounce and

---

[1] Vohl and Eulenburg, Arch. Phar. Chem. 147.
[2] *E. g.*, in M. Beau's cases of angina pectoris from this cause.
[3] Bull. de Thérapeutique, Juin 15, 1869.
[4] Quoted by Taylor on Poisons.
[5] Taylor, op. cit.

a half of crude tobacco, after keeping it some time in his mouth. He became suddenly insensible and motionless, with all the muscles relaxed, very feeble respiration and pulse, strong contraction of the pupils, and (later on) violent tetanic convulsions and profuse purging. After a temporary amendment, consequent on the use of the stomach-pump, the symptoms returned. There was profuse purging, with blood and mucus, and the patient uttered loud cries; the convulsions returned with brief intermissions, the limbs being at intervals rigidly flexed upon the body. There was also grinding of the teeth; death following in syncope, seven or eight hours after the taking of the tobacco. On post-mortem examination, forty hours after death, there was strong cadaveric rigidity, with congestion of the brain, the medulla oblongata, and the pons Varolii; the heart was empty, small, and contracted; the liver and kidneys were much congested; the gastric and intestinal mucous membranes were inflamed and partially abraded; the mesenteric veins were filled with dark blood; the bladder was contracted and empty; the blood was everywhere dark and fluid.

It would appear from a comparison of different reports that, in the most swiftly fatal cases, the action of the poison has been almost entirely expended upon the nervous system and the heart; while patients who survive longer, suffer from severe inflammatory affections of the alimentary canal.

The physiological effects of nicotine are nearly those of tobacco solution, but in a more powerful form, more especially as regards the action upon the heart. An example of its most decided action is afforded by the case of the victim of Count Bocarné, who was poisoned with nicotine, and was believed to have died in less than five minutes; and by a case of suicide recorded by Taylor, in which the alkaloid proved fatal in from three to five minutes.

The most exact recent researches are those of Traube,[1] Rosenthal,[2] Krocker,[3] Erlenmeyer,[4] Schroff,[5] Van Praag,[6] Kölliker,[7] etc., and from these may be derived the following *résumé:* Nicotine paralyzes the brain, producing loss of consciousness and of voluntary movement, after a more or less brief interval of excitement.[8] The primary action on the spinal cord is exciting, and produces clonic and tonic convulsions; this is followed by paralysis, the cord becomes insensible to direct and to reflex irritation, apparently from affection of the gray matter of the anterior cornua. The motor nerve-trunks are but little affected, but their terminals in the muscular substance are at first strongly irritated, and then paralyzed. Small doses affect the heart in a double manner, the vagus branches being in comparison slightly, the cardiac motor nerves much more powerfully affected; the excitement is followed by paralysis. If the vagi be divided, nicotine nevertheless acts on the peripheral vagus-twigs, and the ganglionic apparatus in connection with them. The cardiac muscular structure does not appear to be directly affected; respiration is at first quickened, and there may even be tetanus of the inspiratory muscles;

---

[1] Allg. med. Centralzeitung, 38, 1862.
[2] Med. Centralblatt, 47, 1863.
[3] Ueber die Wirkung des Nicotins, Berlin, 1868.
[4] Corresp.-blatt. Psych., 16, 1864.
[5] Mat. Med. d. reinen Pflanzenstoffe.
[6] Arch. path. Anat., 1855.
[7] Ibid.
[8] Headland and some others style tobacco an inebriant.

apparently this action is exerted upon the nerve termini among the muscles, for section of the vagi does not prevent it; the respiratory excitement is followed by paralysis. A very constant lowering of superficial temperature has been observed by Tscheschichin, which probably depends on rapid cooling of the body in consequence of vaso-motor paralysis. Intestinal peristalsis is hurried, and there may even be intestinal tetanic spasms; while in women there are vigorous contractions of the uterus. The action upon the pupil has been much disputed, but the most recent and reliable researches seem to prove that nicotine (in opposition to the mydriatic group of Solanaceæ) produces contraction of the circular fibres of the iris.

Finally, it may be said that the administration of very large doses seems to produce death by direct paralysis of the heart, without the preliminary so-called phenomena of excitement.

Of the more chronic forms of mischief which are said to be produced by smoking, it is more difficult to speak with confidence. Among adults, beyond the mere local change which is known as granular inflammation of the fauces and pharynx, the injury which has apparently been proved to occur with some frequency is amaurosis, from atrophy of the retina. Chronic dyspepsia, also, which is thought to be caused (or at least greatly aggravated) by the waste of saliva which should have gone to assist digestion, is an undoubtedly frequent result of prolonged excesses in tobacco. The occurrence of angina pectoris as the result of the prostrating influence of great and prolonged excess upon the heart, has been noted by M. Beau and others, but is fortunately an extreme and rare event. General nervous depression has far more frequently been produced, showing itself in restlessness, insomnia, and a tremulous condition of the limbs, not very unlike the phenomena of chronic alcoholism. About all the above forms of chronic mischief from tobacco there is, however, great dispute; that is to say, as to the proportion of cases in which they occur, and the possibility of their production at all, except by very great excess. But as to one matter, now to be mentioned, there can be no doubt, viz., the unmixedly evil effects of smoking in early youth. The use of tobacco in the years before and immediately succeeding the development of puberty has a most prejudicial influence. There can be no question that more than anything, except drinking, it hinders the growth and development of the higher nervous centres, and that both intellect and moral character are capable of being most seriously damaged by this depressing agency continuously exerted during the developmental period. And not merely does this result, but by its interference with digestion and assimilation the use of tobacco by the young impairs the whole process of the consolidation of the bodily frame. There seems to be good evidence that this especially tends to produce amaurosis, from atrophy of the retina.

THERAPEUTIC ACTION.—By far the most important of the purposes to which tobacco or nicotine has been applied in medicine is the rapid and complete relaxation of muscular spasm. The most familiar instance of its application for this purpose is (or rather was, for chloroform has superseded it) the employment of tobacco enemata for the treatment of strangulated hernia. There can be no doubt that this proved successful in many instances, but, on the other hand, the remedy was found to be very uncertain, and occasionally dangerous: in fact, several fatal accidents occurred. More recently, chiefly upon the authority of Mr. Curling, tobacco has been put forward as a remedy for tetanus, and there is really much evidence in favor of its being more potent than most of the remedies that have yet been tried. It is not desirable, however, that tobacco should be used any longer even in so desperate a disease as tetanus, in the objectionable form of watery solution. The subcutaneous injection of nicotine offers every facility for handling this potent agent with precision, as well as effectiveness. A convenient solution for its use is as follows: —Nicotine, ♏x.; mucilag. acaciæ, ♏xl.; distilled water, ♏ccccl. Of this solution five minims may be injected every four hours.

A less important, but still very distressing spasmodic affection, for which tobacco has often proved very useful, is nervous asthma. The late Dr. Hyde Salter placed great value upon it. He, however, employed it solely to cut short actually commencing paroxysms, and pushed its influence quite to the poisonously paralyzing stage, not being content until he had rendered the patient faint and sick. It is probable, however, that asthmatic patients, like sufferers from many other forms of nervous irritation, can derive considerable benefit from a less pronounced action of tobacco than this, and that a mild and stimulant dose of tobacco smoke may not infrequently, by at once bracing and calming the respiratory nervous system, avert an otherwise impending attack. And although tobacco cannot be called a direct hypnotic, yet, by soothing the minor forms of neuralgic pain and general restlessness, it often helps to make sleep possible. To aged persons, provided they have learned to overcome the difficulties of smoking in early life, there can be no doubt tobacco often proves most useful in quieting that nervous restlessness which produces a variety of discomforts, including wakefulness at night. Tobacco is one, and one of the least harmful, of those agents by which soldiers and others, who are compelled to undergo strenuous exertion at times, with a sadly insufficient supply of food, are enabled to hold out, and do their work effectively. What opium is to the Tartar courier, tobacco is to the British soldier or sportsman, in supporting him under severe and continuous physical efforts, when rest and sufficient food are alike beyond his reach. Finally, tobacco seems to be an antagonist to strychnia. The Rev. Professor Haughton published a very interesting account of experiments on the antagonism of nicotia and strychnia. And Dr. Smyley publishes a case of a boy who, having taken about four grains of strychnia, recovered under the influence of an infusion of tobacco. These and many other published cases tend to prove the curative power of tobacco and its alkaloid over strychnine-poisoning.

PREPARATIONS AND DOSE.—Infusum Tabaci (enema), ℥ss.— ℥ij. (15.—60.); Vinum Tabaci, ♏v.—xxx. (.30—2.); Ungt. Tabaci; tobacco mixed with linseed meal is frequently employed as a poultice, especially in epididymitis. If the proportion of tobacco is large, the practice is not altogether safe.

MAGNOLIACEÆ.

DRYMIS WINTERI (Winter's Bark).

Active Ingredients.—Winter's bark, according to the analysis of Henry, contains volatile oil, resin, coloring matter, tannin, acetate and sulphate of potash, chloride of potassium, oxalate of lime, and oxide of iron. Of these the really active ingredients are probably the tannin and the volatile oil. The latter is divisible into two substances: one is a thin, greenish yellow fluid, the other a fatty body of acrid and burning taste.

Therapeutic Action.—The union of the styptic tannin and the stimulating volatile oil renders Winter's bark a tonic, with qualities that one would suppose to be very valuable in various disorders. It was employed for much the same purpose as Canella, which superseded it, and which has itself, in turn, fallen nearly out of use. Upon the whole, it seems to be a fairly active stomachic tonic, well applicable to chronic atonic dyspepsia; it has also been used in paralysis, scurvy, and other conditions of debility.

Preparations and Dose.—None officinal. The dose of the powdered bark is ℨ ss.— ℨ j. (2.—4.), which may be given in infusion.

ILLICIUM ANISATUM (Star-anise).

Active Ingredients.—The fruit of the *Illicium anisatum* contains a substance, anithol (*Anisi Camphor*), which is chemically, though not physically, identical with the nitrogen-containing element of the volatile oils of common anise, of *Anethum janiculum*, and *Artemisia Dracunculus*. Its formula is $C^{10}H^{12}O$.

Therapeutic Action.—The Chinese employ the fruit as a carminative.

Preparations and Dose.—The fruit may be made into a confection, or the volatile oil may be prepared in the same way as true anise oil. The dose is much the same as that of true anise and of anise-oil.

It is probable that other members of the Magnoliaceæ really possess more positive therapeutic powers than the so-called Winteraceæ. Perhaps the most widely esteemed for tonic and also diaphoretic powers (on the continent of America, for it has not been used in Europe), is the *Magnolia glauca.*

This tree is common in the morasses of the middle and southern United States, and bears the vernacular names of "beaver-tree" and "swamp-sassafras." The bark has an aromatic odor, diminishing, however, after it has been kept for some time; and a warm, pungent, and rather bitter flavor.

The bark of the root corresponds, as to properties, with that of the stem or trunk, and, for pharmaceutical purposes, is considered preferable.

The Pharmacopœia of the United States likewise includes the *Magnolia acuminata*, called in the vernacular the "cucumber-tree," and the *M. tripetala*, or "umbrella-tree." The bark of both these trees is reputed to possess the same properties as that of the *glauca*.

All three species being cultivated in England, it might be worth while to inquire, by experiment, if the difference of climate in any degree impairs the medicinal value.

It should be added that the tincture of the fruits has been found equally valuable with that of the bark; and that an infusion of the fruits, while green, in whiskey or brandy, is extensively employed in the United States against intermittent fevers and for rheumatism.

MENISPERMACEÆ.

CALUMBA (Cocculus palmatus).

ACTIVE INGREDIENTS.—The two elements which, probably, are alone efficient, are berberine and starch; for the alkaloid calumbine, $C^{11}H^{22}O^1$, is most likely inert.[1] Berberine, $C^{20}H^{17}NO^4$, exists in the root in combination with calumbic acid; its action will be discussed below. It forms, out of an aqueous solution, small glittering yellow crystals, which have a bitter taste and a neutral reaction. At a temperature of 120° C. it melts into a reddish-brown resinous mass. It is little soluble in cold, much more soluble in boiling water; very soluble in alcohol; little so in benzol; insoluble in ether and petroleum-ether. Charcoal precipitates it from a solution, but alcohol will set it free. Cold water precipitates it from its solution in alcohol. Its solutions have no influence on the plane of polarization. The starch is a most important element, and exists in the root to the amount of 33 per cent.

PHYSIOLOGICAL ACTION.—Although calumba itself can hardly be spoken of as having any toxic effect, since it would probably be impossible to administer the drug in sufficient quantities to produce such action, there is at least one ingredient of it—berberine—which is possessed of poisonous properties. These, however, are weak and variable.

Berberine, it will be remembered, has been mentioned as a constituent of several plants of the ranunculaceous family (Podophyllum, Hydrastis, etc.). Its actions have been investigated by Falck[2] and Guenste,[3] and no doubt is left that, upon certain animals at any rate, it can act as a fatal poison. Rabbits died in from eight to forty hours after a subcutaneous injection of 7½ to 15 grains. Dogs, on the other hand, resisted all at-

---

[1] This opinion is based on the absolutely negative results of the researches of Schroff, and of those of Falck and Guenste; but other authors have conjectured, probably on the sole ground of its bitterness, that calumbine is the active principle of calumba root.

[2] Deutsche Klinik, 14, 15, 1854.

[3] De Columbino et Berberino Observationes. Inaug. Diss. Marburg, 1851.

tempts to kill them, even by the largest doses injected into the blood. Given to them by the stomach, berberine produced some temporary tremor, restlessness, and thirst, and a few watery evacuations. Injected into the veins, it produced salivation, watery evacuations, and (after a second injection) convulsive shudderings followed by paralysis (especially of the hind limbs) and difficult and frequent breathing; but all these symptoms gradually wore off, the tremor being the last to disappear. In pigeons and fowls repeated injections of berberine solution into the crop caused vomiting, watery evacuations, quick breathing, and loss of appetite; given in pills amounting to a quantum of four to eight grains with each day's food, the drug produced progressive loss of appetite, to the extent of causing marked inanition.

On man, berberine has never been known to produce actively poisonous effects; but several observers record pain in the belly and watery diarrhœa, as the result of swallowing doses of 7 to 15 grains—effects which at least are remarkable, and different from those which result either from the employment of small doses of berberine itself, or from the ordinary medicinal employment of ounce doses of the infusion of calumba twice or three times daily.

THERAPEUTIC ACTION.—Calumba is powerfully tonic and antiseptic, and when resorted to for medicinal use has the great recommendation of being free from nauseous flavor. Another excellent character of the drug is that, while free from acridity and astringency, there exists in it so considerable an amount of starch that it becomes demulcent. Of all known tonics, moreover, calumba is the least likely to disagree with the stomach. It is anti-emetic; it promotes the appetite, and assists digestion, yet is not stimulant. The excellent tonic and stomachic effects are produced without any accompanying nausea, sickness, or headache—symptoms which so frequently follow the exhibition of tonics—and hence, it can be usefully employed when other remedies of its class would be instantly rejected. So far, indeed, is calumba from ever being a cause of sickness, that its qualities may be pronounced to be exactly the reverse; in a word, as just now said, it is *anti*-emetic. Not only does it arrest the vomiting produced by tartar-emetic and ipecacuanha, but, when combined with these medicines, their operation is rendered milder.

Calumba is employed with great benefit to the patient in all those affections of the stomach and bowels which are accompanied by an increased secretion of vitiated bile. In a languid state of the stomach also, especially when nausea and flatulence become troublesome, experience has fully justified its reputation. So, again, in the "bilious attacks," as they are commonly termed, to which delicate females are often subject, the conjoined use of infusion of calumba and of effervescing draughts is in most cases exceedingly beneficial; while in dysentery, when the inflammatory symptoms have subsided, it has proved so serviceable that, in Germany, calumba goes by the name of *Ruhrwurzel*, or "dysentery-root." The utility of the drug is specially observable when the dysentery is chronic, and attended by ulceration of the colon.

In habitual diarrhœa, when tonics are admissible, calumba possesses value similar to that displayed in dysentery.

Vomiting caused by kidney disease and renal calculi has been alleviated by the employment of calumba. The anti-emetic effects have also been found particularly useful in cholera. Combined with soda, powdered culumba-root is serviceable also in pyrosis.

After attacks of fever, when cinchona or disulphate of quinia is about to be exhibited, calumba, in infusion, is an excellent preparation. In hectic fever, and in the last stages of phthisis pulmonalis, calumba has been found an excellent agent for the checking of colliquative diarrhœa, also in allaying irritability, and imparting to the stomach a certain amount of vigor.

Women, during the early period of pregnancy, often suffer from a distressing amount of vomiting. This is frequently allayed by the use of calumba; and if, after confinement, puerperal fever should supervene, calumba is, according to Denman, preferable to cinchona.

The vomiting and diarrhœa which often accompany dentition, are likewise amenable to the influence of this excellent medicine.

Sydney Ringer (p. 410) speaks of calumba in the following words: "Calumba is used as a tonic to increase both appetite and digestion. It has a slight irritant action upon the stomach, as indeed most bitter substances have, and is considered by this property to be able to set aside slight changes in the mucous coat of the stomach, and, in this indirect way, to assist appetite and digestion. It is easily tolerated by the stomach, and is thus employed when this organ is in a weak state, as during the convalescence from an acute disease, when it is often found that calumba is borne with benefit, while stronger tonics may upset the stomach."

PREPARATIONS AND DOSE.—Extract. Calumbæ Fluidum, ♏v.—xxx. (.30—2.); Infus. Calumbæ, ℨss.—ij. (15.—60.); Tinct. Calumb., ℨi—ij. (4.—8.)

PAREIRA BRAVA (Cissampelos Pareira).

ACTIVE INGREDIENTS.—The active ingredient of pareira is now pretty well ascertained to be nothing more or less than buxine, $C^{16}H^{21}NO^3$, an alkaloid which will be found mentioned as entering into the composition of plants of two other families (Lauraceæ and Euphorbiaceæ), besides being present in various species of Cissampelos, and, probably, some other plants of the order Menispermaceæ. Buxine, discovered by Fauré in boxbark, is substantially identical with the bibirine which was found by Rodie, Maclagan, and Tilly in the "Bibiru" tree, and the pelosin, discovered by Wiggers in pareira.

Buxine, when pure, is a white amorphous substance, which loses 8.2 per cent. of water when heated to 212° Fahr. It has a strong and clinging bitter taste, and a marked alkaline reaction; is very insoluble in water, freely soluble in absolute alcohol and in ether, remarkably soluble in chloroform and acetone. It completely neutralizes acids, and forms uncrystallizable salts with them. Concentrated sulphuric acid changes it to a brownish-yellow resinoid body.

PHYSIOLOGICAL ACTION.—The root of *C. Pareira* is a well-known tonic and diuretic, exerting a specific influence over the mucous membrane which lines the various passages. What are the effects produced upon the system in general does not appear to be well known, though, in large doses, it would seem certain that pareira is aperient.

THERAPEUTIC ACTION.—The use of pareira at the present day is almost limited to the cure of discharges from the urino-genital mucous membrane. Gonorrhœa, leucorrhœa, and chronic inflammation of the bladder are successfully treated with it. In the last-named disease Sir B. Brodie considers pareira superior to uva-ursi. It lessens, he tells us, the secretion of the ropy mucus, and appears, at the same time, to diminish the inflammation and the irritability of the organ.

PREPARATIONS AND DOSE.—The best form [Phillips] for the administration of pareira is the decoction [B. Ph.], to which may be added, as occasion renders desirable, alkaloids, acids, or narcotics, especially hyoscyamus. The officinal preparations are: Extract. Pareiræ Fluidum, ʒss.—j. (2.—4); Infus. Pareiræ, ʒj.—ij. (30.—60.)

COCCULUS INDICUS (Anamirta Cocculus).

ACTIVE INGREDIENT.—The properties of cocculus seem to be fully represented by its alkaloid cocculine, or picrotoxine, $C^{12}H^{14}O^5$. This substance crystallizes out of a clear solution in stellate bundles of colorless shining needles, free from water of crystallization. It is odorless, neutral in reaction, and bitter in taste; very highly soluble in alcohol; soluble in ether, amylic alcohol, and chloroform; also soluble in aqueous solutions of alkalies and of ammonia, and in concentrated acetic acid; but not very soluble in water.

PHYSIOLOGICAL ACTION.—The shell of the Anamirta seed is bitter and acrid; the seed itself is also very bitter. The substance of the seeds introduced into the stomachs of quadrupeds operates upon the spinocerebral system in such a way as to cause trembling, tetanic convulsions, and insensibility, and in large doses becomes fatal. Deadly effects of a similar kind are induced by cocculus indicus in *fish*, whence it is employed as a fish-poison.

When cocculus is swallowed by man, its action appears to be exerted chiefly upon the muscles of volition. If used in the brewing of beer, as a substitute for hops (which is done by dishonest brewers, though prohibited by statute), it gives the drink more "bottom," and renders it more intoxicating. The intellectual faculties appear to be less influenced by it than the muscular powers. No chemical antidote is known, but, if taken in excessive quantity, relief appears to be given by acetic acid.

The poisonous action of picrotoxine is remarkable and important; leading to the belief that the abandonment of cocculus as a remedy by many physicians has been a singularly unwise step. There is a general resemblance to the phenomena of strychnine, but this is not so accurate as has been represented. Falck observed that fish made twisting and boring movements, with intervals of quiet swimming. Roeber found that frogs had violent tetanic convulsions, with intervals of stupor, and a notable protrusion of the belly from overfilling of the lung with air. Small doses, $\frac{1}{30}$ to $\frac{1}{15}$ of a grain, produced restlessness, weakness of movements, sinking in of the eyes, somnolence, occasionally a temporary loss of reflex irritability, which afterwards became active again, and, in fifteen minutes, attacks of opisthotonos (with distention of the belly), at intervals of half a minute, semicircular swimming movements, winding up with violent tonic spasms, and violent noisy expulsion of air from the mouth. Then came an interval of exhaustion, followed by emprosthotonos, and most singular movements of the hinder limbs. The attacks increased in number and severity, but death was not induced for several hours, sometimes not till after several days. The heart's action is slowed, or even stopped; during the tetanic spasms the capillaries are gorged. In experiments on mammalia there has been the same alternation of tonic and clonic convulsion, the latter frequently of the most singular but definite kind, as swimming, thrusting backwards and forwards, rolling over on the axis of the body, accompanied by great slowness of respiration and of heart-action. The pupils seem to be at first contracted, and afterwards dilated. After death, in mammalia, the meninges are always found engorged with blood, while the nervous centres themselves are either normal or anæmic; the venæ cavæ and the flaccid heart (especially the right side) are very full of dark blood; the lungs are also engorged, and present apoplectic spots, and there is a profusion of mucus in the mouth, throat, trachea, and bronchi. The main action of picrotoxine seems to be that of an excitant to the centres, especially the medulla oblongata. It is obvious that the motor centres, the vagus centres, and the centres for restraint of reflex movements, are powerfully affected. It has been experimentally proved that the convulsions can be excited in an animal from which the brain has been removed.

THERAPEUTIC ACTION.—In medicine cocculus indicus has long been considered useful for the destruction of pediculi; also in certain skin diseases, such as porrigo, scabies, and ring-worm of the scalp; for all of which purposes it is best employed in the form of ointment. The last-named disease (tinea tonsurans) and ringworm of the body (or tinea circinata) are both advantageously treated also with decoction of cocculus, the skin being well washed with common brown soap and hot water every night and morning, before the employment of it. If the ointment be preferred it should be prepared from the kernels alone.

In Vomiting, etc.—Cocculus indicus likewise possesses considerable value as a medicine for internal exhibition,—in certain forms of vomiting, for example, such as those which are accompanied by dull and heavy pain in the head, giddiness, and intolerance of light and sound. Vomitings of this class are sometimes very severe and persistent. In other patients there are alternations of constant nausea, with violent but ineffectual efforts to evacuate the stomach, the latter becoming so exceedingly irritable that no description of food can be endured. In either case cocculus is an efficient agent in allaying the symptoms, and seldom fails to give the relief so much desired. It is when the part primarily affected is the head (as indicated in the pains there felt, and in the giddiness), and stomach affections of the character spoken of ensue, that cocculus, by relieving the cephalic symptoms, removes the stomach affections which follow them.

In Dyspepsia.—Cocculus is likewise a valuable medicine in various forms of dyspepsia, and specially for employment in particular stages of that disorder. In cases where there is severe epigastric pain, aggravated by pressure, or by taking food,—and which are accompanied by flatulent distention of the stomach, and perhaps of the intestines generally, with nausea, giddiness, headache, dryness of the mouth, and a feeling of hunger, yet a repugnance towards food,—the same good results attend the exhibition. If not invariably successful, there is at all events great probability of relief being experienced.

Again, when the colon is distended with flatus; when the bowels are constipated, and the motions are more or less hard and lumpy, cocculus proves, in many cases, of singular service.

The same result ensues in those milder cases of tympanitis so often met with in the later stages of peritonitis, and frequently encountered also in enteric fever; here, as in the preceding, a few doses of the tincture of cocculus will often remove the pain, and relieve the distention. Colic is similarly amenable to the influence of cocculus, as are most other spasmodic affections of the stomach and bowels depending on flatulent distention of any of the abdominal viscera.

The good effects of cocculus are likewise seen during pregnancy. At this period the intestines are often much distended with flatus, and the patient suffers from frequent desire to urinate, referable in part to flatulent pressure on the bladder.

In Menstruation, etc.—Other affections to which women are subject, and which cocculus is able to mitigate, if not to subdue, are found among those which occur during the menstrual period. In females of nervous temperament, and of thin and delicate fabric of body, the menses are often preceded by certain paroxysmal pains of a colicky nature, felt in the hypogastric region, and accompanied by more or less pain in the back and hips. These pains not only precede the arrival of the catamenia, but accompany them for the first day or two. They are of a twisting, griping, or colicky character, and are attended by a scanty discharge, or by a profuse one, in either case somewhat paler perhaps than customary, and attended not uncommonly by clots or shreds. The administration of cocculus in these cases, commencing a few days prior to the expected flow, and continued during the first two or three days of its progress, will frequently ward off the pains, and render the discharge more natural.

Leucorrhœa is also treated advantageously with cocculus, especially when the discharge is of a sero-purulent character, and accompanied by pain in the lumbar region.

Chlorosis too, where the menses disappear for months together; or where the discharge takes place only at irregular and long-separated intervals, and is pale and scanty. Patients so conditioned often suffer from profuse and exhausting leucorrhœa between the periods; but by the employment of cocculus they may receive great benefit, and have the menses re-established and regulated.

Nervous Affections.—Cocculus is a very valuable medicine in various nervous affections, such as certain forms of hemiplegia, paraplegia, and paralytic stiffness, accompanied by a sense of heaviness, and by a loss of motor power in the lower limbs, with giddiness, and a feeling of lightness in the head. I have recently witnessed its good effects upon a patient who had suffered for several months from loss of power in the lower limbs, moving them only with considerable difficulty, and who, when standing erect, became giddy, disposed to be sick, and light-headed. This patient quite recovered through a few weeks' perseverance with the tincture.

Nervous affections, such as hysterical hemiplegia, choreic hemiplegia, and epileptic hemiplegia may likewise be usefully treated with cocculus. I have seen well-marked cases of hysterical paralysis—where the sensibility and the muscular power were both impaired—yield quickly to this drug; and cases, in particular, which have been accompanied by menstruous irregularities, and by spasms attacking different organs. Some of those epileptic cases which are attributable to onanism likewise derive benefit from cocculus.

Reil employed tincture of cocculus seeds in chorea; in hemiplegia arising from cold; and in paralysis of the sphincter vesicæ; and in every instance with advantage to the patient.

Gubler recommends picrotoxine in chorea.

Tschudi recommends cocculus also in paralysis of the extremities, and of the sphincters; resorting, however, to picrotoxine.

PREPARATIONS AND DOSE.—None officinal. A tincture may be made of the usual strength (1—8), of which the dose would be from ♏ij.—x. (.12—.60). The dose of picrotoxine is from gr. $\frac{1}{66}$—$\frac{1}{13}$ (.001—.005).

PAPAVERACEÆ.

OPIUM (PAPAVER SOMNIFERUM).

ACTIVE INGREDIENTS.—A large number of substances enter into the composition of opium, and of some of these the physiological action is by no means settled. The most practically important are the alkaloids morphia and codeia, which exist in combination with meconic acid, an inert ingredient of which nothing further need be said. A third alkaloid, more or less resembling these in properties, but the action of which is not so accurately known, is narceia. Somewhat analogous to them, but rather intermediate in properties between these three and the group to be next mentioned, is papaverine. In strong contrast to the alkaloids, on which depend the properties of opium that are most familiarly known, stand narcotine and thebaia; the former of these being an antispasmodic tonic in small doses, and producing convulsions and death when given in large quantities; while the latter is apparently useless in very small doses, and a violent tetanizing poison when given in about the same quantity as would constitute a fatal dose of strychnia. Lastly, there is meconine, or opianyl, which has lately attracted some notice as a hypnotic.

As less important constituents of opium may be mentioned opianine, porphyroxine (or opine), cryptopine, rhœadine, papaverosine, and thebolactic acid.

Morphia, $C^{17}H^{19}NO^3 + H^2O$ (that is, with one atom of water of crystallization) forms small white silky needles; or (if slowly crystallized out of alcohol) colorless, semi-transparent, six-sided, oblique, rhombic prisms. It has a distinctly alkaline reaction, and a bitter taste, but no smell. It melts at about 130° C., and a higher degree of heat destroys it. It is soluble in about 1,000 parts of cold, or 400 parts of boiling water; much more soluble in chloroform and in alcohol; insoluble in ether. It completely neutralizes the acids, forming salts which are, for the most part, crystallizable, odorless, bitter, soluble in water and in alcohol, but insoluble in ether, chloroform, and amylic alcohol. Nitric acid dissolves morphia with effervescence, producing a rich orange color; perchloride of iron strikes a deep indigo blue, turning to green when the reagent is added to excess; strong sulphuric acid and bichromate of potash produce a green tint, owing to reduction of oxide of chromium. These reactions are due to the deoxidizing power of morphia. Morphia itself is not employed medicinally.

Of the salts of morphia four are used in medicine—the acetate, muriate, sulphate, and bimeconate. The acetate, when carefully crystallized, presents bushy groups of silky needles; it is highly soluble in water (1 in 24), and evolves an acetous odor. The muriate occurs either in the form of groups of silky needles, or of large transparent prisms; it dissolves in 20 parts of cold water, 80 of cold alcohol, and 19 of glycerine. The sulphate forms bushy groups of needles that are soluble in 2 parts of water. The bimeconate (the form under which morphia exists in opium) is a resinoid body.

Codeia, $C^{18}H^{21}NO^3$, is a powerful base, capable of precipitating the solutions of the salts of iron, lead, nickel, copper, and cobalt. It crystallizes out of a pure ethereal solution in small colorless crystals, which are without water; they melt at 150° C., and crystallize again on cooling. From a watery solution codeia separates in crystals which contain water; and, if the process be slow, forms large, regular, transparent octahedra, and right rhombic prisms; these part with their water of crytallization at 100° C.; in boiling water they melt into an oily fluid, odorless, and of a weakly bitter taste. Codeia is soluble in 17 parts of boiling and 80 parts of cold water; easily soluble in alcohol, chloroform, and ether. Codeia salts are mostly crystalline, have an exceedingly bitter taste, and are almost insoluble in ether. Two are used in medicine besides the alkaloid itself—viz., the muriate and the nitrate. The muriate crystallizes out of dilute solutions, in star-shaped groups of short needles; it is soluble in 20 parts of cold, and less than one part of boiling water. The nitrate crystallizes out of boiling water in small prisms, which melt under heat, and cool again into a resinoid mass.

Narceine, $C^{23}H^{29}NO^9$, crystallizes from its solutions in water, alcohol, or dilute acetic acid, in long, white four-sided rhombic prisms, or clusters of fine needles, which have a taste at first bitter and then styptic; these melt at a high temperature and solidify amorphously on cooling. They are very insoluble in water, slightly soluble in cold, and easily so in boiling alcohol. Narceine dissolves in weak mineral and vegetable acids, if pure, without discoloration; if impure, often with a blue tinge. Although a weak base, it forms salts with these acids. Concentrated sulphuric acid first colors narceine brown, and then forms a bright yellow solution; concentrated nitric acid forms with it a yellow solution.

Papaverine,[1] $C^{20}H^{21}NO^4$, crystallizes out of alcohol in white, confusedly heaped-up needles or scales. It is very weakly alkaline in reaction. It is almost insoluble in water, and but little soluble in cold alcohol or ether; boiling alcohol takes it up freely; chloroform extracts it with facility either from an acid or an alkaline watery solution. Concentrated sulphuric acid changes papaverine instantly to a deep violet blue, and forms a violet solution, the color of which fades very slowly. Papaverine salts are mostly difficult of solution in water.

Narcotine, $C^{23}H^{25}NO^7$, crystallizes out of an alcoholic or ethereal solution in colorless, transparent, pearly prisms, or grouped needles on the right-rhombic system: precipitated by an alkali from an acid solution, it presents the form of light white flakes. It is odorless, tasteless, and neutral in reaction. Cold water hardly dissolves it; boiling water only in very small proportion. It is moderately soluble in alcohol and in ether;

---

[1] This alkaloid is not to be confounded with another substance also (most unwisely) named papaverine, which exists in the unripe poppy capsule. The latter is a resinoid body.

chloroform dissolves it with great facility. Its neutral solutions rotate the plane of polarization to the left; its acid solutions rotate the plane to the right. With acids, narcotine forms salts of acid reaction; usually uncrystallizable; soluble in water, alcohol, and ether; and extremely bitter to the taste. Concentrated nitric acid converts it into a red resinoid body, red vapors being also given off; this resin, on boiling with potash, develops methylamine.

Thebaine, $C^{19}H^{21}NO^4$, crystallizes in silvery, four-sided scales, or in needles, or in small and horny crystals. It is strongly alkaline in reaction, and strongly electro-negative on friction. It is almost entirely insoluble in water, and in aqueous solutions of ammonia or alkalies; freely soluble in alcohol and ether, and rather less so in benzol. Dilute acids dissolve it easily, and form salts which will crystallize out of an alcoholic or an ethereal solution, but will not crystallize from a watery one. Concentrated acids, and also chlorine and bromine, have each the power to change thebaine into a resinoid body. Concentrated sulphuric acid dissolves it blood-red, the color afterward changing to yellow. The taste of thebaine is sharp and styptic rather than bitter.

Opianyl or meconine, $C^{10}H^{10}O^4$, forms tasteless, white, shining needles, not easily soluble in cold water; soluble in ether, alcohol, and acetic acid. It dissolves without color in pure sulphuric acid, but, on heating the solution, it becomes purple.

The above-named substances are the really important components of opium, so far as present knowledge goes, described with just sufficient detail to allow of their physical and chemical identification. So much seemed absolutely necessary, but to go farther into the vast subject of the chemistry of opium would be foreign to the scope of the present work.

PHYSIOLOGICAL ACTION (A) OF OPIUM, AND (B) OF ITS INDIVIDUAL INGREDIENTS.—It is necessary to consider these divisions separately. (A.) The physiological action of opium given in poisonous doses presents somewhat different aspects according to circumstances. Where the quantity is very large, and the forms such as to permit rapid absorption of the whole, the course of symptoms is as follows:—the patient, if an adult, quickly becomes conscious of a sense of fulness in the head, which seems to commence in the nape of the neck, and to spread therefrom; and in the course of a few minutes feels great and increasing drowsiness, and a sensation of general heat, which increases to an almost intolerable degree, and is then accompanied by sweating. This sweating, which generally bathes the body in moisture, tends to suspend all the other secretions, and is probably nature's medium for eliminating the poison. If the dose given has been very large, it reduces the surface to a clammy coldness. The drowsiness passes into semi-coma; the patient, unless aroused, lies unconscious, and is heedless of everything around him, but if roughly spoken to and shaken, can still be induced— though the speech is thick and hesitating—to answer questions, and even to get up and walk about; on leaving him, however, he immediately subsides into his former stupor. The pupils are generally contracted to the size of a pin's point, and become insensible to light, and there is buzzing in the ears. The pulse, which in the first or hot stage was rapid and somewhat full, becomes feeble and irregular, and falls to the normal rate, or even below it; simultaneously with this alteration, the features become pinched and ghastly, and flushing gives place to livid pallor, while the muscles of the limbs are affected with spasmodic jerkings.

The mouth and fauces are dry, and there is commonly nausea, and often vomiting. The general depression of the system is marked. The respiration becomes more and more embarrassed; in the later stages the presence of a quantity of mucous secretion in the tubes makes itself known by the rattling sound of the breathing. The respiration comes to a standstill usually some minutes before the entire cessation of the heart's action. Death, that is to say, takes place by apnœa.

The course of events differs considerably in young children, especially in young infants. Here there is a more rapid passage into profound stupor; but what is specially characteristic is the much greater frequency with which convulsions occur. The convulsive movements vary from mere twitchings of the facial muscles to rhythmical startings of the limbs; to severe clonic convulsions, which may be hemiplegic, or may affect both sides of the body alike; and even to tetanic spasms.

I have seen three adults thrown into a state of apparently complete tetanic rigidity of the whole body, the mouth open and squared, all the facial muscles highly tetanized, the spine strongly curved, resembling the shape of a well-drawn bow; and this state of opisthotonos continue in each case, without any abatement, from twelve to forty-eight hours, and then only become relaxed to become rigid again for a longer or shorter time, according as the effect of the opium was kept up.

One of the patients remained more or less in this state for four or five weeks, opium having been given in the first instance to relieve toothache, and continued for the convulsive spasms which followed. As soon as the opium was discontinued, the patient recovered; but some months afterwards, having accidentally again taken opium, the same results quickly reappeared.

It is very interesting to note that this same tendency to convulsive movements under opium-poisoning, which in the European human race is almost limited to the first few years of life, is, in many of the lower animals, a regular feature of profound opium-poisoning.

Among the other exceptional poisonous effects of opium must be reckoned a singular action which it exerts upon the skin of certain patients, especially that of elderly ones. In such individuals an intense irritation of the skin, with intolerable itching, and sometimes even the production of a sudaminous rash, has been observed; these effects resulting both from the external and from the internal use of opium.

B. *Physiological Action of Individual Opium Salts.*

(*a*) NARCOTIC ACTION OF MORPHIA.—Of the various constituents of opium, there can be little doubt that morphia most nearly represents the aggregate toxic powers of the drug. In poisonous doses it congests the cerebral vessels, producing coma; it paralyzes common sensation, and, in a less degree, voluntary motion; it paralyzes the muscular coats of the intestines and the bladder, and generally produces death by paralysis of respiration, or, in a word, by apnœa. It possesses, moreover, in a high degree, the less conspicuous power of opium (usually referred exclusively to the action of thebaine), of producing convulsions, either clonic or tetanic. In this last respect even the illustrious Claude Bernard does not appear to have been fully aware of the activity of morphia. In estimating the relative power of the different opium-salts as tetanizers, he places morphia only fifth; in a position inferior, that is, not only to thebaine, but to papaverine, narcotine, and codeia. There is the best possible reason to think that in this view Bernard was mistaken. Very numerous records of accidental and criminal morphia-poisoning show that tetanic spasm is really quite as frequent a symptom in the human adult as clonic convulsions are in the child. Dr. Anstie has also shown,[1] by experiments on such various animals as dogs, cats, rats, rabbits, guinea-pigs, and frogs, that every variety of abnormal muscular action can be produced by morphia; the most characteristic, however, being cataleptic rigidity, shown in the rat, and tetanus, shown in the frog. But indeed there are other strange anomalies in this comparative table of Bernard's. In particular, he actually reckons narceine as superior to morphia in hypnotic power. This is so exceedingly incorrect (as will presently be seen) as to shake our confidence in these experiments, great as is Bernard's reputation.

(*b*) **NARCOTIC ACTION OF CODEIA.**—The action of poisonous doses of codeia has been mainly established by the researches of Wachs,[2] Baxt,[3] Fraser, and Crum-Brown[4] upon brute animals. On men there has been no opportunity of observing any but those minor poisonous effects which are produced by a somewhat excessive medicinal dose. The general result seems to be that the narcotic influence falls most severely on the cerebellum and medulla oblongata, producing convulsions. Wachs, who experimented on a large series of animals (under the direction of Falck), observed, in rabbits and dogs, the following general train of symptoms: muscular tremors, a sudden shriek, and stretching out of the limbs; twitching of the lips, and jerking of the eyeballs; (occasionally spasms of the jaw-muscles, and backward locomotory movements, or running in a circle;) weakness of the limbs, restlessness, projection of the eyeballs, immediately followed by opisthotonos and inspiratory spasm, which was rapidly fatal if the dose was very large, but with rather smaller doses, repeated at intervals, was varied by clonic spasms, ending in exhaustion and death. The whole suite of phenomena much resembled those produced by picrotoxine. After death, Wachs usually found very pronounced hyperæmia of the cerebral membranes; and in dogs there was also much fluid collected in the ventricles. The substance of the brain seems to have been always uniformly free, both from hyperæmia and from local hæmorrhage. The heart was always greatly distended by dark blood; the lungs were greatly congested, as were the liver, spleen, and kidneys; while the alimentary canal, the urinary bladder, and the pancreas were anæmic; the gall-bladder was excessively full.

The minor poisonous effects, as noted in human beings, are semi-coma instead of sleep, nausea, vomiting, severe pain in the stomach, sometimes tinnitus aurium, slight salivation, feeling of pressure in the temples, weakness of sight, and a somewhat remarkable slowing of the pulse.

(*c*) **NARCOTIC ACTION OF NARCEINE.**—Scarcely any salt that exists has given occasion for such disputes and contradictions, respecting the extent of its physiological activity, as narceine. By Claude Bernard it was pronounced the first among the opium-salts as a hypnotic, but as regards tetanic action the least active of them all, and as regards general

---

[1] "Stimulants and Narcotics." London, 1864.
[2] Das Codeia. Eine Monographie. Marburg, 1868.
[3] Virchow's Archiv, 1869.
[4] "On the Connection between Chemical Constitution and Physiological Action." Edinburgh, 1868.

narcotic action only fourth, but still more active in this way than either morphia or narcotine. At the other extreme, Fronmüller declared, as the result of his experiments, that narceine has no activity whatever; that out of 22 cases in which he gave it, either subcutaneously or by the mouth, in one only was there a doubtful appearance of hypnotic effect; and, on the whole (since there was not a single instance of characteristic narcotic action upon the pulse, the temperature, the pupils, etc.), Dr. John Harley also expressed himself [1] as entirely sceptical as to the poisonous activity of narceine when pure; he had invariably found that any such power which might appear to belong to a particular specimen, was really due to its being contaminated with some of the other alkaloids of opium. Eulenburg, who is in general a very trustworthy witness, states, as the result of his experiments with narceine (prepared by Merk), that this alkaloid must be given in doses twice as large as those of morphia, in order to produce narcotic effects corresponding with those of the latter. Bouchardat (fils) considers [2] that the most important physiological effects are dryness of the mouth, nausea, and vomiting; (diarrhœa when as much as 1 to 1½ grains is subcutaneously injected;) subcutaneous twitchings, and slowing of the pulse and respiration. Reviewing the whole matter, it must be allowed that much doubt hangs over the action of narceine, and must continue to do so until the final clearing up of the essential question whether a pure specimen of the alkaloid has ever produced the special effects which are supposed to characterize this agent. On the whole, it seems most probable that although mistakes have been made by certain observers, through employing an impure specimen, this cannot explain such results as those obtained by Eulenburg, who experimented with the carefully prepared narceine of Merk.

(*d*) **NARCOTIC ACTION OF PAPAVERINE.**—Respecting this alkaloid also there have been great disputes, but the researches of Baxt [3] have placed it beyond doubt that papaverine is a narcotic of considerable power in paralyzing the heart (independently of any action of the vagus), and in paralyzing the function of reflex irritability, besides possessing a hypnotic action, which will be considered under the heading "Therapeutics."

(*e*) **NARCOTIC ACTION OF NARCOTINE.**—This alkaloid likewise possesses decided narcotic qualities when given in large doses. The latest and most trustworthy researches are those of Baxt and of Kauzmann. The former concludes that the poisonous action of narcotine is very analogous to that of thebaine, but attended by less convulsion, and by more of absolute paralysis. Kauzmann destroyed a cat with about five grains. Symptoms commenced in four hours with tremors, tetanic rigidity of the extremities, contraction of the pupil, clonic spasms of half a minute to a minute's duration, succeeded by coma and death in 36 hours. Claude Bernard, while placing narcotine among those opium-alkaloids which are exciters of convulsion, considered it the least poisonous of all opium-salts; and this opinion must probably be correct, considering the quantities of narcotine which have been given in India as anti-periodic medicines by O'Shaughnessy and others.

---

[1] Practitioner, 1868.
[2] De la Narceïne. Thèse de Concours. Paris, 1865.
[3] Virchow's Archiv, lxx., 1869.

(*f*) **Poisonous Action of Thebaine.**—This alkaloid, which is the most poisonous of the opium-salts, has an action much resembling that of strychnia. It is a powerfully tetanizing narcotic. The most complete observations are those of F. W. Müller,[1] who experimented upon frogs, dogs, rabbits, and pigeons. The last-named were found to be more resistant than the others. In dogs and rabbits tetanic spasms set in within a few minutes (preceded in rabbits by general restlessness), and this was followed by paralysis. In pigeons, with fatal doses, there was general tremor of the whole body, and then a strong tetanic attack, followed by isolated spasms and a paralytic condition. Where recovery took place there was vomiting, and either general tetanus or only partial tonic spasms. From experiments made upon frogs' hearts removed from the body, Müller concluded that thebaine acts partly as a direct poison to the heart; if the heart be immersed in a weak aqueous solution of thebaine its beats diminish, and it comes to a stop more rapidly than if plunged in pure distilled water. Müller decided a question of much interest in regard to the explanation of the physiological action of opium itself. He showed that morphia and thebaine do not antagonize each other; for a mixture of the two, subcutaneously injected into a rabbit, caused death of a kind similar to that produced by thebaine, and with great rapidity. No therapeutic application of thebaine has yet been made.

Therapeutic Action of Opium.—As a Hypnotic, opium is the best and most trustworthy of drugs, with the solitary exception of hydrate of chloral, the only medicine which in this respect approaches it. In former times there was remarkable failure of appreciation of the circumstances under which its soporific influence should be sought, and of the doses most appropriate for the purpose. When opium is given in such a manner as to produce sleep (in a previously healthy person) in the best way, its action is very peculiar. The patient, in about fifteen to thirty minutes after taking the medicine, perceives first a slight sense of muscular languor and relaxation in all parts of the body; and to this is soon added a not unpleasant sense of slight fulness and weight at the back of the head. The vague discomfort which had before pervaded the system, now gives way to a feeling of perfect lightness and freedom; and the mental state is that of agreeable indifference to everything, without any approach to stupor. After this condition has lasted for a variable time, drowsiness, such as precedes healthy sleep, comes on, and the patient soon sinks into slumber. On awakening he feels refreshed, and there is no nausea, and only a very little furring of the tongue, nor is there any headache. Very often, however, the next action of the bowels is slightly delayed, and, when it occurs, is somewhat constipated from extra dryness of the fæces. Such effects are commonly produced by half a grain to a grain of opium, or its equivalent in tincture, etc.

Very different is the result when marked wakefulness is treated by the administration of a heavy dose, such as two to five grains; then, as above remarked, the patient is truly narcotized. The early symptoms are those of semi-delirium; if sleep occurs, it is opium-coma rather than genuine sleep, and on waking, the patient experiences the nausea, headache, and constipation (with absence of bile from the stools) which indicate the receding stages of a real (though slight) opium-poisoning.

---

[1] "Das Thebain; eine Monographie." Marburg, 1868.

There are curious differences, as yet not at all fully explained, between the hypnotic effects of different preparations of opium. Laudanum fairly agrees in action with opium *per se;* the only differences depending, apparently, on slowness of solution of the latter in the stomach and consequent slowness of its absorption. On the other hand, the well-known "liquor opii sedativus" of Battley, and the preparations by which that medicine is most successfully imitated, contain about 50 per cent. more opium than simple laudanum,[1] but are generally admitted to produce a much more tranquil hypnotic effect, with less of brain-excitement at the commencement, and less of unpleasant after-consequences. So also the "black drop" and the "acetum opii" of the old Dublin Pharmacopœia. These, however, were much stronger preparations than laudanum; and are now nearly gone out of use. The "extractum liquidum" of the present [British] Pharmacopœia is the officinal representative of the kind of preparation which was sought for in Battley's liquor, but is one-seventh stronger than the tincture.

As an Anodyne, opium is without an equal. It will, at least temporarily, relieve pain of any kind; whether it be what is called merely functional, or such as depends on inflammation, or such as is the result of wounds, or of destructive organic processes, cancer for instance. It will be best, however, to defer the discussion of the employment of opium as an anodyne under circumstances of inflammation, to a separate paragraph.

The most characteristic example of the anodyne influence of opium is seen in its effect upon simple nerve-pain, such as neuralgia; or upon the pain of a severe wound, in which of course nerves must be cut, or torn, or bruised. Here its influence is direct and speedy, but there is a point that deserves attention. When the pain, however severe, is merely felt as a local distress, and does not throw the general nervous system into commotion, it will, I believe, be invariably found that the smaller doses of opium act fully as well as, if not better than, the large. On the other hand, where the secondary general distress is great, and especially where the emotions of the patient are excited and kept on the rack by the pain, there is no choice but to place him in a state of tolerably deep narcotic stupor, so as to get rid of the influence of the mind. The smaller doses of opium may, however, be effectively used as prophylactics, even in cases where the pain (then only just commencing) would otherwise be of the severest type; as, for example, in severe neuralgia.

As a local application for pain, opium is often exceedingly useful, though often inferior to aconite, belladonna, and some other drugs. It may be applied to the unbroken skin in the form of the pharmacopœial liniment [B. Ph.]; or of laudanum applied on cotton wool, or—much better—sprinkled on the surface of a hot linseed poultice. As an application to painful ulcers it sometimes has the highest value; but there are singular anomalies in this respect, and it occasionally happens that the local use of opium severely aggravates the pain of ulcers, setting up indeed a smart inflammation, with great heat, swelling, and throbbing. The ointment usually employed is not pharmaceutical, but consists of one part of soft extract of opium to nine parts of simple ointment. A much weaker preparation than this would probably be found to act better in a considerable number of cases. More will be said as to the irritative action of opium upon raw surfaces under the head of Morphia.

As a Relaxer of Spasm, opium has enjoyed for many centuries the

---

[1] Squire's "Companion to the British Pharmacopœia."

highest reputation. As regards the reduction of local spasms in muscles that are accessible, almost everything useful that can be said may now be included under the heading of local hypodermic injection of morphia, to be presently discussed. Where, however, we are unable to obtain a proper hypodermic syringe and morphia solution, we may effect some good by the local application of hot opium fomentations, or poultices. And when the localized spasm is in a deep-seated part, for instance, in ordinary intestinal colic, we may afford relief by the internal use of opium (two to four grains solid opium, or thirty to sixty minims of laudanum). Sometimes this is usefully combined (especially in lead colic) with a moderate dose of castor-oil; and the effect may be kept up afterwards by $\frac{1}{31}$ grain doses of morphia, given from time to time. But there is an especially important use of opium in the more formidable varieties of acute obstruction of the bowels. Formerly these affections[1] were universally treated with medicines intended to increase the peristaltic action of the intestines, and the results were most disastrous. Whether the obstruction depended upon a purely functional spasm, or on local paralysis of a portion of bowel, or on the accidental snaring of a piece of intestine by a band of lymph, the effect of heightened peristalsis was simply to drive down a portion of gut from above, and increase the obstruction, besides in all likelihood producing adhesive inflammation. To the late Dr. Brinton was chiefly due the method of treatment by large and repeated doses of opium, the singular results of which are well described in his little treatise on Intestinal Obstruction.[2] Not merely does the steady persistence in the use of such quantities as $\frac{1}{2}$ grain, given every four hours for two or three days or longer, arrest the most dangerous symptoms, such as stercoraceous vomiting, but it positively brings about a painless purgation, large quantities of semi-fluid and dark-colored fæces being expelled, to the immense relief of the patient. It will be remembered that belladonna may often be successfully employed in the treatment of these cases of intestinal obstruction.

The effect of opium in allaying irritation and hypersecretion of mucous membranes is often remarkable. The commonest example of this is found in the case of catarrhal irritation of the mucous membrane of the nares, the larynx, and the bronchial tubes. It ought carefully to be remembered, however, that it is to the early stages of these affections that opium is specially applicable, and that its employment at more advanced periods is often hurtful. This is particularly the case in fully developed bronchitis, more especially when the secretion is copious, and the expulsory power is feeble. Here it is quite possible to do even fatal mischief by the use of opium. And in general it may be laid down as a maxim, that moderate doses only are useful in the irritative affections of the air-passages. True spasmodic asthma stands in a somewhat peculiar position, and full doses of opium have sometimes in this complaint produced good effects; but it is now known that other narcotics, especially belladonna and stramonium, are more efficacious and less dangerous. [Fothergill[3] in reference to this use of opium and its derivatives says: "It is useful in the hacking cough of phthisis, when cough is excited by the presence of

---

[1] I leave out of the question the treatment of external strangulated hernia; here the remedies are taxis, under chloroform, extraction of gas and fluids with the aspirator, and, failing these things, operation.

[2] Second edit., edited by Dr. Buzzard.

[3] The Antagonism of Therapeutic Agents, London, 1878.

diseased masses in the lungs; and where the cough is distressing and yet useless, and incapable of getting rid of the source of irritation. Here it is necessary to stop the reflex mechanism of cough. This morphia does most effectually. It is noted, however, that morphia is not an unalloyed good in such cases. In addition to its action in destroying the appetite and locking up the bowels, it lowers the respiration while checking the cough. When under the influence of morphia, the patient sweats profusely, and to the extent of producing much exhaustion. I have combined belladonna with morphia, and found that this checks the sweats, while the morphia allays the cough: . . . . The effect of the belladonna is not confined to the skin, but it influences the respiration favorably."]

In catarrhal irritation of the gastric mucous membrane opium sometimes proves valuable; but it is here a two-edged weapon, for its constipating effects may more than undo any momentary advantage which it produces. However, in chronic gastric catarrh, attended with much pyrosis, opium in small doses, perhaps combined with the mineral acids, is sometimes very useful: and constipation may be met by the use of mild aperient waters, such as Pullna or Friedrichshall, if necessary.

In catarrhal affections of the intestines, especially in autumnal diarrhœa, and in English cholera (early stage), the good effects of opium, combined or not, according to circumstances, either with mineral acids, or with chalk and aromatic confection, are well known. One-sixth to one-fourth or half a grain, or an equivalent dose of tincture, should be given every two, four, or six hours. In the early stages even of Asiatic cholera, notwithstanding all that has been said against it, the preponderating voice of those practitioners who have had the largest experience still declares that opium is one of the most decidedly valuable agents. But in the stage of collapse, and in that of reaction, it is worse than useless. In irritable conditions of the mucous membrane of the genito-urinary organs, small doses of opium are often most valuable; but it is necessary always to bear in mind the possible existence of kidney disease, which often renders the employment of opium very dangerous.

As a Sudorific, opium is occasionally very useful: its effect is greatly heightened when it is combined with ipecacuanha, as in Dover's powder; ten grains of the latter will often give the greatest relief in febrile catarrh, with hot dry skin, and in various inflammatory diseases.

In Conditions of Acute Delirium opium is occasionally very useful, but it requires to be employed with great circumspection. The use of it (formerly frequent) in acute mania and in *délire aigu* is nearly abandoned in favor of hydrate of chloral, combined with the careful and assiduous administration of food, and, in some cases, of stimulants.

In Delirium Tremens also, opium may be said to be practically superseded by chloral, and by bromide of potassium. In the delirium of zymotic fevers opium has always been regarded with suspicion and dislike by some of the best practical physicians. Graves and Billing, however, have endeavored to show that it may often be employed usefully in such cases, in combination with small doses of tartarized antimony; and even the danger of opium pure and simple (in typhus, for example) has probably been a good deal exaggerated. One caution it is absolutely necessary to remember; viz., if kidney complication be present, opium will probably increase the danger very seriously. [Prof. A. L. Loomis, of New York, has introduced the practice of administering morphia in *acute* uræmia. In an oral communication to the editor, he stated that he first used it in a case of uræmic convulsions in 1868, with such satisfactory results, that he has frequently employed it since. The commencing dose should be small, not exceeding ten minims of Magendie's Sol. In *chronic* uræmia, however, where there is danger of œdema of the brain, he believes that morphia should be used with very great caution.

Ulcers.—The use of opium in forwarding reparative processes in certain kinds of ulcers and wounds is one of the most interesting facts in therapeutics. First in importance under this head is its action when boldly and promptly given in the phagedenic ulcer; here its employment, in doses of a grain or two grains every three or four hours, has frequently exerted a surprising influence, not only in easing pain, but in arresting the destructive processes. Again, in the indolent ulcer, with large flabby granulations, the internal administration of opium (three to four grains daily), combined with the external use of opium in lotion or ointment, has often entirely changed the character of the sore, and induced rapid healing. The use of opium, locally in particular, is by no means so commendable in the so-called "irritable" ulcer, although here it might have been thought especially applicable; these ulcers on the contrary are often aggravated by opium, and are best treated in quite a different manner. [Skey was a prominent advocate of this practice, and the editor has frequently verified the above statements, though giving the opium in smaller doses, gr. j.—ij., daily.]

The above curative actions of opium are established on indisputable evidence; but there is much more difference of opinion among authorities respecting its value under the following circumstances.

In General Convulsions the action of opium is very uncertain, and the instances in which it is useful are rare, though sometimes striking.

It has been employed very largely in tetanus; and there was the more reason to hope for good results from its use because, in this disease, the system shows an extraordinary tolerance of large quantities of opium. The enormous daily quantum of 20 or 30 grains has not infrequently been given without producing any decided stupor, and with the simple result of relieving the spasms to a greater or less extent. At present, however, I believe I am justified in saying that no high authority regards opium as a remedy possessed of any considerable curative power in tetanus; it would seem rather that the cases which have recovered under its use belonged to the class, not inconsiderable in number, which recover spontaneously.

In Epilepsy the highest trust has, by some authorities, been reposed in opium, but the day has gone by for this kind of practice, and few believe that any effect, except a quite temporary one, can be thus produced. Indeed, there is the best reason to suppose that serious mischief would follow the use of opium in many such cases.

Puerperal Convulsions form, perhaps, an exception: the high authority of Scanzoni may at all events be cited in favor of the beneficial influence of morphia in this affection. But it is probable that far better results may be obtained by the employment of chloroform, bromide of potassium, aconite, or hydrate of chloral.

In the Convulsions of Infancy, it is probable that opium, in the majority of cases, is not merely useless, but actively hurtful, unless given in extremely minute doses.

The Value of Opium in Inflammation was formerly one of the most disputed topics in therapeutic literature; but the observation of Brochet (1828) probably represents what is now the general opinion of experienced practitioners, viz., that the drug is far more useful in inflammations of membranous than of parenchymatous structures. In many cases of peritonitis, pleuritis, enteritis, and gastritis its use cannot be too highly extolled. In puerperal peritonitis it has undoubtedly often proved beneficial; but it is very much to be regretted that too many medical men employ it as a matter of routine in this disease, instead of selecting other remedies (aconite, for instance), which are decidedly more suited to individual cases. In diffuse inflammation of cellular tissue, especially where the periosteum is involved, opium, in tolerably full repeated doses, often exerts a most powerful antiphlogistic influence.

The employment of **Opium in Insanity** is at present regulated by very different principles from those which formerly prevailed. I have already alluded to the general tendency to abstain from the use of opium in acute delirium, and it may now be said that the old plan of attempting to narcotize acutely maniacal patients with this drug is very generally abandoned.

On the other hand, the great value of opium in insanity attended with depression, particularly in typical cases of melancholia, has been established beyond doubt, especially by the authority of Dr. Maudsley.[1] It may be given in repeated and gradually increasing doses, several times daily; and the results of this treatment, if early adopted, are often strikingly beneficial. How far any employment of opium is ever justifiable in circumstances of blood-poisoning, where there is notable embarrassment of the excreting organs, must be considered very doubtful. On the whole, these conditions must be regarded as forming a *primâ facie* and serious counter-indication.

Lastly may be mentioned, as a matter still under dispute, the question of the utility of opium in **Diabetes.** Very striking cases have been adduced in evidence of the power of full doses of opium to reduce the amount of sugar-elimination and diuresis, and to improve the general bodily condition; but, on the other hand, the drug has often been tried without the least good effect. The secret of this discrepancy very probably rests on differences between the causation of the disease in different cases; it is likely, in particular, that disease of the medulla oblongata in some instances counts for much, and in others for very little. The whole subject requires further investigation from the clinical, and also from the physiological and the pathological points of view.

THERAPEUTIC ACTIONS OF MORPHIA.—There are important differences between the remedial action of morphia and that of opium, which depend partly on intrinsic qualities of the latter, owing to the presence of other ingredients, and partly on the superior facility of administration which distinguishes the former.

Under the first head we may establish the following comparative estimate: Morphia is more applicable to the relief of pain, and less potent in the mere production of sleep, than opium; it is less useful, also, than the latter in checking hypersecretion from mucous membranes. On the other hand, it can be much more effectively employed, especially in very small doses, for the relief of certain spasmodic affections. We may here especially mention whooping-cough, for which the treatment mentioned by Dr. Edward Smith,[2] with doses of $\frac{1}{24}$ grain, three or four times daily, is

---

[1] See Practitioner, Jan., 1869.
[2] Art. "Whooping-cough:" Reynolds's System of Medicine, vol. i.

often extremely effective; and spasmodic asthma, in which small doses of morphia (especially if hypodermically injected) often afford a relief which would be sought in vain from opium.

The most practically important advantages which morphia possesses over opium have been revealed by the modern practice of hypodermic injection. First adopted in 1857, by Dr. Alexander Wood, as a mere local remedy for pain, the hypodermic injection has since developed in the hands of Hunter, Eulenburg, Behier, Lewin, and many others, into a method of quite unexpected power and utility, and one adapted for the administration of many substances besides morphia. No drug, however, so fully illustrates the value of hypodermic injection, or the dangers that attend indiscriminate resort to it.

On the mechanical part of the process (instruments, mode of puncture, etc.), it is unnecessary to say anything, as these matters have been fully discussed in the well-known treatises of Hunter and Eulenburg, in the various papers in the Practitioner, and in the Report of the Royal Medico-Chirurgical Committee.[1] As to the solution, it is probable that the acetate of morphia, five grains to the drachm, dissolved in plain distilled water, is the most convenient; and, as regards dose, it may be laid down as a general rule that the potency of injected morphia is at least three times as great as that of morphia swallowed.

The most striking effects of hypodermic injection of morphia are seen in its rapid relief of pain. In the treatment of neuralgias, especially, this remedy has proved most valuable; the severest pain being often quite relieved in a few minutes by the injection of $\frac{1}{16}$ or $\frac{1}{8}$ to $\frac{1}{4}$ grain. Nor is the alleviation always merely temporary; in recent cases it is often possible, by injecting once or twice a day (according as the attacks recur), to completely efface the malady. It is of great consequence not to begin with too large doses, for if the patient be heavily stupefied we soon set up a severe craving for the repetition of the remedy, and a dangerous opium-habit may thus be engendered. The golden rule appears to be to employ just such doses, and no larger, as are sufficient to arrest the pain and produce either no narcotism (*i. e.*, no stupor, no contraction of pupil, nor after-headache and constipation), or as little as may be possible.

Again, in the severe acute pain of serous inflammations, especially pleurisy, may usually be afforded all the relief which formerly was sought through the employment of the lancet or of leeches, by the hypodermic injection of $\frac{1}{16}$ to $\frac{1}{4}$ grain of acetate of morphia. One or two such injections daily, during the early stages, will usually suffice to keep the patient free from any considerable suffering.

Indeed I may probably go further, and assert that in all febrile diseases (fevers, acute inflammation, pyæmia, etc.), if it be necessary to use opium for any purpose, it is highly advisable to employ hypodermic morphia. The reason is that morphia, given in the manner referred to, produces much less disturbing effect on the stomach than any opiate administered by the mouth, provided it be not given in excessive doses; and it need hardly be said that, in acute diseases, it is often of the greatest importance not to interfere with the power of digesting food. The hypnotic power of hypodermic morphia is not materially different (except as regards the dose employed, and the rapidity with which the brain is influenced) from that of morphia swallowed; but, for the reason just men-

---

[1] Transactions, vol. li.

tioned, it is often highly expedient to adopt the former method of administration.

Much difference of opinion exists in regard to the relief of localized pains; viz., as to whether the injection should be made near to the painful part or in an indifferent situation. In favor of the former plan we may cite the authority of Wood, Eulenburg, Behier, Lawson, etc.; and in favor of the latter, Hunter, Anstie, the committee of the Medico-Chirurgical Society, and others. Probably it is only in a minority of cases that the local injection presents any advantage over injection into a convenient indifferent part; and, in fact, absorption takes place, under ordinary circumstances, so rapidly, that the drug must reach the general circulation before it can act upon the site of pain.

Of precautions against sudden fatal accidents from hypodermic injection of morphia it is unnecessary to speak at length. It would seem that the only serious danger of such an occurrence lies in the possibility of injecting by mishap directly into a vein. It becomes, therefore, the duty of the injector always to avoid such localities as are evidently or probably supplied with large superficial veins.

THERAPEUTIC ACTIONS OF CODEIA.—As a hypnotic, codeia has received a great deal of commendation from trustworthy authorities, though Garrod and some others somewhat unaccountably depreciate its powers. Berthé and Aran specially recommend it for procuring sleep in cases of tormenting cough, of gouty pains, and of pains from cancer; and Krebel praised it highly, more especially in cases of nervous insomnia, and sleeplessness from rheumatic pains. The latter authority particularly states that the sleep it causes is light and refreshing, and not followed by the disagreeable after-effects of opiates. Reissner employed codeia subcutaneously, and found it analogous to morphia both in its beneficial and in its inconvenient effects. In France, codeia is largely employed, more especially as a hypnotic, and Claude Bernard placed it (along with morphia and narceia) in the soporific group of opium-alkaloids; nor can it be supposed to be much weaker than morphia in this respect, since the highest dose employed by Krebel was about one grain; and he recommended only the $\frac{1}{15}$ or $\frac{1}{10}$ grain for sensitive subjects. The nitrate of codeia, which was recommended by Magendie because of its solubility, must on that account be given in somewhat smaller doses than the alkaloid itself.

As a remedy for pain, codeia on the whole is doubtless greatly inferior to morphia. Erlenmeyer employed it subcutaneously for neuralgia, but without any effect; and for the relief of any very severe pain it is probably, according to English experience, nugatory. Yet it has been recommended warmly by Berthé, Aran, Krebel, and others, in abdominal neuralgias, and Krebel also recommended it in rheumatic sciatica; but assuredly it cannot compete with hypodermic morphia in this or in any other capacity.

In irritation and hypersecretion of the bronchial mucous membrane, there is reason to think that codeia may often play a valuable part when morphia or other opiates are not well borne. It was strongly recommended for this purpose by Vigla and Aran; and Dr. Anstie informs me that he has employed it with satisfactory results in such cases in the dose of $\frac{1}{4}$ grain every three to six hours.

THERAPEUTIC ACTION OF NARCEIA.—Here again we meet with great conflict of opinion among authorities, the various views ranging from absolute denial of any medicinal virtues in narceia, to the statement that it is the best promoter of sleep and soother of pain that exists. There can be little doubt that each of these extreme views is quite incorrect. The substantial agreement of all the best French authorities (Claude Bernard, Behier, Delpech, Bouchardat, fils) with two of the best German experimenters, Erlenmeyer and Eulenburg, completely disposes of the idea that narceia is an inert or nearly inert substance; on the contrary, it is plainly a hypnotic of considerable power, though it fails to affect some individuals, and is comparatively free from the tendency of opiates to produce after-headache. It is probably an exceedingly good remedy in irritative cough, and especially in phthisis, for which Behier has particularly recommended it.[1] Unfortunately the alkaloid itself cannot be conveniently injected under the skin, since it causes violent local irritation. This seems to depend in part on its high insolubility, and may possibly be hereafter remedied, as Husemann suggests, by the use of the very soluble lactate. The dose, as a hypnotic, given in powder or pill, is half a grain to a grain. To me it seems sudorific as well as hypnotic and anodyne.

THERAPEUTIC ACTIONS OF PAPAVERINE.—This alkaloid has also been the subject of the most conflicting statements; but the conjoint weight of the clinical researches of Baxt and those of Leidesdorf has probably convinced most persons who are impartial that the powers of genuine papaverine are very real. Baxt and Sander reckon it equal in strength to morphia, but this seems doubtful; on the other hand, the trials of Hofmann and of Reissner, which gave quite negative results, must surely have been made with bad specimens of the alkaloid. On the whole, it seems probable that for cases of insomnia with nervous excitement, papaverine is an effective hypnotic in doses of about a grain. The phosphate is not adapted for subcutaneous injection on account of its irritating properties. The muriate was successfully employed in this way by Leidesdorf, but does not appear to be more effective when injected than when swallowed.

THERAPEUTIC ACTION OF MECONINE OR OPIANYL.—This may be referred to for completeness' sake, since there is some prospect, from John Harley's experiments on animals, that opianyl may prove a valuable tranquillizing hypnotic, without unpleasant after-effects.

I must conclude these remarks on the therapeutic use of opium and its alkaloids with the offer of a few cautions. The most important of these is the lesson, which ought never to be forgotten, as to the extreme susceptibility of young children—at any rate during the first two or three years of life—to the poisonous influence of opium and also of morphia. Instances are by no means uncommon in which so small a dose as a single drop of laudanum, given to a child three or four months old, has produced profound stupor lasting for many hours; and, in one case within my knowledge, the $\frac{1}{16}$th grain of acetate of morphia produced coma and convulsions in an infant of seven weeks. These are doubtless extreme cases, but they serve to show what the possibilities are, and, at any rate, there is always need for the greatest caution in administering any kind of opiate to children. Another caution may be useful in regard to the possible formation of the opium-habit. It is a familiar fact that the continued use of opium sets up in the system, in the first place, a tolerance of the drug,

---

[1] See also Bouchut: Bull. de thérapeutique, i., p. 87, 1865.

and finally a violent craving for it. It is much easier to point out this danger than to say how it can be avoided, since many patients, suffering from severely painful and incurable diseases, must and will have opium daily. All that can be said is that we may greatly postpone and mitigate the otherwise inevitably large and constant increase in the doses required to relieve pain, by commencing with the smallest quantities, and never sanctioning a needlessly large advance. As a rule, the hypodermic injection secures the greatest economy in keeping down the doses, but it is often impossible to trust the discretion of the patient's attendants in the management of so potent an agent, and equally impossible for the medical man always to administer it himself. On the subject of the hypodermic injection it is indeed necessary to speak very strongly; for, unfortunately, this remedy has passed with wonderful rapidity into the hands of that portion of the public who persist in being their own doctors, and great mischief and many serious and even fatal accidents have occurred. If ever it should become necessary to entrust the performance of hypodermic injection to the patient, or to an attendant, it is the practitioner's duty to satisfy himself, by the evidence of his own eyes, that the person undertaking it thoroughly understands what he has to do, and will most implicitly obey orders, and never increase the dose without medical authority.

PREPARATIONS AND DOSE.—The following preparations are officinal:

Pulvis Opii, gr. ss.—ij. (.03—.12);
Acetum Opii,
Vinum Opii,
Tinctura Opii, ⎫ ℥x.—xx. (.60—1.20);
Tinct. Opii Acetata,
Tinct. Opii Deodorata, ⎭
Tinct. Opii Camphorata, ʒ i.—ij. (4.—8.);
Extract. Opii, gr. ss.—i. (.03—.06);
Pilulæ Opii, No. 1-2;
Pilula Saponis Composita, gr. v. (.32);
Pulv. Ipecac. co., gr. x. (.65);
Confectio Opii, ʒ ss. (2.);
Trochisci Glycyrrhizæ et Opii;
Suppos. Opii, No. 1;
Suppos. Opii et Plumbi, No. 1;
Emplastrum Opii;
Morphiæ Acetas, ⎫
Morphiæ Murias, ⎬ gr. ⅙—½ (.01—.03);
Morphiæ Sulphas, ⎭
Liquor Morphiæ Sulph., ʒ i. (4.);
Trochisci Morphiæ et Ipecac;
Suppos. Morphiæ, No. 1;
Codeiæ Sulphas, gr. ⅙—½ (.01—.03).

APOMORPHIA.

As an appendix to the article on opium and its alkaloids, it is necessary to give a description of the new substance apomorphia, which, though not a natural constituent of opium, has been obtained by treating morphia in a particular manner. In 1869 it was discovered by Matthiessen and Wright that, by heating morphia for two or three hours in a closed tube, with large excess of hydrochloric acid, the result was the formation of water and a new substance—Apomorphia, $\overset{\text{Morphia.}}{C^{17}H^{19}NO^3} = H^2O + \overset{\text{Apomorphia.}}{C^{17}H^{17}NO^2}$. The new substance is precipitated with bicarbonate of soda, and the precipitate removed with chloroform or ether. The chloroform or ether solution is treated with concentrated hydrochloric acid, and chloride of apomorphia becomes deposited on the sides of the vessel; it is afterwards precipitated by bicarbonate of soda. Pure apomorphia is a snow-white substance, rapidly changing to green upon contact with the atmosphere: when it has become green it is partially soluble in water and in alcohol, forming in either case a beautifully colored solution; the ethereal and chloroformic solutions, on the other hand, are respectively purple-red and violet. The chloride of apomorphia also assumes a green color upon contact with atmospheric air or on heating; probably from oxidation.

The therapeutic action of this new substance proves to be widely different from that of any of the original constituents of opium. The investigations of Dr. Gee,[1] from which I have copiously extracted, show that it is an extremely prompt and certain provoker of vomiting. The experiments were made with the chloride, since the base itself is exceedingly unstable. One-fifth of a grain taken by a healthy man produced dizziness and depression in twenty minutes, an uncomfortable sensation in the head, and nausea, without the least somnolence. The patient then became very pale and salivated; in twenty-five minutes after the dose he vomited freely; after vomiting twice he felt much relieved, drank half a glass of wine, and in half an hour regained his usual condition. Subcutaneous injection of $\frac{1}{10}$ grain of the chloride in a healthy adult produced similar sensations, and free vomiting in ten minutes. Subsequently Dr. Hensley produced vomiting in a drunken man in three minutes, by injecting $\frac{1}{8}$ grain; he also produced vomiting in a drunken woman in three minutes by injecting $\frac{1}{10}$ of a grain, and in a drunken man in six minutes by $\frac{1}{10}$ grain.

There are reasons for thinking that, over and above the production of vomiting, apomorphia may act as a contra-stimulant or antiphlogistic sedative.

Given in large doses to brute creatures, chloride of apomorphia was found to produce grave nervous symptoms. Two grains were injected into a dog; the animal vomited, in two or three minutes began to run round and round the room in a curiously persistent and methodical manner, and in the end recovered completely. Three grains injected into a cat produced greater excitement. Besides the running round, there were occasional high leaps and somersaults; the pupils were dilated to the extreme, and became insensible. Two more grains produced epileptiform convulsions; and upon the injection of yet two more, making seven in all, the convulsions were followed by perfect relaxation, the heart beat somewhat forcibly, and next morning the creature was found dead. All the organs were found in a perfectly natural condition: there was no hyperæmia of any one of them. Only once has anything similar been observed in the human subject. This occurred with a man who was under treatment for chronic *morbus Brightii*, and who had on each of two succes-

---

[1] Clinical Trans., vol. ii.

sive days taken the emetic draught of the hospital pharmacopœia without effect. A day or two after he was injected with $\frac{1}{16}$ gr. of apomorphia. In four minutes he vomited freely. The vomiting continued at intervals for half an hour; he then passed into a mildly delirious state for half an hour, after which he went to sleep for an hour; when he awoke he felt, according to his own account, better than before the injection.

It is interesting to remark that, although the action of small doses of apomorphia on man is so extremely unlike that of small doses of morphia, the poisonous action of large doses of these alkaloids upon cats is very similar. Dr. Anstie has observed[1] that wide, fixed dilatation of the pupil is an invariable and early-produced symptom of morphia-poisoning in the cat; and he informs me that, with the exception of the vomiting, every symptom described by Dr. Gee is produced in cats by morphia in very large doses. I have already mentioned that morphia produces epilepsy in dogs. There are sundry expectations, as yet not sufficiently grounded on fact, that apomorphia may prove antagonistic to tetanus. Whether this hope will prove well founded or not, it is certain that in apomorphia a very valuable medicine has been secured, its emetic powers being extraordinarily active, and the emesis itself being followed by no nausea, and only by very transient depression. The entire freedom from any tendency to irritate the subcutaneous tissues gives to apomorphia much additional value.

SANGUINARIA (Sanguinaria Canadensis).

Active Ingredients.—The powers of sanguinaria chiefly depend on the presence of the alkaloid sanguinarine or chelerythrine, $C^{19}H^{17}NO^4$. It forms colorless crystalline needles, which cling together in lumps, or in stellate groups: when dry they are opaque. If a dry crystal be nibbled, it seems tasteless: but the alcoholic solution has an acrid and bitter flavor. The powder provokes sneezing if it reaches the nostrils. It is insoluble in water; soluble in ether, alcohol, and the fixed and volatile oils.

Sanguinaria contains also an alkaloid, named puccine, about the action of which there is no certain information. [The so-called "Sanguinarin" of the Eclectics is a mixture of several principles, and is said (Coe) to contain a resin, a resinoid, an alkaloid, and a neutral body.]

Physiological Action.—Sanguinarina has been recently experimented on by Weyland. Frogs injected with .0156 grain died in less than two hours: the symptoms were clonic convulsions and an early cessation of the heart's action; on opening the animal, the heart would sometimes give a spontaneous movement or two, and at any rate retained a weak electric irritability. The muscles of the extremities, on the contrary, retained a strong electric irritability, and it was plain that the peculiar action of veratria as a muscle-poison (which was pointed out by J. L. Prévost, and by Bezold, Guttmann, and others) is not possessed by sanguinarina. Drs. Tully, Eberle, and other authorities state that in human beings the effect of large doses of sanguinaria is to slow and weaken the pulse, and to produce nausea, vomiting, burning at the stomach, great thirst, faint feelings, vertigo, anæsthesia, irregular heart-action and palpitation, with great prostration, and sometimes convulsive rigidity of the

---

[1] Stimulants and Narcotics.

limbs. At the Bellevue Hospital, New York, four deaths occurred from taking large doses. Sanguinarina, in short, is a powerful acro-narcotic, and as a poison may be said, perhaps, to combine many of the powers of veratrum with those of digitalis. [From an extended series of observations and experiments, Dr. R. M. Smith (*Am. Jour. Med. Sciences,* October, 1876) concludes as follows;

"1. Sanguinarina destroys life through paralysis of the respiratory centre.

"2. It causes clonic convulsions of spinal origin.

"3. It has no effect on either motor or sensory nerves.

"4. It causes marked adynamia and prostration from its depressing action on the spinal ganglia and muscles.

"5. It decreases reflex excitability through irritation of Setschenow's centre, and by ultimate paralysis of the spinal ganglia, from large doses.

"6. It produces in cats, dogs, and rabbits a fall of pulse and blood pressure, the fall of the latter being preceded by a temporary rise after the administration of proportionably small doses.

"7. The fall of blood tension is caused by a paralysis of the vaso-motor centre, and by a paralysis of the heart itself, probably of its muscular structure.

"8. The temporary rise in blood pressure is due to irritation of the vaso-motor centre, previous to its paralysis, by small doses.

"9. The reduction in the pulse is due to direct action of the poison on the heart through paralysis of its motor power.

"10. Sanguinarina has no action on the liver.

"11. It causes marked salivation.

"12. It slows the respiratory movement, by prolonging the pause after expiration.

"13. This reduction is caused by loss of tonus of the respiratory centre.

"14. Small doses cause an irritation of the respiratory centre, and consequently in increase in the number of respiratory movements.

"15. Applied locally, sanguinarina soon causes complete paralysis of striped muscular fibre.

"16. It always causes dilatation of the pupil.

"17. It is an emetic.

"18. It always lowers the temperature.

"19. When introduced into the circulation, it diminishes muscular contractility."

Rutherford and Vignal, in the series of experiments with cholagogues, already alluded to, employed the resinoid "sanguinarin" and found that: "In one experiment three grains, in another one grain of sanguinarin, when mixed with a small quantity of bile and water and placed in the duodenum, powerfully stimulated the liver. 2. It rendered the bile more watery; nevertheless, it caused the liver to secrete more biliary matter in a given time. 3. The secretion of the intestinal glands was slightly increased by these doses. The results show that the statements of Tully and Mothershead ought not to be treated with indifference and neglect, as they at present appear to be in practical medicine."]

THERAPEUTIC ACTION.—Sanguinaria, though little employed in England, is much esteemed by American physicians: the disuse of it here is due probably, in part, to the discredit thrown on it, some years back, by an absurd attempt to ascribe to the drug the power of curing cancer. In an inaugural dissertation (New York, 1822) Bird describes it as one of the best acro-narcotics contained in the Papaveraceæ. The following summary of the medical virtues of sanguinaria is given by Dr. V. Vander Espt, of Courtrai,[1] as a *résumé* of the experience of physicians in America, and of his own trials of the drug in Belgium:

"The irritant and escharotic properties of sanguinaria give it an important place in therapeutics, and render it suitable to a variety of uses. It has thus been employed in the treatment of mucous polypi of the nasal fossæ. Dr. Smith, of Hanover, first suggested its use in this affection, in the form of snuff. By virtue of the same action it has come to be used for repressing the fungous granulations of indolent ulcers, when it frequently produces a rapid cure. The ulcers are powdered with it daily. We have successfully employed the following compound: Glycerine, 80 parts; alcoholic extract of sanguinaria (the so-called sanguinarin, to be presently mentioned), 1 part. A small piece of charpie is smeared with this and applied to the ulcerated surface. This same preparation has rendered service in cases of hospital gangrene. We have since had occasion to employ sanguinaria in injection for anal fistulas; in one case a cure resulted in a fortnight.

"As to the external and internal use of sanguinaria in tinea, we have not had occasion to test it; and the recorded facts are too few to allow of conviction; time and experience must decide. The stimulant effect of sanguinaria on the alimentary canal, and on all the other organs of the body, has occasioned its employment in various gastric affections accompanied by general debility. Given in small doses internally, it revives strength during convalescence from protracted adynamic diseases. It is of use whenever the stomach needs moderate stimulation. We have thus employed it in dyspepsia; and we have seen migraine yield to a few doses of sanguinaria: this is in cases where headache was due to a disturbance of stomach functions. Under the influence of the remedy the stomach contracts, the secretions are re-established, and digestion resumes its natural course.

"It is especially in various affections of the respiratory organs that sanguinaria has been employed successfully. In acute bronchitis it is given in doses small enough not to produce vomiting, but frequently repeated, until the rapidity of the pulse diminishes; in this manner the inflammatory irritation is checked, and expectoration becomes easy. Combined with hyoscyamus, conium, camphor, or opium, sanguinaria may be very useful in chronic bronchitis. The Americans have found it a very efficacious remedy in typhoid pneumonia, when mercurials, and especially calomel in small repeated doses, only did harm. We frequently employ it in pneumonia and pleuro-pneumonia, as a contra-stimulant, after having first reduced the hard resisting pulse, the frequent, anxious, shallow and painful breathing, by phlebotomy; in short, when the fever has somewhat abated and the graver symptoms have amended. Many American practitioners employ sanguinaria as an expectorant, or as an emetic in the treatment of whooping-cough. We believe that, combined with henbane, it may do much good in nervous spasmodic cough.

"Preparations of sanguinaria have been suggested for use in phthisis, with an idea of specific action, in which, spite of the assertions of the American practitioners, we put no faith. We believe that they may do much good; for besides helping expectoration they otherwise benefit the disease by reviving the enfeebled powers of the stomach.

---

[1] Journal de médecine de Bruxelles, Juillet, 1868.

"Many remedies have been lauded for the treatment of croup or pseudo-membranous laryngitis. Diphtheria being a general malady with local manifestations, emetics, joined with other agents, are reckoned among the most useful remedies. The choice of a suitable emetic in this disease is an important matter; and the best emetic, so far known, is undoubtedly sanguinaria. Many American authors consider it the specific for croup; and they count on the alterative action of the remedy in this disease. For the last five years we have employed sanguinaria in different cases of croup. Always, when the disease had not made too formidable progress, we had occasion to praise its efficacy. In many cases we combined sanguinaria with ipecac.: for instance, syrup of ipecac., 2 ounces; powdered sanguinaria, 20 grains; powdered ipecac., 5 grains: a teaspoonful every quarter of an hour till vomiting was produced; afterwards, a half teaspoonful every hour.

"Sanguinaria never produces diarrhœa, as tartar emetic does, with the very dangerous prostration which ensues. Dr. A. Allen, of Middleburg, U. S., says that the administration of finely powdered sanguinaria in the first stage of croup, in doses sufficient to cause vomiting, will strangle the disease. If traces of it remain, he gives a solution of acetate of sanguinarine, in as strong doses as are possible, without renewing the vomiting; he repeats the same dose every two, three, or four hours. According to Dr. Grover Coe, sanguinaria is very useful in secondary and tertiary syphilis. It is easy to understand that its stimulant action may relieve the stupor in which the whole organism is plunged under the combined influence of syphilis and the specific treatment. Sanguinaria has been recommended as a specific in gangrene; further experience is needed to decide in what cases it is likely to do good. Sanguinaria is also one of the favorite remedies of American physicians in hepatic engorgement from accumulation of secretion in the canaliculi without organic disease. A dose of $\frac{3}{4}$ grain to $\frac{1}{2}$ grain, every four hours, will cause the bile to resume its flow. Dr. Lee recommends the use of equal parts sanguinaria and aloes: this remedy may continue to be given for a long time without the inconveniences which attend the use of mercurials. We have used powdered sanguinaria, either alone or combined with podophyllin, in many cases of hysteria due either to profound disturbance of the nervous system from pain or from moral causes, or to chronic hepatitis: we have always obtained a speedy cure. We usually employed this formula:—Podophyllin, 3 grains; sanguinaria, 8 grains; soap, 8 grains; extract of hyoscyamus, 3 grains; make 20 pills: from two to four to be taken every day.

"Finally, Dr. Coe declares that sanguinaria holds the first rank as an emmenagogue: in cases of weakness he recommends it to be combined with tonics. We have given it in several cases of amenorrhœa, and the flux has returned. The Americans almost always combine sanguinaria with other remedies having analogous properties."[1]

Enough has now been said to show that the merits of sanguinaria are very positive, and that the neglect of it in English practice is a remarkable instance either of short-sightedness or of prejudice. This neglect is the more singular, because, as long ago as 1839, the researches of Probst and Reuling had shown the powerful physiological properties of chelerythrine.

[[1] The writer of the above seems to have obtained most of his information from eclectic sources.]

PREPARATIONS AND DOSE.—
Tinctura Sanguinariæ,
Acetum Sanguinariæ, } ♏ xv.— 3 ss.-(1.—2.).
"Sanguinarin" (not officinal), gr. $\frac{1}{6}$—$\frac{1}{2}$ (.01—.03).

CELANDINE (Chelidonium Majus).

ACTIVE INGREDIENTS.—Chelidonium root contains two alkaloids. One of these is chelidonine, $C^{19}H^{17}N^{3}O^{5}$, which crystallizes in colorless, glittering tables, with two atoms of water that can be driven off by a heat of 212° F. It has a bitter taste, leaving an after-feeling of irritation and alkaloid reaction. It is insoluble in water; in the crystalline form it is also nearly insoluble in ether and in alcohol; it is more soluble in fixed and volatile oils. Solutions of its salts, when tested with alkalies, let fall the chelidonine as a voluminous cheesy deposit, which gradually becomes horny—an exceedingly characteristic test. It is not quite certain how far this alkaloid can be called an active ingredient of Chelidonium. The other alkaloid is sanguinarin, or chelerythrine (already described under Sanguinaria), as to the activity of which there is no doubt whatever.

PHYSIOLOGICAL ACTION.—Chelidonium undoubtedly possesses the qualities of a narcotic irritant, but exact experiments are altogether lacking as to the physiological action of the juice itself. Of sanguinarine we have already spoken; of chelidonine the only exact account is given by Probst and Reuling. These observers agree that up to the dose of five grains it produces no poisonous effect either on animals or on man; while the sanguinarine found in the same plant proved fatal to rabbits in doses of $\frac{1}{8}$ grain. The symptoms were like those obtained with the tincture in Orfila's experiments, and death occurred in ten minutes.

THERAPEUTIC ACTION.—The popular repute which celandine formerly enjoyed was chiefly as an aperient, a diuretic, and a sudorific. It was also considered a powerful deobstruent, and employed with that view in jaundice, acute and chronic hepatitis, gall-stones, and other hepatic affections; also in hæmorrhoids, and in pneumonia with hepatic complications. It has been supposed useful in opacity of the cornea. As a nervine remedy it has been employed in paralysis, for spasmodic coughs, and neuralgia (Dr. Buchmann). It has also been recommended for eczema and herpes. All this, however, must be deemed vague; and although celandine might in all probability be made a useful remedy, further exact researches are required for this purpose. [From personal experience we would be disposed to place the drug by the side of podophyllum and iris versicolor, as capable of energetically affecting the liver.]

PREPARATIONS AND DOSE.—No officinal preparation; we have always employed a tincture made by macerating the fresh root in twice its weight of alcohol, dose ♏ v.—xx. (.30—1.20).

CRUCIFERÆ.

SINAPIS (Mustard).

Active Ingredients.—For so familiar a substance as it is, mustard has a rather complex chemistry.

(*a*) The fixed oil, which forms about 25 per cent., can be obtained from either white or black mustard by expression of the ground and sifted seeds. It is almost devoid of acridity, but has been used as a purgative and vermifuge. It would possess little interest, but for the fact that it contains—

(*b*) Erucic or brassic acid, $C^{22}H^{42}O^4$. This is a crystalline body, forming long, slender needles, insoluble in water, but highly soluble in alcohol and ether. At present it is not known that erucic acid possesses any marked physiological properties.

(*c*) Myronic acid, $C^{10}H^{19}NS^2O^{10}$, exists in black mustard-seeds in the form of myronate of potash, $C^{10}H^{18}KNS^2O^{10}$. This myronate of potash is easily soluble in water, insoluble in ether, chloroform, and benzol, somewhat soluble in dilute alcohol: it has a cool, bitter taste and neutral reaction; when heated, it gives out a pungent smell.

(*d*) Myrosine is an albuminoid substance analogous to emulsin, and acts as a ferment to the myronic acid in mustard-seeds, as soon as water is mixed with the flour. By this reaction is produced—

(*e*) The volatile oil of mustard, which has no existence in the dry seeds. It is colorless or light yellow; has a pungent odor and taste, and a neutral reaction. It largely consists of the sulpho-cyanide of allyl, C^4H^5NS, mixed with some cyanide of allyl. It is almost, if not quite, identical with a similar oil obtained from horse-radish.

White mustard-flour, like black, also undergoes a fermentation process, when mixed with water; but the oil developed, which is of a pungent character, is not the volatile oil of black mustard.

(*f*) Sinapine, $C^{16}H^{23}NO^5$, exists in both white and black mustard-seeds, under the form of the sulpho-cyanide. Sinapine is not to be obtained pure in a solid form, but can be procured as a watery solution, indirectly, through the medium of its bisulphate, which, again, can be got by adding concentrated sulphuric acid to hot alcoholic solution of the sulpho-cyanide. The watery solution of sinapine has a marked alkaline reaction, and a light-yellow color; but, during evaporation, it changes through green and red, and at last leaves a residue which is of a brown color and non-crystalline.

Physiological Action.—The well-known irritant and acrid qualities both of white and black mustard-seeds are dependent on the fermentation products, already described, which are the result of the mixture of mustard with water. The volatile oil of black mustard, ascending in the steam of hot water which is poured on mustard-flour (as in a mustard foot-bath), acts either as an irritant (causing sneezing, watering of the eyes, etc.), or, if the vapors be not too strong, as a soporific. The local irritant effects of mustard and water on the stomach and bowels are well known; carried to excess, they end in gastro-enteritis. Mustard-flour mixed with water and applied to the skin produces effects of the same kind as cantharides; proceeding to vesication, and even to ulceration and gangrene, if pushed to an extreme. The main difference of mustard irritation from cantharides irritation is that the former subsides more quickly.

The irritant qualities of the volatile oil have been thoroughly tested. It is one of the most poisonous of the ethereal oils: rabbits have been killed in two hours by a drachm, and in fifteen minutes by half an ounce; there was violent gastro-enteritis, and the odor of the oil was perceptible in the blood, the urine, and the breath. Upon the human skin the volatile oil acts with excessive severity, causing great burning pain, with redness and rapid vesication, even when much diluted.

THERAPEUTIC ACTION.—Mustard has been administered and applied in many forms. As to its internal use, we may first mention the old practice of administering the uncrushed seeds, which has received support in recent years. It has been shown [1] that mustard-seeds by no means act (as had been asserted) merely as mechanical irritants. On the contrary, in their somewhat slow passage along the alimentary canal they swell up and become mucilaginous (especially *white* mustard-seeds, of which the external coating is thin), and then give out the true acrid products of the fermentation of myronic acid and myrosine. It is urged that from the slow and equable manner in which this takes place, we have, in uncrushed mustard, a singularly effective and manageable stimulant to all the functions of the alimentary canal. But the objection is that the seeds are apt to accumulate in some one place, as the appendix vermiformis, and then produce dangerous inflammatory effects. Otherwise, there can be no doubt that both laxative and diuretic effects can be produced in this way.

As an emetic, a tablespoonful of mustard-flour in warm water is one of the readiest and best; it is especially indicated in cases where it is desired to stimulate the action of a failing heart, whether in ordinary diseases, in narcotic poisoning, in the collapse of cholera, or in certain forms of paralysis.

It is probable that, in future, mustard, when administered internally as a tonic stimulant, in cases of dyspepsia with torpid liver and loss of appetite, would be best employed in the form of alcoholic solution of the volatile oil. In the Austrian army they have even substituted the oil for mustard advantageously for dietetic purposes. The solution of 24 drops of the oil to the ounce of spirit, as originally recommended by Meyer and Wolff, is an excellent and powerful medicine both in chronic catarrhal dyspepsia and chronic bronchial catarrh (in doses of three to five drops in some emulsion): it has also received much praise as a diuretic in dropsy, but the special indications for its use in the latter way have not been defined.

The distilled water of mustard-seed has been employed in cases of itch.

The principal use of mustard is, however, in the form of poultice, or cataplasm. In various affections of the brain, such as the stupor and delirium of low fever, apoplexy, and in cases of poisoning by opium, or other narcotics, it is most valuable, and should then be applied to the feet and ankles; operating on the principle of a blister, its speedy operation giving it a great advantage over the blister similarly formed by canthar-

---

[1] Journal de méd. de Bruxelles, tome 50, 1870, p. 260. Paper by M. Commailhe.

ides. So, too, in pulmonary and cardiac affections, the poultice may be applied to the chest of the patient with a certainty of the most beneficial effects resulting. Mustard poultice is excellent again as a counter-irritant in inflammation, in neuralgic pains, and in spasms, as well as in pleurodynia, and arrested catamenia.

PREPARATIONS AND DOSE.—As an emetic, from a teaspoonful to a tablespoonful of the ground seed mixed with warm water. It is, however, seldom used for this purpose except in emergencies or in cases of narcotic poisoning.

COCHLEARIA ARMORACIA (HORSE-RADISH).

ACTIVE INGREDIENTS.—The root has a very pungent odor, and an acrid taste, depending on a volatile oil, C^4H^4NS, produced by the action of myrosin or myronic acid in the presence of water; it is identical with oil of mustard, and is readily dissipated by heat or exposure to the air. It is very volatile; one drop is sufficient to odorize a whole room. It imparts its properties to water, vinegar, and alcohol; is very easily soluble in alcohol, less so in water. It is of a pale-yellow color, and heavier than water.

PHYSIOLOGICAL ACTION.—Horse-radish has a very pungent odor, very irritating to the nostrils, and causing the eyes to flow profusely with tears. Its taste is sweetish, hot, and acrid. Taken with food, it produces a sense of warmth in the stomach, and is said to assist the digestion, especially of animal food. Freely administered in the form of infusion, it produces a sensation of soreness and relaxation in the throat, with a burning heat in the stomach, soon followed by nausea and vomiting; the urine is increased in quantity, and often acquires the peculiar odor of the plant.

Tiedemann injected an ounce of the juice into the crural vein of a dog, and immediately after detected its odor in the animal's breath. On the skin it produces perspiration when administered internally; and when the scraped root is applied externally, it causes redness of the surface, which often leads to vesication.

THERAPEUTIC ACTION.—Sydenham highly valued horse-radish in cases of dropsy supervening upon intermittent fever. In modern practice it is not much employed. Its effects are similar to, though more energetic than mustard seed. It has been employed with good results in atonic dyspepsia, chronic rheumatism, and paralytic affections, especially of the tongue, and in this last it often helps mastication by its powerful sialagogue property. A syrup prepared from a concentrated infusion of the root, and swallowed leisurely, a teaspoonful at a time, removes hoarseness arising from relaxation of the throat. Hoarseness may also be relieved by employing horse-radish as a masticatory, or as a gargle.

The infusion, taken with draughts of warm water, readily produces vomiting, and may be employed by itself, or to assist the operation of other emetics. As an antiscorbutic it is much praised by Cullen and others.

PREPARATION AND DOSE.—Spiritus Armoraciæ compositus, (B. Ph.), ʒ i.—ij. (4.—8.).

VIOLACEÆ.

VIOLA ODORATA (The common Sweet Violet).

Active Ingredients.—The active element of the *V. odorata* is the alkaloid violine, discovered in the root by Boullay (1828). It is a pale-yellow, bitter-tasting powder, which, on the application of heat, first melts and then burns like a resin. It is still uncertain whether violine is or is not identical with emetine, the active principle of ipecacuanha, but it appears to be more soluble in water, and less soluble in alcohol and ether than the latter substance. Pereira affirmed the operation of the two substances to be similar, and states that the violet-root may be employed as a substitute for ipecacuanha; but the experiments of Orfila and Chomel throw much doubt upon this, and show at least that violine is very inconstant in its action. Possibly violine may be an impure form of emetine, as suggested by Husemann.

Physiological Action.—Violine is distinctly irritant to the alimentary canal; though the degree in which it will produce vomiting or purging in any individual man or animal is always uncertain. Its extreme effects are seen in such experiments as those of Orfila, when he placed 5½ grains in the stomach of a dog, and ligatured the œsophagus: in 12 hours the animal became exceedingly depressed, with rapid pulse, and in 48 hours died in convulsions; gangrenous inflammation of the stomach was found on dissection. The same quantity of violine subcutaneously injected, proved fatal to a dog in 10 hours; a corresponding dose killed a dog when injected into the jugular vein; but the half of this quantity, when thrown into the circulation, produced no effect. Chomel's experiments upon men proved that the power of violine to produce vomiting or diarrhœa varies very much. Garrod says that the root of the *Viola odorata* is purgative and emetic in doses of 30 to 60 grains.

Therapeutic Action.—In the present state of uncertainty as to whether violine is a really independent substance, or merely an impure emetine, it seems impossible to accept as satisfactory any statement respecting the medicinal uses of the plant before us. Those species of Violaceæ which contain emetine, require, on the other hand, no special mention here, since their possible actions will be tacitly included in the remarks to be made under the head of the therapeutic action of ipecacuanha. It may be mentioned, however, that syrup of violets has been found a very useful remedy in coughs of a nervous spasmodic character, attended by much dyspnœa and little or no expectoration. In some cases of whooping-cough the employment of this medicine lessens the spasms considerably.

In **Hysteria**, attended by depression of spirits and constant weeping, it is also good. A mixture of equal parts of oil of almonds and syrup of violets, administered in a dose of one to two teaspoonfuls, often serves as a nice laxative for new-born infants.

The *Viola canina* is said not only to have emetic roots, but to be useful as a depurative, and to have been recommended for the cure of cutaneous disorders; while, in Italy, the herbage of the common pansy, *Viola tricolor*, is said to be employed in cases of tinea capitis.

PREPARATIONS AND DOSE.—None officinal. The shops keep a syrup of violets, dose ℥i.—ij. (5.—10.)

[WILD PANSY (Viola Tricolor).

ACTIVE INGREDIENTS.—Little known. Boullay failed to find Violine in this plant (Gubler).

PHYSIOLOGICAL ACTION.—A strong infusion made from ℥ ss.—℥ i. of the herb without the root, does not give rise to any suspicion that it contains violine or emetine. Its action is exceedingly mild, sometimes proving slightly laxative, at other times diuretic; as a rule giving rise to very little disturbance.

THERAPEUTIC ACTION.—Viola tricolor has long been a favorite in France in the treatment of eczema capitis et faciei, and the editor has employed it for many years with great satisfaction in chronic cases of this affection. The watery preparations have appeared to answer better than the alcoholic, and our usual procedure is to give it in infusion, combined with purgative doses of senna for the first few days. Afterward the violet is continued alone. To make the compound infusion, take one ounce of the herb and half an ounce of senna leaves. Pour on them a pint of boiling water, and when cold strain. Of this a tumblerful (for an adult) is taken at night. The dose should be regulated so as to produce three or four evacuations daily for a week or so; after that the senna is omitted and the viola continued in somewhat smaller quantity. For children, Cazin macerates ℥ i.—ij. in half a pint of cold water for twelve hours, then boils the infusion and adds a little milk and sugar. This amount to be taken daily. Viola tricolor is not found wild in this country, and the cultivated plant is not to be relied on. The imported herb should be employed, and care taken to procure a good quality. Most of it in the market is inferior.]

POLYGALACEÆ.

SENEGA (Polygala Senega).

ACTIVE INGREDIENTS.—The only portion of senega employed in medicine is the root, which, although nearly inodorous, possesses a peculiar, bitter, and pungent flavor. If chewed, it leaves a sensation of acrimony, which is deepened if the saliva be swallowed. The interior is nearly or quite inert, the virtues residing exclusively in the bark. These are brought out by water, but more completely by spirit; but it is said that neither the decoction nor the infusion possesses value equal to that of the simple powdered root. The most important ingredient is the alkaloid polygaline or senegine, called also saponine, struthiine, quillagine, githagine, monninine, and monesine, according to the source whence de rived, since it is present in many different plants. This alkaloid is an amorphous white powder, neutral and odorless, but capable of exciting severe sneezing if smelt; the taste is sweet at first, but has a secondary sharpness and pungency. Senegine is freely soluble in water, very little soluble in cold strong spirit, and not at all soluble in ether. Concentrated sulphuric acid dissolves it with a reddish-yellow color, that changes to bright red. The watery solution is peculiar, from its lathering like soap-suds.

There is probably a second active ingredient in senega, specially soluble in alcohol, and which does not exist to any large extent in infusions or decoctions.

PHYSIOLOGICAL ACTION.—The action of senegine, and the similar substances extracted from other plants, has been carefully studied by a number of experimenters. Schroff[1] found that in man, in doses of $\frac{1}{4}$ grain to 3 grains, it produced a certain amount of nausea, with a bitter taste and prickling in the mouth. From $2\frac{1}{2}$ to 3 grains also produced irritative cough, and secretion of mucus, lasting for several hours, in the bronchial tubes, but no effect was manifested upon either the kidneys or the skin. Applied locally to any breach of skin or mucous membrane, senegine and the kindred substances (especially monesine) produce severe pain and irritation, followed by ulcers which give out a plastic exudation, and become covered with a more or less gray-colored membrane.

There is ample proof, however, that senegine is, in large quantities, a poison to the voluntary muscles; in this respect githagine, and one or two of the other representatives, are more powerful even than senegine itself.

THERAPEUTIC ACTION.—The earliest use of senega appears to have been with a tribe of North American Indians called Senegaroos, who esteemed it an antidote to the bite of the rattlesnake. They applied it both externally and internally.

Whether superior, as a specific, to many other plants which are similarly employed, appears still to be unsettled. That it is beneficial in cases of pneumonia and other chest affections appears to be well established. In the advanced stages of pneumonia, when the cough is obstinately dry, irritating, and painful, with a sense of tightness and oppression across the chest, and when other remedies, in appearance more markedly indicated, have failed to give relief, I have found this medicine remove the tightness and oppression, relieve the cough, and rapidly promote expectoration.

In the bronchitis of old people, and especially when complicated with emphysema, where the cough is harsh, dry, and irritating, the breathing oppressed, and pains are felt more or less throughout the chest, senega will often act more beneficially than any other drug.

I have also employed it in humoral asthma with much success. It not only diminishes the secretion, but promotes easy expectoration and relieves the oppression. In such cases it has been highly commended by many physicians. The American Dr. Archer introduced it as a remedy of great power in *croup;* while Dr. Barton and many other celebrated American practitioners regarded it as a valuable auxiliary to medicines ordinarily employed in this disease. It is very desirable that experimental inquiries should be instituted in regard to the action of this drug in croup. It is sometimes used also in whooping-cough.

---

[1] Lehrbuch der Pharmakologie.

Dr. Percival and others have praised senega as a diuretic that may be usefully employed in dropsies dependent upon kidney disease; but many recent writers have denied its action on the kidneys or skin. It unquestionably exerts a powerful influence on the nervous system, somewhat resembling that of arnica.

Senega-root has further been employed with advantage in amenorrhœa, a saturated decoction being given to the extent of a pint in twenty-four hours, during the fortnight preceding the monthly function. On the other hand, it is interesting to note that senegine (or rather monesine) has been successfully employed by Martin St. Ange, in 2-grain doses, as a remedy for uterine hæmorrhage.

Lastly, it is said that in consequence of its stimulant and diaphoretic effects, senega has proved itself a powerful help in the treatment of chronic rheumatism and rheumatic paralysis.

PREPARATIONS AND DOSE.—Extractum Senegæ, gr. i.—v. (.06—.30). Syrupus Senegæ, ʒ i.—ij. (5.—10.). Decoctum Senegæ, ℥ ss.—i. (15.—30.)

RHATANY (Krameria triandra).

ACTIVE INGREDIENTS.—Rhatany-root owes its powerful styptic qualities to the presence of krameric acid, $C^{24}H^{22}O^{11}$, a dark-red, shining, amorphous substance, little soluble in cold water; a large quantity of tannin; and a substance called rhatanine, which is dissolved out in the watery extract, and crystallizes in coherent masses of slender and fragile crystals; formula $C^{10}H^{13}NO^3$. It is said to be identical with angeline, the resinous alkaloid of the leguminous plant *Fereira spectabilis*.

PHYSIOLOGICAL ACTION.—This has not yet been investigated with any exactness. [According to Trousseau and Pidoux, the extract of rhatany given in doses of eight to sixteen grains causes a painful sensation of weight in the stomach, with sharp pricking pains; digestion is interfered with and constipation induced. After some hours general malaise with embarrassed respiration.]

THERAPEUTIC ACTION.—Rhatany is a very valuable tonic in cases of indigestion arising from direct debility, and is particularly serviceable when the habit is flaccid or leuco-phlegmatic. Dr. Percival, the celebrated Manchester physician, strongly recommended it in advanced stages of typhus fever, stating that it possesses all the good qualities of port wine without the disadvantage of containing alcohol. Sir Henry Halford was accustomed to prescribe it for fluor albus, and with marked success. He gave it also for passive uterine and other hæmorrhages.

Rhatany is an excellent tonic also to associate with diuretics, cathartics, and absorbent stimulants, in cases of dropsy arising from debility. By reason, too, of its tonic qualities, when Peruvian bark, in any of its preparations, disagrees with the stomach, it may be substituted for that drug with the most beneficial results. Unfortunately, it is itself apt to be rejected by the stomach until after three or four doses have been taken, but in such cases of course it may be ordered in the shape of pills instead of extract. Administered in diabetes, rhatany diminishes the quantity of urine; but the secretion of sugar, and other morbid phenomena, remain unaltered. Ipecacuanha and its preparations are incompatible with rhatany, as with other vegetable astringents.

Externally employed, the extract of rhatany is styptic. It operates powerfully upon certain kinds of tumors, resolving the parts, and restoring tone to them. It also corrects and cures many kinds of ulcers when applied in plasters. [Boyer introduced the use of rhatany in treatment of fissure of the anus. Trousseau recommends that it be used in this disease in the following manner: let the patient take an ordinary enema in the morning; then, after the bowels have been moved inject a mixture composed of extract of rhatany, fifteen to forty grains, tincture of the same fifteen minims, water two ounces. This to be retained for a moment and the injection repeated night and morning.]

PREPARATIONS AND DOSE.—Tinct. Krameriæ, ℨ ss.— ℨ ij. (2. —8.); Ext. Krameriæ, gr. ij.—v. (.12—.30); Ext. Kram. Fl., ♏v.—xx. (.30—1.20); Syrup. Kram., ℨ ss.— ℨ ij. (2.50—10.); Infus. Kram. ℥ ss.—ij. (15.—60.).

LINACEÆ.

COMMON FLAX (Linum usitatissimum).

ACTIVE INGREDIENTS.—The albumen of the seeds yields to boiling water an inodorous and almost tasteless mucilage. By expression the kernels (or cotyledons) yield a bland, inodorous, and sweetish oil, the specific gravity of which is ·939. It is much more soluble in alcohol than olive-oil, and, as it is one of the "drying oils," after proper preparation it loses the original unctuousness. Linseed-oil is not congealed except by a cold below 0° of Fahrenheit, and boils at 600° of the same scale.

The seeds, before being submitted to any process, contain about one-fifth of mucilage and one-sixth of oil. [In some qualities of seed the proportion of oil is much larger.] When the oil has been expressed, the refuse is used for fattening cattle, under the name of "oil-cake."

PHYSIOLOGICAL ACTION.—Linseed is emollient and demulcent, and the oil is a mild laxative. It has been thought available for human food; but when used as an article of diet, it relaxes the digestive organs, and produces a viscid, slimy mucus, and a morbid acid in the primæ viæ. These effects may be obviated by the addition of bitter extractive. Linseed, in any case, affords but little nourishment, and is found to impair the stomach, as long ago noticed by Galen. [This does not agree with the recent observations of Sherwell (*Trans. Am. Dermatological Assoc.*, 1878), who found both the whole and the ground seed nutritious, and not a disturber of digestion.]

THERAPEUTIC ACTION.—When it can be obtained good,—that is to say, as the result of simple expression, as ordered by the Pharmacopœia, instead of by means of heat, which gives it a disagreeable taste and smell,—linseed-oil is an excellent corrective of habitual costiveness. If a little tincture of rhubarb be added, the most fastidious stomach will not be discomposed.

Linseed, put to stand for a short time in boiling water, constitutes an admirable poultice.

A decoction of the seeds contains not only the mucilage, but a portion of the oil. Hence it becomes a useful material for injections, when there is abrasion or ulceration of the mucous membrane of the intestines.

The infusion, called "linseed-tea," is, for the same reason, a valuable drink for persons who are suffering from irritation of the fauces.

Linseed-tea is also much employed for diseases of the urinary organs; but the wisdom of this use may be questioned.

Mixed with lime-water, linseed-oil has always been a favorite application for burns.

[Linseed has been largely used, and with benefit, by Sherwell, in the treatment of certain obstinate cutaneous affections. We have personally confirmed his observations, sometimes using it mixed with wheat-flour, or corn-meal, and made into bread or cakes in the usual way. As linseed contains no starch, we have suggested it as an ingredient in diabetic biscuits. The linseed employed is not the native variety, but that imported from India.]

PREPARATIONS AND DOSE.—Lini Farina, ℨj.—ij. (30.—60.); Infus. Lini Compos., ad libitum; Linimentum Calcis, external use.

PURGING FLAX (LINUM CATHARTICUM).

ACTIVE INGREDIENTS.—The active principle is said to be the neutral body linine, discovered by Pagenstecher. Linine occurs in shining, silky white crystals, which are very soluble in alcohol, ether, and chloroform, but exceedingly insoluble in water; yet they have the same persistently bitter taste which is said to belong to the watery infusion of the plant. It is changed to a deep violet blue both by concentrated sulphuric and by concentrated phosphoric acid.

PHYSIOLOGICAL ACTION.—The experiments of Schröder have put it beyond doubt that linine will produce in exaggeration the drastic irritant effects on the intestinal canal which are characteristic of purging flax when administered in other forms.

THERAPEUTIC ACTION.—Two ounces infused in a pint of water constitute a medicine which has often been administered to delicate subjects as a valuable tonic purgative. The medicine certainly has the recommendation of being obtained from an indigenous British plant.

PREPARATIONS AND DOSE.—A wineglassful of the above-mentioned infusion, twice a day, generally keeps the bowels open. If more decided effects are desired, it may be given oftener, or a little rhubarb or some neutral salt may be added.

MALVACEÆ.

MARSH-MALLOW (ALTHÆA OFFICINALIS).

ACTIVE INGREDIENTS.—These are two. At any rate there is good reason to suppose that not only the mucilage, but also the asparagine (an alkaloid, which will be described with the liliaceous plant, asparagus) is a really active substance.

PHYSIOLOGICAL ACTION.—Scarcely to be spoken of in the case of althæa, since it could only be produced by enormous and unmanageable quantities of the plant.

THERAPEUTIC ACTION.—The mucilage which is copiously extracted from althæa by aqueous decoction has certain undoubted uses. In inflammatory irritation of the throat, stomach, and intestines, it can unquestionably exercise that mechanically protective power which is probably the true function of all mucilaginous demulcents in so far as they are anything more than mild nutrients. When, however, it is assumed that by internal employment they can exert the same kind of action upon the trachea and bronchi, or upon the genito-urinary tract, it is obvious that the assertion stands on a different basis; for we should have to suppose that mucilaginous matter has a similar action after absorption into the blood as upon immediate contact with tissues—a very improbable idea. Still, it would be rash to assert that no effect is produced on distant organs by althæa, since, as already noticed, it contains asparagine, which alkaloid, though not a powerful poison, so operates as to support the belief that, in moderate doses, it may prove an antiphlogistic remedy of some importance.

The practical applications of althæa are chiefly to those irritations of the alimentary and urinary tracts which have been mentioned. Internally, it is given in one- or two-ounce doses of the decoction. Externally, the decoction can be employed to foment inflamed surfaces, or a mucilaginous cataplasm made from the roots may be applied. The decoction is useful as an enema in tenesmus; and, according to Montgomery, as a vaginal injection in cases of difficult labor. The French prepare pectoral lozenges, called *pâte de guimauve*, with althæa. They are supposed to relieve pectoral and laryngeal irritation. A decoction of the root and stem has been resorted to as a remedy in flatulent colic; and a lukewarm decoction of the flowers has been used to relieve irritant cough.

PREPARATIONS AND DOSE.—None officinal. The decoction may be used *ad libitum*.

COTTON (GOSSYPIUM HERBACEUM).

ACTIVE INGREDIENTS.—Cotton is a modification of lignine or cellulose, $C^{12}H^{10}O^{10}$, and in all its ordinary chemical properties corresponds with woody fibre. It is insoluble in water, alcohol, ether, oils, and vegetable acids. It is dissolved by strong alkaline leys, and decomposed by strong mineral acids.

THERAPEUTIC ACTION.—The *root* of the cotton-plant in decoction has long been employed in the East Indies, according to Ainslie, as a demulcent in cases of strangury and gravel. In the southern United States, according to the reports of Drs. Bouchelle and Shaw,[1] a similar preparation is resorted to, because of its power of intensifying uterine contraction, and even of originating it. To this end it was much valued by the female slaves. From some of the statements there would seem to be good reason for regarding it as more energetic even than ergot.

An emulsion prepared from the *seeds* is in the West Indies considered useful in dysentery. The same preparation is believed to be pectoral; and in India it is said to be employed as a galactagogue. The oil expressed from the seeds is very bland, and may be applied to all the same purposes as almond oil.

Cotton-wool itself is employed in the treatment of burns and scalds, for allaying pain, diminishing inflammation, and excluding the atmospheric air; the last-named function is a very important one. It has also been used as a topical application in cases of erysipelas, and for the encircling of joints that are inflamed by gout or rheumatism. In these cases it should be overlaid with oiled-silk or gutta-percha paper, so that the parts may be kept in a state of perspiration. ["Absorbent cotton," that is, cotton thoroughly cleaned and deprived of all oily matters, is a new and admirable surgical convenience. It readily absorbs moisture and discharges from wounds, etc., and makes an elastic padding for splints.]

PREPARATIONS AND DOSE.—The decoction of the root is made by boiling 4 ounces of the inner bark in a quart of water, until reduced to one-half. The dose recommended is a wineglassful every 20 or 30 minutes.

PYROXYLON, or GUN-COTTON.

When cotton is immersed for three minutes in a mixture of nitric and sulphuric acids, five fluid ounces of each, then well washed, and dried upon a water-bath, it is converted into *pyroxylon*. The fibre does not alter in appearance, but is found to have acquired highly explosive properties, while its weight has increased 70 per cent. At a temperature of about 300°, pyroxylon inflames, explodes, and disappears. Yet by immersing in water, and being kept wet, the explosive power of the cotton is held in abeyance, and in this condition it may be carried any distance. The formula is $C^{12}H^9(NO^4)+HO$.

COLLODIUM.

Collodium is a solution of pyroxylon in ether or alcohol. It is prepared by taking pyroxylon, 1 ounce; ether, 36 fluid ounces; and rectified spirit, 12 fluid ounces; mixing the two latter, and allowing the pyroxylon to dissolve in them.

---

[1] American Journal of Med. Science, 1841; Charleston Med. Journ., xi. 118.

The formula is $C^{18}H^{23}(NO^{14})O^{19}$.

Collodion is a colorless, inflammable fluid, having an ethereal odor, and rapidly drying upon exposure to the atmosphere. When the drying is complete, pyroxylon is left as a transparent film, insoluble both in water and rectified spirit.

This property allows it to be usefully employed in cases of cracked nipple, fissures of the anus, superficial ulcers of various kinds, and superficial burns and wounds, to all of which it forms a protective covering. The same with erysipelas; and as a stopping, applied upon cotton, for carious teeth: it forms likewise a very excellent application to slight cuts, such as are made by a razor while shaving.

PREPARATIONS.—Collodium flexile; collodium cum cantharide.

AURANTIACEÆ.

THE SWEET ORANGE (CITRUS AURANTIUM).

ACTIVE INGREDIENTS.—The flowers contain volatile oil (called oil of Neroli), bitter extractive, and other substances. The water prepared from them should be colorless and fragrant. The rudimentary and, of course, unripe fruits contain volatile oil, a bitter extractive (hesperidine), bitter astringent matter, and citric and malic acid. The rind of the ripe fruit contains volatile oil, isomeric with oil of turpentine ($C^{10}H^{14}$), hesperidine, and a little gallic acid. The juice of the ripe fruit contains citric acid, malic acid, mucilage, albumen, sugar, citrate of lime, and water.

PHYSIOLOGICAL ACTION.—M. Imbert-Gourbeyre states that in the south of France, where the orange is largely cultivated for the sake of the rind, the persons who are engaged in removing it are affected in a singular manner. Their hands become inflamed with an erythematous, a papular, or a vesicular eruption; they suffer from headache, dizziness, tinnitus aurium, deafness, neuralgia, oppression in breathing, constriction of the throat, nausea, pyrosis, irritation, and thirst. They are disturbed by dreams; they experience cramps and twitchings of the muscles; and, occasionally, convulsions of an epileptic character. These symptoms continue only so long as the occupation is pursued. Like other substances of an aromatic nature, orange-peel produces in the stomach a grateful sense of warmth.

THERAPEUTIC ACTION.—Orange-peel stimulates the digestive system, and is valuable as qualifying the action of other bitters in the treatment of dyspepsia. It is usually employed as a cover for the taste of quinine, and as an associate with purgative medicines of a griping character, or when the bowels are distended with flatus. M. Imbert-Gourbeyre states that he has successfully employed the essential oil for hysterical and other nervous affections. In certain febrile inflammatory complaints, orange-juice allays thirst and diminishes preternatural heat.

PREPARATIONS.—Aqua Aurantii Florum; Syrupus Aurant. Flor.; Syr. Aurant. Corticis; Confectio Aurant. Cort.

THE SEVILLE OR BITTER ORANGE (Citrus vulgaris).

This is a smaller tree than the preceding; the leaves are more elliptical; the flowers are more decidedly white, and yield a stronger perfume; the rind of the fruit is not yellow, but deep orange color, and the pulp is acid and bitter. A good distinctive character is found also in the remarkable unevenness of the fruit.

The rind of the fruit, which is bitter, contains a volatile oil, and is employed as a constituent of the compound infusion, the mixture, and the compound tincture of gentian. This and other parts of the plant are used for the same purposes as the sweet orange.

PREPARATIONS AND DOSE.—Aurantii amari cortex; Tinct. Aurantii, ℥ i.—ij. (4.—8.).

THE LEMON (Citrus Limonum).

ACTIVE INGREDIENTS.—The recent and outermost part of the rind of the fruit has an aromatic and bitter taste; the odor is strong and peculiar. These qualities depend upon the presence of a volatile oil, which, when pure, is colorless or pale yellow, limpid, fragrant, and possessed of a warm and bitter taste. It is soluble in anhydrous alcohol; less so in rectified spirit. The composition is $C^{10}H^{16}$—that is to say, it is isomeric with the oils of turpentine, savin, juniper, and copaiva. The juice of the fruit consists of citric acid ($H^3C^6H^3O^3H^2O$, sp. gr. 1:039), malic acid, gum, bitter extractive, and water. Of these substances the citric acid, of which there are about thirty-two grains in every ounce of lemon-juice, is the most important. It occurs in colorless crystals, of which the right rhombic prism is the primary form, and which are very soluble in water, less soluble in rectified spirit, and insoluble in pure ether.

PHYSIOLOGICAL ACTION.—Oil of lemon-peel is stimulant, carminative, and diaphoretic, and hence becomes a grateful stomachic. Applied externally, it is stimulant and rubefacient. The juice of the ripe fruit is refrigerant and antiscorbutic; and similar but much feebler properties pertain to citric acid. The juice also, dissolved in water, is employed as a beverage. It allays thirst, diminishes preternatural heat, abates undue perspiration, and quickens the action of the kidneys.

THERAPEUTIC ACTION.—The chief employment of oil of lemon-peel is as a carminative, but it is generally used in connection with medicines of more energy, and appears to have its chief value in dyspepsia.

Rheumatism.—Lemon-juice has been recommended as a remedy in acute rheumatism and in gout, and has been successfully employed in England, France, Italy, and the United States. At present, however, it is not looked upon with the same confidence with which it was regarded by its introducer, Dr. Owen Rees. Its effects are about the same as those of a mild alkaline treatment.

Narcotic poisoning.—Lemon-juice is valuable also in cases of poisoning by narcotic substances such as opium. In these, after the poison has been removed from the stomach, the effects may be partly counteracted by the free use, either of the juice of the fruit, or of citric acid in solution.

· Scurvy.—By far the most important use of lemon-juice is its employment in scurvy. Lemon-juice and lime-juice, the latter more especially, have been adopted, as the result of frequent and striking experience, as the most convenient and perfect prophylactic and curative in sea scurvy; and, as such, lime-juice is ordered to be constantly carried in stock by the ships both of the Royal Navy and of the merchant service.

But it has long been known that no substance peculiar to the lemon is needed for the prevention and cure of scurvy: and especially, it has been completely proved that citric acid is by no means the only antiscorbutic. Given alone, citric acid has proved but feebly curative. It has also been found that a number of other fruits and vegetables[1] are antiscorbutic, always provided that their juices are in a fresh state. So markedly is this the case that a theory was once set up, the very contrary of that which placed the cure to the score of the acid, attributing the antiscorbutic action to the citrates and malates of potash which the juices contain. This theory, however, has proved no more stable than the other; and at present it can only be said that various fruits and vegetables are both prophylactic and curative in scurvy, and that lemon-juice is a very convenient form in which to administer this kind of corrective, because easily preserved by the addition of a small proportion of spirit (10 per cent.).

As a perfume, oil of lemon is an exceedingly useful adjunct to sulphur ointment and to evaporating lotions.

Lemon-juice, converted into "lemonade," is of great service in febrile and inflammatory complaints, and in hæmorrhages.

Combined with bicarbonate of potash, it forms a citrate of potash, which is a mild diaphoretic and diuretic, and often allays restlessness and watchfulness in fever.

It is adapted, also, for lithic acid deposits, but is sometimes objectionable in phosphatic ones.

PREPARATIONS AND DOSE.—Limonis succus, ℨss.—ℨi. (15.—30.); Syrupus Limonis, as a vehicle; Mistura Potassii Citratis, ℨss.—ij. (15.—60.).

BAEL, OR INDIAN BAEL (ÆGLË MARMELOS).

ACTIVE INGREDIENTS.—No accurate analysis of the composition of bael has yet been made. It contains an astringent principle allied more or less to tannic acid, and to which the active properties appear to be due.

PHYSIOLOGICAL ACTION.—Little is yet known on this point. The liquid extract of the fruit acts as a mild aperient when given in moderate doses, but when in large ones as an active cathartic.

THERAPEUTIC ACTION.—The physicians of Malabar consider the root, bark, and leaves to be refrigerant, and prescribe a decoction in hy-

---

[1] Dr. Buzzard (Article "Scurvy," Reynolds' System of Medicine, vol. i.) mentions oranges, lemons, limes, cabbage, lettuce, potatoes, onions, mustard and cress, dandelions, sorrel, scurvy-grass, cactus, grapes, apples, sauer-kraut, etc.

pochondriasis, melancholia, and disease of the heart. They also employ a decoction of the leaves in asthmatic complaints.

In India the fruit is declared to be a valuable and efficacious remedy in dysentery and diarrhœa, and all atonic affections of the bowels; also in irritation of the mucous membrane of the stomach. It relieves diarrhœa without constipating the bowels, and is recommended also for the relief of habitual constipation: I have myself administered it in some obstinate cases of English dysentery and diarrhœa with the best results.

PREPARATION AND DOSE.—Extractum Belæ liquidum [B. Ph.], dose, 1 fluid drachm, to ½ fluid ounce. (4.—8.)

POLYGONACEÆ.

RHUBARB (RHEUM).

ACTIVE INGREDIENTS.—There is still a considerable degree of uncertainty as to the part taken by the individual constituents of rhubarb in its action on the body. The first ingredient that requires notice is Chrysophanic acid, $C^{14}H^{10}O^4$, now known to be the same substance which, under various impure forms, has received the names of Rhabarber-bitter, Rhabarberine, Rheumnine, Rhabarbergelb, Rhabarbergelbsäure, Rheine, Rhaponticine, Lapathine, and Rumicine (Peckolt). Chrysophanic acid, as such, does not perhaps exist to any large extent in rhubarb, but is formed by the splitting up of Chrysophane (to be presently described). It crystallizes out of alcohol in small orange-yellow prismatic and lustrous needles of a somewhat bitter taste. It is friable by heat, slightly soluble in water, freely soluble in ether and hot alcohol. From its solution in benzol, on slow evaporation, it is deposited in rather paler yellow crystals which are oblique rhombic tables. Concentrated sulphuric acid dissolves it red; water added to this solution separates the acid in yellow masses unchanged. Aqueous solutions of fixed alkalies and ammonia change it to a beautiful red; the ammonia solution, being evaporated, leaves unchanged chrysophanic acid, but the potash solution, if evaporated, leaves red chrysophanate of potash. [Chrysophanic acid is chiefly obtained from a vegetable powder, called Goa or Poh di Bahia, which is the product of some unknown Brazilian plant. The powder contains from 70% to 80% of acid.]

The other vegetable acid present in rhubarb is a peculiar tannic acid, $C^{26}H^{26}O^{14}$, a yellowish brown powder which attracts water very readily; is easily soluble in water and spirit, but not in ether. The brown colored aqueous solution has an acid reaction, precipitates perchloride of iron blackish green, reduces gold and silver salts, precipitates lime and albumen, but does not give a precipitate in solution of tartar emetic.

The neutral body Chrysophane, $C^{16}H^{18}O^9$, is a glucoside, which, when boiled with dilute sulphuric or hydrochloric acid, splits up into sugar and Chrysophanic acid. Chrysophane is obtained in microscopic orange-yellow prisms; it is present in the proportion of about 9 or 10 grains to the pound of rhubarb-root, is highly soluble in water, insoluble in ether.

Concentrated sulphuric acid gives it a brownish color; from water it subsides in green flocky masses.

Emodine, $C^{10}H^{30}O^{13}$, is an extra product of the (De La Rue) process for the extraction of chrysophanic acid; it forms deep orange-red crystals varying in size up to oblique rhombic prisms of two inches in length; these do not melt and decompose below a temperature of 482° F. Alkaline watery solutions turn it red; ammonia gives a violet-red.

Phaoretine, $C^{16}H^{16}O^{7}$, is an extra product of the process for extracting the rhubarb tannin; it is a dull brown shining mass which pulverizes to yellow brown; is tasteless, melts on heating, and evolves yellow fumes that smell like rhubarb. It is insoluble in water, ether, and chloroform, very soluble in spirit, also in warm acetic acid.

Aporetine and Erythroretine are resinoid bodies which are possibly mere mixtures of other rhubarb constituents; at any rate, nothing accurate is known about them.

PHYSIOLOGICAL ACTION.—Rhubarb furnishes a striking instance of the widely, though not universally operative law, that changes in dosage not only alter the degree in which a medicament will act upon the body, but, when carried beyond a certain point, altogether change the mode of action. Taken in small doses, such as 4 to 8 grains of the powder, not only does it exert no purgative action, but, as will be presently shown, it is efficient in checking chronic diarrhœa; it exercises in addition a remarkable tonic influence upon primary digestion, increasing appetite, and enabling the food to be disposed of without discomfort in cases where previously there had always been sluggish digestion and flatulence. On the contrary, if taken in large doses, such as a scruple to a drachm, rhubarb induces none of these tonic effects upon the stomach, and acts as a direct laxative. The operation commences in from 4 to 8 hours after the medicine has been taken. According to the magnitude of the dose, and conditionally upon its having been repeated or otherwise, there may be more or fewer evacuations; but on the whole the character of the stools is fairly constant. They are of loose, but not of watery consistence, and usually of a markedly yellow-brown color. When the number of evacuations has been considerable, a brief period of constipation usually follows.

When these actions of rhubarb are further inquired into, it is found impossible as yet to decide as to the intimate working of the drug. Is the purgative action produced simply by a direct stimulation of the muscular fibres of the stomach and intestines, pushing their contents more steadily and rapidly onward? Most observers are inclined to adopt this idea; and if it were correct, the same loose character of the stools might be accounted for without the supposition of any unusual outpour of secretion, by the theory that the fæces simply passed onward too rapidly to allow of the normal mucous secretions being reabsorbed, a process which certainly contributes to the dryness of normal, and still more of constipated, dejections. And it is remarkable that, even in the largest doses, rhubarb does not appear, like jalap, scammony, and other resinous purgatives, to cause inflammation of the mucous membrane; at least such a result has never to my knowledge been recorded. The usual results of an unnecessarily large purgative dose are limited to the production of more or less violent colicky pains, such as do not attend the milder laxative operation.

On the other hand, it has been asserted by several writers that rhubarb specially affects the mucous secretions of the duodenum as well as its peristaltic action, and also that it increases the flow of bile. The former assertion must be incapable of proof, and the latter very probably rested partly on the absurd and effete doctrine of signatures, and partly on the fact that the stools contain a large quantity of yellow coloring matter. As to this coloring matter, however, it is in all likelihood only the same which makes its appearance in the urine, the sweat, and the milk of persons who have taken rhubarb.

Another question which as yet remains a puzzle, relates to the respective shares in the general physiological action which are taken by the different ingredients of rhubarb. So lately as 1871, Nothnagel,[1] a very high authority, considered that the purgative action was proved, by the experiments of Schroff, to depend chiefly on the chrysophanic acid. Theodor and August Husemann[2] take the opposite view, and think that the crysophanic acid has little or nothing to do with the purgative effect. It is worth while to review the evidence upon which these opinions respectively rest.

Schroff[3] administered chrysophanic acid (obtained from the lichen called *Purmelia parietina*) in a dose of 8 grains; this was followed by eructation of wind and repeated semi-liquid stools, which commenced twenty-four hours after the dose was swallowed, and continued to recur for five days, during which period there were also observed loss of appetite, fulness of the head, giddiness, and dull depression. On the other hand, there is the concurrent evidence of Schlossberger, Buchheim, Meykow, and v. Auer to the effect that chrysophanic acid produces no result worth mentioning on the intestinal tract, even in the same dose (8 grains) that was employed by Schroff. Then there is a contradiction as to the cause of the color-change in the urine; Schlossberger ascribing it to the elimination of phaoretine and erythroretine, while Schroff and Buchheim always found the same coloring of the urine in their experiments with chrysophanic acid. Meykow also believes this acid always to be the cause of the color, and says that in experiments where the resinoid ingredients have seemed to produce it, this was because they contained an admixture of chrysophanic acid.

[Dr. J. Ashburton Thompson has recently made an extended series of observations with crude chrysophanic acid (goa-powder). After trial in ninety cases he concludes that in a dose of 25 grains for adults, or of six or more for children, it is an emeto-cathartic, the action of which is unattended with any inconvenient symptoms. He also made observations on a hundred and sixteen persons with pure chrysophanic acid. The action of the latter is similar to the former, with the difference that while in a suitable dose each will cause vomiting and purging, if the dose is too small, goa-powder is most likely to purge only, while chrysophanic acid is most likely to cause vomiting only.]

The great difficulty in supposing chrysophanic acid to be the cause of laxative action in rhubarb is the small proportion (2 per cent. ?) in which it exists in the root. It might be imagined that fresh chrysophanic acid is generated from the chrysophane as soon as that body becomes dissolved in the gastric juice; but chrysophane itself is only present in the proportion of 1 in 1,000 of the root; it therefore appears impossible that the

---

[1] Arzneimittellehre.
[2] Die Pflanzenstoffe.
[3] Wien. ärztl. Woch. 1856.

purgative action can be materially aided by it. It is certainly almost impossible to understand how there can be a sufficiency of the acid from any source in (say) a 30-grain dose of rhubarb to produce any decided action on the bowels, seeing that no less a quantity than 2 or 3 grains can be supposed necessary for the purpose. Nothnagel thinks it likely that the action of chrysophanic acid is increased by some of the salts present in rhubarb root; this brings to mind the old opinion that a test of good rhubarb was its containing a large number of oxalate of lime crystals, the section receiving therefrom a peculiar appearance; but this idea is now generally abandoned, although the best Russian rhubarb is undoubtedly richer in lime than are the English sorts. According to Michaelis,[1] the cathartic effect is produced by the resinoid matters together with the oxalate of lime, while the chrysophanic ("rhabarberic") acid is purely tonic. Kubly[2] believes that phaqretine causes a part, but only a part of the purgative action; in opposition to the idea that the remainder of that action might be due to the oxalate of lime. I may cite, moreover, the experience of the London Hospital in Pereira's time. English rhubarb was there habitually used, and found to answer well, although it contains far less oxalate of lime than the so-called Turkey sort.

If the reader will refer to the description of the various ingredients of rhubarb, he will observe that very few of them are freely soluble in water. The rhubarb tannin, to which it is impossible to attribute the purgative action, is almost the only one so soluble, except chrysophane; and this latter, as already said, is trifling in amount. On the whole, therefore, it would appear that the purgative action of rhubarb must be a compound result of several stimuli, any one of which singly would be inadequate. This must especially be the case where such a preparation as the infusion is employed; here, some 24 grains (sliced, not powdered) are infused with 2 ounces of boiling water, and virtue enough is extracted to form a draught which is mildly but decidedly laxative. Where, on the other hand, powdered rhubarb is introduced (either in pill, or partly dissolved and partly suspended in water) into the stomach, it would be more possible to imagine that some one or two of the ingredients, being thoroughly dissolved by the gastric juice, might suffice for the production of the effect. As yet, it must be confessed, we are very much in the dark. [Rutherford & Vignal, in their experiments determined that: "1. An infusion of 17 grains of Indian rhubarb when placed in the duodenum never failed to increase the secretion of bile. 2. The bile, although secreted in increased quantity, had the composition of normal bile as regards the biliary constituents proper. 3. The doses which so powerfully excited the liver, had in one case no marked purgative effect, but in two cases the purgative effect was considerable."—*Brit. Med. Jour.*, Nov. 6, 1875.]

THERAPEUTIC ACTION.—If the physiology of the action of rhubarb be still obscure, there is no doubt that the popular opinion as to the great value of the drug for several purposes is correct.

As a Laxative, in doses of 20 to 30 grains, it acts with certainty and ease; but such doses as these should not be frequently repeated, since an increasingly constipating after-effect is apt to follow; and when it becomes needful to take a mild aperient rather frequently, rhubarb should only be given in small doses, combined with other purgatives, as in Pil. Rhei

---

[1] Journal Mens. de la Soc. Chim. de Paris. 1868.
[2] Neues Rep. f. Pharm. xvii. 214.

comp. For a single purgative dose for children, the best plan is to combine it with one-third its weight of carbonate of soda or magnesia and a little ginger, or other carminatives, as in Gregory's powder. Unfortunately the gritty sensation of the powdered rhubarb is not very easily to be obviated. For adults, a pill of 6 grains of the extract with 2 grains of ginger, followed, if needs be, by a so-called "Seidlitz" draught, may answer the purpose. As a laxative in hæmorrhoids, rhubarb is often very useful.

As a Stomachic.—The stomachic uses of rhubarb are continually recurring in practice. Combined with gentian or chamomile, rhubarb, in small doses, furnishes a most valuable pill in atonic dyspepsia; an excellent stomachic may also be procured by the combination of drachm doses of the tincture with some bitter infusion, three times a day. The popular idea that rhubarb is useful in relieving pain in the bowels too often leads to mischief; for though it may be doubted if rhubarb ever originates inflammatory mischief, it can aggravate it when already existing, as in the case of chronic gastro-intestinal catarrh. When the simple union of a tonic and an antacid is required, rhubarb is often effectively combined with soda; hence 3 or 4 grains of rhubarb, with an equal quantity of the exsiccated carbonate of soda, forms a good pill for meal-times.

In Diarrhœa.—It has already been mentioned that rhubarb (like various other Polygonaceæ) contains a large proportion of tannin, and this circumstance confers on it the property of restraining excessive diarrhœa with peculiar effectiveness. More especially where the diarrhœa is wholly or partly to be attributed to the operations of undigested food or other irritant matters, rhubarb acts in a twofold manner, by gently removing the offending substances, and afterwards mildly constringing the intestinal vessels, and restraining the flow of mucous secretion. On the other hand, as already mentioned, it does not appear suitable to those cases in which diarrhœa or other troublesome symptoms depend upon an inflammatory condition of the mucous membrane.

A general review of the therapeutic action of rhubarb shows that it holds a peculiar position. As a purgative it perhaps more nearly resembles aloes than anything else, but the action is milder, and it has none of that specific tendency to act upon the large intestines which distinguishes aloes. Its very remarkable tonic properties, apparently limited in their action to the stomach and intestines, stamp it as a member of a not very large class of remedies. It is not often, in the study of drugs, that one and the same medicine is found specially affecting a particular tract of the body in two entirely different modes, each of which most usefully supplements the other; but such is the case with rhubarb. A great portion of the action of rhubarb is undoubtedly exerted after its ingredients have entered the circulation. Yet it is likely that a portion of the stimulant effects upon the stomach and intestines is locally exerted before absorption. This is rendered probable from experience of its use.

As a Local Stimulant.—Sir Everard Home advised the application of pulverized rhubarb as a local stimulant to indolent ulcers; and though this plan has not been carried out to any considerable extent, the neglect arises, probably, not from its intrinsic inefficiency, but from the circumstance of there being more convenient methods of treatment.

[Crude chrysophanic acid has been in use for some time in India for the treatment of ringworm (*trichophytosis*). As such it was a few years since introduced to the notice of European physicians. Squire, of London, however, has recently discovered that it is one of the most effective agents we possess in the treatment of psoriasis. It is used in ointment (ʒ ss.— ʒ i. to the ounce). Both in this disease and in parasitic affections we have found it exceedingly useful.]

PREPARATIONS AND DOSE.—Pulvis Rhei, gr. v.—xxx. (.30—2.); Ext. Rhei, gr. v.—xv. (.30—1.); Pil. Rhei, No. 1—5; Pil. Rhei Co., No. 1—5; Pulv. Rhei Co., gr. x.—xxx. (.65—2.); Tinct. Rhei, ʒ i.—iv. (4.—15.); Tinct. Rhei et Sennæ, ʒ ss.— ʒ i. (15.—30.); Vin. Rhei, ʒ i.—iv. (4.—15.); Ext. Rhei Fluid., ♏v.—xxx. (.30—2.); Syr. Rhei, ʒ ss.—i. (20.—40.); Syr. Rhei Aromat., ʒ ss.—i. (20.—40.); Infus. Rhei, ʒ ij.—iv. (60.—120.).

BYTTNERIACEÆ.

COCOA (THEOBROMA CACAO).

ACTIVE INGREDIENTS.—The Oleum Theobromæ is a concrete oil obtained from the ground seeds by expression and heat, and appears to be composed chiefly of stearine with a little oleine. The consistency is that of tallow; in color it is yellowish; the fracture is clean, presenting no indications of an admixture of foreign matter; exposure to the air does not render it rancid; in water it is insoluble, but in alcohol, ether, and oil of turpentine it dissolves; it melts at a temperature of 122°. To these qualities are added a bland and agreeable flavor, and an odor resembling that of chocolate. Good shelled cocoa-beans contain about 52 per cent. of this oil.

The oil or butter is the only ingredient of cacao which is officinally recognized, but the seeds also contain a substance of very different character, which must be taken into account in estimating the effects of cocoa and chocolate used as a beverage, viz., Theobromine. This alkaloid, $C^1H^nN^1O^2$, forms microscopic white rhombic needles with a strong, bitter taste, which develops but slowly on the tongue; it is tolerably soluble in water, less so in alcohol, very little soluble in ether. Hot solutions become opaline as they cool. The percentage of theobromine in cacao seeds has been very differently stated, but apparently the quantity must be very small, since, as we shall see, it is a powerful poison.

PHYSIOLOGICAL ACTION.—The oil of cacao can only be considered as a bland oleaginous food; it serves, however, a mechanical purpose as a good vehicle for the making up of soluble pessaries, suppositories, &c.

Theobromine is a very powerful poison, the action of which has been especially investigated by Mitscherlich. In rabbits the early symptoms were grinding of the teeth, slowing of respiration (often as much as to the third or the fourth of the normal rate), and increased rapidity of pulse, with weakened heart-pulsation. The further symptoms varied according to the rapidity of absorption; when this was great, there were convulsions of spinal origin; but when absorption was slow, there was gradual loss of voluntary power. The secretions, appetite, throat, &c., did not seem affected. Retching and vomiting were occasional symptoms. In frogs there was great distention of the lungs and of the bladder.

Death occurred with phenomena of vagus and spinal cord paralysis; the spinal convulsions in cases of rapid absorption hastened the end. Dissection showed appearances corresponding to the symptoms: when death occurred very rapidly, there was long-continued irritability of the voluntary muscles, and persistence of peristaltic movements; when the poisoning had been slower, the heart and muscles were found quite inirritable, the ventricles were contracted, the auricles, great venous trunks, and small vessels, in all parts of the body (especially in the mucous and serous membranes, the brain, liver, and kidneys), were very hyperæmic; there were extravasations of blood almost always in the lungs, and frequently in the mucous membranes and under the peritoneal covering of the bladder; the blood was dark-red, and speedily became oxidized on exposure to air. Mitscherlich concluded that theobromine was a poison of the same general kind as caffeine, but fatal in much smaller doses. It can be absorbed either from the stomach or from the subcutaneous tissues.[1]

THERAPEUTIC ACTION.—The present employment of the oleum theobromæ is in the preparation of suppositories which have for their ingredients tannic acid, mercury, morphia, or lead. Theobromine has not yet been applied to any therapeutic purpose, but it could probably be used with advantage in nervous affections of the same kind as those which are relieved by caffeine, only in smaller doses.

In regard to the dietetic use of cocoa and chocolate as substitutes for tea and coffee, it may be said that the chief advantage of the two former is the predominance of nutrient over stimulating ingredients in their composition. Cocoa and chocolate are really foods, with only so small a proportion of alkaloidal stimulant as probably just serves to render digestion more active.

RHAMNACEÆ.

PURGING BUCKTHORN (RHAMNUS CATHARTICUS).

ACTIVE INGREDIENTS.—The purgative properties of buckthorn are probably due, for the most part, to an uncrystallizable bitter substance called Rhamnocathartine, the descriptions of which are still somewhat vague. It has been questioned whether this principle may not be identical with rhamnine, another ingredient of buckthorn berries; but the latter is almost or quite tasteless, and can be shown by the microscope to present a regular crystalline structure. Rhamnocathartine must not be confounded with cathartic acid, $C^{190}H^{96}N^2O^{61}$, the active principle of senna, and found also in the bark of the *Rhamnus Frangula*. Rhamnocathartine seems to be a sure purgative, in two or three grain doses, for children, but is exceedingly disagreeable to take, and is slow in action.

PHYSIOLOGICAL ACTION.—The juice of the berries causes great dryness of the mouth and throat, with intolerable thirst, nausea, in cer-

---

[1] Mitscherlich, Der Cacao und die Chocolade. 1859. Quoted in Husemann, Pflanzenstoffe, p. 98.

tain cases vomiting, griping pains that extend throughout the abdomen, and violent purging, the character of the evacuations being watery.

THERAPEUTIC ACTION.—The juice spoken of was at one time prescribed in dropsy, also in gout and rheumatism, but now is seldom employed except in domestic practice, which regards it as a laxative suitable especially to children, in the form of syrup.

PREPARATIONS AND DOSE.—Syrupus: [B. Ph.] dose, ½ to 1 oz.; for young children, ½ drachm to 2 drachms. [Little used in this country.]

[ALDER BUCKTHORN (Rhamnus frangula).

ACTIVE INGREDIENTS.—The bark of the R. frangula contains a substance named "frangulin" by Casselman, who first obtained it pure. Its composition is said to be $C^6H^4O^3$. It has been obtained in the form of citron yellow, satiny, tasteless, odorless crystals. It is insoluble in water, but soluble in 160 parts of hot alcohol.

PHYSIOLOGICAL ACTION.—It is a hydragogue cathartic and vermifuge (sometimes emetic), somewhat similar to, but milder in its action than R. catharticus.

THERAPEUTIC ACTION.—The preparations of this bark seem latterly to be gaining in favor as mild cathartics, well adapted to some cases of habitual constipation. Its vermifuge properties should not be overlooked.

PREPARATIONS AND DOSE.—None officinal. The purgative dose of the powdered bark in decoction or otherwise is about 20 grains. Squibb prepares a fluid extract.]

CONIFERÆ.

TURPENTINE (Terebinthina).

ACTIVE INGREDIENTS.—The colorless turpentine of pharmacy is procured by distillation from the common yellow. The formula is $C^{10}H^{16}$. Specific gravity ·84. It is miscible with absolute alcohol, benzol, sulphide of carbon, chloroform, and ether, and is insoluble in water. The penetrating odor and burning taste are sufficiently well known.

PHYSIOLOGICAL ACTION.—In large doses turpentine is a powerful narcotic irritant poison, exerting its effects especially upon the alimentary and the genito-urinary tract, the whole of which is influenced by it; and acting also as a deliriant narcotic, producing symptoms of brain-intoxication not very unlike those of alcoholic tipsiness. A dose of two or three teaspoonfuls produces great heat at the stomach, and afterwards all over the body, with rapid and tense pulse, slight giddiness, and mental confusion: the breath, the sweat, and the urine smell slightly of turpentine, or, more accurately speaking, the last smells not of turpentine, but of violets. If the dose be larger, nausea and vomiting set in, with delirium; purging sometimes occurs, but not always; if, however, the dose be very large, purging and tenesmus are nearly sure to take place. In the latter case there are burning pains in the abdomen; intelligence becomes torpid; instead of diuresis (which at first appeared), there is strangury, with passage of small quantities of bloody urine, often followed by complete suppression. These effects pass off in from twenty-four to forty-eight hours, or they may lead on to a fatal result, with coma, or convulsions, or both. It is remarkable, that in the not very numerous fatal cases recorded, comparatively slight organic traces of the irritative process were found. The poison evidently acts chiefly through the nervous system.

Locally applied to the unbroken skin, turpentine produces a sense of burning, followed by inflammatory redness and stinging pain; if the application be carried far enough there is vesication.

THERAPEUTIC ACTION.—Turpentine has been applied to a large number of remedial purposes, and its value, though often exaggerated, is very great.

As a Stimulant Narcotic in neuralgias, turpentine sometimes proves of wonderful service; but there is a marked difference in the frequency with which its benefits are manifested in different forms of the disease. Sciatica is pre-eminently the neuralgia in which success may be expected from the use of turpentine; yet, as Nothnagel remarks, it is to be regretted that, although nothing can be more certain than its occasional great value therein, scientific indications for its use are entirely lacking. Turpentine is far from being a specific for sciatica; and all we can say is that it is usually in cases where other remedies have failed, and possibly in those in which a rheumatic taint is particularly distinct, that this medicine cures. When it does succeed, the cure is often astonishingly rapid and complete. Of late years turpentine has been naturally a good deal pushed out of view as an anti-neuralgic medicine by the many new and powerful remedies for pain which have been introduced; but its powers in this respect should not be forgotten, as they may sometimes prove highly serviceable.

In Chronic Rheumatism, threatening to become inveterate, and having already produced considerable deformities in the joints, turpentine is one of the few remedies which will occasionally arrest the downward progress of the case, and frequently give relief to the pains.

In cases of Excessive and Unhealthy Discharges from mucous membranes, turpentine is a remedy that frequently acts with great effect. In bronchorrhœa, especially when the discharge becomes fœtid, and more particularly in gangrene of the lung, turpentine has been used with the best effect, both in small doses by the stomach, and by inhalation from hot water. It has been employed, with occasional success, in chronic blennorrhœa, and in chronic cystitis, with ropy secretion.

For Hæmorrhage of various kinds there can be no doubt that turpentine often proves very efficacious. Inhaled from hot water, or taken in repeated half-drachm doses by the stomach, it has not unfrequently checked serious hæmoptysis. But its most undoubtedly beneficial action in this way is shown in typhoid fever; here it often proves invaluable in cases where there is a disposition at once to hæmorrhage and to extreme tympanitis. It is then best given by enema, 30 to 60 minims in starch-mucilage, either alone or (if there be much discharge and pain) with 10 minims of liq. opii.

Tympanitic Conditions of the large intestines, occurring under many other circumstances besides typhoid fever, are often greatly benefited by turpentine.

In cases of Ulceration of the Bowels, when the tongue parts with its fur in large flakes, and when the surface appears smooth and varnished, turpentine is strongly to be recommended. It not only moistens the tongue and covers it with a healthy fur, but all the other ulcerative symptoms quickly abate.

In Pure Atonic Constipation, with gaseous distention of the large bowel, the persistent use of this medicine has not unfrequently triumphed when all other remedies have failed. In the melanismus which occurs in a certain class of cases often reckoned among the puerperal fevers, the stimulant influence of turpentine again often leads to the happiest results; and it may be suspected that the warm praises bestowed upon turpentine by Ramsbottom, Marshall Hall, and others, as a remedy in puerperal fever, really referred to cases of this sort, and not to instances of the genuine puerperal utero-peritonitis.

As a Simple Purgative for any one particular occasion, turpentine is best given in enema. Half an ounce, with an equal quantity of castor-oil, and half a pint to a pint of gruel, usually acts very promptly and conveniently.

As a Vermifuge, turpentine once enjoyed much approval, but may now be said to be superseded. Its chief efficacy is in the treatment of tape-worm, but it is so decidedly inferior to Filix-mas and Kousso for this purpose as to be scarcely worth consideration.

In Iritis of the so-called "rheumatic" variety, turpentine has been very successfully used by Carmichael and others, in small repeated doses.

Dropsy, with albuminous urine, depending upon non-desquamative disease of the kidney, yields in a remarkable way to drop or even to half-drop doses of turpentine every two to four hours.

Poisoning by Phosphorus.—Oil of turpentine, given in 30-minim doses in mucilage, every quarter of an hour or thereabouts, is an excellent antidote for the poisonous effects of phosphorus. [Ordinary turpentine cannot be relied on as an antidote in acute phosphorus poisoning; but if the turpentine be thoroughly impregnated with oxygen it is probably the most efficient agent for the purpose yet known, a fact first brought to light by Köhler,[1] since confirmed by others.]

External Uses.—As an external remedy, oil of turpentine has numerous modes of application; the principle of its employment being, in every instance, that of counter-irritation. As a liniment, either cold or hot, it is valuable in chronic rheumatism, sprains, sore throat, and various neuralgic affections; also, as a fomentation, in puerperal peritonitis, pleuropneumonia, and all inflammations of serous membranes. Severe and dangerous burns and scalds, especially when the local injury is accompanied by great constitutional depression, are likewise treated with it very successfully; and occasionally it is found useful in those dry and chronic forms of gangrene which are not preceded by inflammation.

PREPARATIONS AND DOSE.—Oleum Terebinthinæ, ♏ x.—℥ ij. (.65—8.); Linimentum Terebinthinæ.

---

[1] Ueb. Werth u. Bedeutung des sauerstoffhaltigen Terpenthinöls, u. s. w. Halle 1872.

TAR (Pix Liquida).

(From the Pinus Sylvestris.)

PHYSIOLOGICAL ACTION.—Large doses of tar cause vomiting purging, violent pains in the bowels and across the kidneys, with diarrhœa and extreme exhaustion. When large doses have been taken, the pulse appears to be depressed, but not increased in frequency, while tar-water in small doses is said to quicken the pulse, and to act both as a diuretic and a sudorific. When the vapor of tar is inhaled, it increases the secretion from the bronchial mucous membrane. Taylor (on Poisons) reports some fatal cases of poisoning by tar.

THERAPEUTIC ACTION.—Tar may be used internally in chronic catarrhal affections, especially in chronic bronchitis. It diminishes the secretion and allays the cough. It is also valuable in tubercular phthisis, acting as an expectorant, soothing the cough, and relieving the oppression; at the same time it helps the digestive system, increasing the appetite: and if colliquative diarrhœa be troublesome it tends to remove it. On the other hand, tar frequently overcomes constipation. It is very valuable in complaints of the urinary passages, such as chronic vesical catarrh; this latter affection is greatly improved by daily injections of tar-water into the bladder. Its use in certain chronic diseases of the skin has long been known. The tar-ointment quickly relieves the itching of scabies, and cures the disease in ten or twelve days. It is also used in psoriasis, eczema, and herpes circinatus (ringworm) with good effect. Trousseau recommends it strongly for prurigo. In unhealthy ulcers tar-water dressing has been extensively and successfully resorted to. As a disinfectant and deodorizer it has been most beneficially employed. Inhaled, as before stated, the vapor is useful in alleviating chronic bronchitis and other pulmonary affections; in acute inflammations and hectic fever its use should be avoided. In pulmonary affections the steam of tar-water from a vaporizer is the most useful mode of inhalation. It not only increases the bronchial secretions where the mucous membrane is dry and inactive, but, in other cases, diminishes it where the secretion is already too profuse.

PREPARATIONS AND DOSE.—Infusum Picis Liq., ℥ ss.—i. (15.—30.); Glyceritum Picis Liq., ℨ ss.—ij. (2.—8.) ; Ungt. Picis Liquidæ.

BURGUNDY PITCH [ABIES EXCELSA.]

PHYSIOLOGICAL ACTION.—The local action of Burgundy pitch is that of a mild irritant. Certain constitutions are affected by it in a very remarkable manner, a troublesome inflammation of a vesicular and pustular character ensuing upon its extensive employment.

THERAPEUTIC ACTION.—Burgundy pitch-plaster is usefully applied to the chest in chronic pulmonary affections; also to the loins in cases of lumbago, to the joints in chronic articular affections, or to any external part of the body where there is local pain of a rheumatic character, and also where (as in chronic pleurisy) it is desirable to afford firm mechanical support and pressure.

PREPARATIONS.—Emplastrum Picis Burgundicæ; Empl. Picis cum Cantharide; Emplast. Antimonii; Emplast. Ferri; Emplast. Galbani comp.; Emplast. Opii.

CANADA BALSAM [Abies Balsamea].

Active Ingredients.—These appear to be much the same as those of turpentine.

Physiological Action.—Similar to that of turpentine and copaiva, but more irritating to the genito-urinary passages.

Therapeutic Action.—Canada balsam may be resorted to in mucous discharges from the urino-genital organs, such as gleet, gonorrhœa, leucorrhœa, and chronic cystirrhœa. It is useful also in certain forms of chronic rheumatism, especially lumbago and sciatica; and in chronic catarrh, whether pituitous or mucous, occurring in old persons of lax fibre and lymphatic temperament.

Preparations and Dose.—Canada balsam may be given in doses of five to twenty grains in pills or emulsion. The only officinal preparations are Charta Cantharidis; Collodium cum cantharide; and Collodium Flexile.

JUNIPER (Juniperus communis).

The juniper and its dark purple fruits are well known, and need no description, especially as the medicinal uses are very limited.

As regards its employment in pharmacy, very little that is useful can be said. It was formerly given extensively as a diuretic, sudorific, and carminative, but is by no means so innocent a diuretic that it may be used recklessly. In acute and subacute nephritis, and in the nephritis of scarlatina, juniper has often aggravated the mischief. Given in excess, it has also produced much irritation of the bladder and urethra. As a diaphoretic, it has occasionally been substituted for guaiacum.

The active ingredient is an ethereal oil, obtained from the berries and the young tops, that from the latter being considered the best. [A tar obtained from juniper wood by distillation and called in pharmacy oil of cade (*Oleum Cadini*), is much used as a local application in cutaneous diseases. It may also be given internally.]

SAVIN (Juniperus Sabina).

Active Ingredients.—The leaves, the tops, and the berries have a fœtid and offensive odor, and a bitter, hot, and acrid taste. Distilled with water, they yield a considerable quantity of an essential oil, which is the active element of the plant. This oil is a yellowish fluid, becoming colorless when rectified; it has a penetrating odor of the fresh plant, and a burning taste. Specific gravity, .89—.94. It dissolves perfectly in absolute alcohol, and in twice its weight of alcohol sp. gr. .85.

Physiological Action.—Savin in large doses is a powerful irritant poison, resembling many ethereal oils in the general character of its effects. It produces vomiting and purging, and other signs of gastro-intestinal inflammation; the experiments of Mitscherlich and Hillefeld have shown, moreover, that its irritant action on the kidneys produces hæmaturia; about two drachms sufficed to kill middle-sized rabbits in a few hours. A considerable number of instances of fatal poisoning in the human subject have been caused by the practice of administering savin (the oil or decoction of the tops) with a view to produce abortion. Christison records two such cases, and Taylor (on Poisons) expresses the opinion that death from savin, given for this purpose, is far more common than is generally supposed. It is somewhat curious that Orfila is silent respecting the operation of savin, except as regards dogs; which seems to show that, at his date, it had attracted comparatively little attention on the Continent. It is only by the production of such violent irritation of the abdominal and pelvic organs as generally endangers life, that the pregnant uterus can be stimulated to expel its contents.[1]

Therapeutic Action.—The value of savin has been the subject of much dispute, except with regard to its well-known uses as a local irritant.

As an Emmenagogue, savin has been declared by Van der Kolk, and some other writers, to be useless; but there is a very large body of good evidence[2] which must be considered to fully establish its worth in this respect; and the failures which have attended its use must have been due either to adulteration of the oil, or insufficiency of the dose. Cullen expressly states that he had often been deterred from giving it in sufficient doses to act as an emmenagogue, on account of its acrid and heating qualities. I consider that savin is one of the most certain and powerful emmenagogues in the Pharmacopœia, with the additional advantage that it can be given with perfect freedom from risk of doing harm.

In Menorrhagia, Leucorrhœa, and Uterine Hæmorrhage, singularly enough, savin has proved useful in many hands. I have myself given the tincture with the best effects, in doses of from five to ten drops, in a tablespoonful of cold water, every half-hour to every three hours; and Aran considered savin to be one of the most powerful medicines which we possess for combating the various atonic conditions of the uterus which may give rise to the above-mentioned disorders. Aran gave the powdered leaves.

In Chronic Gout, and in rigidity of the joints from chronic rheumatism, savin was formerly much employed. Dr. Chapman (of America) thought highly of it in the latter class of affections. Nothnagel not only denies these actions of savin, but speaks even of its action in atonic amenorrhœa, with a scepticism which has every appearance of being based on theory untested by practical experience.

As an External Irritant, savin is chiefly employed (in the form of ointment) to keep blistered surfaces open and suppurating. The concreted discharges require to be removed from time to time, so that the action of the ointment may be continuous. It is preferable to ointment of cantharides, because less acrid.

---

[1] Fodéré gave 100 drops daily for three weeks, without producing the slightest movement towards abortion.
[2] Pereira, vol. ii. p. 332; Home, Clinical Experiments, p. 387; Tilt, p. 18; Sir Charles Clarke; Sir C. Locock (who gave it in combination with iron and aloes).

The power of savin is useful as a caustic for the destruction of warts, and other excrescences; in combination with verdigris it may be employed for the removal of condylomata and venereal warts. Upon wounds and ulcers its action is that of an acrid (not chemical) caustic.

PREPARATIONS AND DOSE.—Sabina, gr. v.—xv. (.30—1.); Oleum Sabinæ, ♏i.—v. (.06—.30); Ext. Sabinæ Fluid., ♏v.—xv. (.30—1.); Ceratum Sabinæ.

ARBOR VITÆ (Thuja Occidentalis).

[**ACTIVE INGREDIENTS.**—The leaves or rather fronds of Thuja contain Thujin, $C^{20}H^{22}O^{12}$, and Thujigenin, $C^{14}H^{14}O^{7}$, discovered by Rochelder and Kawalier in 1858. Thujin forms citron-yellow shining microscopic crystals, slightly soluble in cold, better in hot water, and freely in alcohol. Thujigenin is amorphous. There is also a camphoraceous oil and wax and a peculiar form of tannin called pini-tannin.

PHYSIOLOGICAL ACTION.—Little has been recorded on this subject, except by the homœopaths, who in one treatise devote a hundred and fifty pages to the subject.]

THERAPEUTIC ACTION.—The common arbor vitæ is considered to partake of many of the properties of savin. It is useful in rheumatic and arthritic pains, and in ulcerated surfaces, especially about the corona glandis. My own experience induces me to recommend it in the highest terms for the cure of warts with a narrow base and pendulous body. These may occur upon any part of the surface, and often have an hereditary history. Many of them are removed by cleanliness and the external application of nitrate of silver, glacial acetic acid, or the acid solution of nitrate of mercury, or by cutting off the growth with a pair of scissors, and dressing the part with an astringent lotion. But these remedies are often inferior to a strong tincture of Thuja occidentalis, applied to the part three times a day for a week or a fortnight. Sometimes the warts fall off in two or three days after this application, leaving the base perfectly healed. Five drops of the tincture should be taken in a wine-glass of water internally every night and morning during the period of the external application.

Condylomata about the anus or pudenda of either sex—whether of a syphilitic character or otherwise—are often rapidly cured by the external application of the same tincture. [Leaming[1] has found Thuja of service in warts, epithelioma, and in bronchial affections. Personally we have employed it for many years, and with the utmost satisfaction in non-syphilitic warts of the penis and vulva (*condylomata acuminata*). In syphilitic lesions (*condylomata lata*), however, we have never seen the slightest benefit from its use. In papillomatous growths in general it will be found valuable, as well as in gleet dependent on granular urethritis, employed internally and by injection.

PREPARATIONS AND DOSE.—None officinal. The tincture or fluid extract should be made from the fresh leaves. We have found preparations from the dry plant very unsatisfactory. Our own experience has

---

[1] N. Y. Jour. of Med., 1855.

been confined to doses not exceeding fifteen minims, but quite recently a case of cure of supposed epithelioma has been reported in which the dose was a drachm.]

GUTTIFERÆ.

GAMBOGE (Cambogia).

ACTIVE INGREDIENTS.—The source of the physiological activity of gamboge is still a puzzle. The only active ingredient which has yet been identified is gambogic acid, $C^{20}H^{24}O^4$, which, according to Christison, exists to the extent of 72 per cent. in the purer (pipe) gamboge, the remainder being made up by 23 per cent. of gum and 5 per cent. of water. But it has been distinctly shown by Christison himself, and by Pabo, that gambogic acid is not so active as gamboge itself; so that there is some yet unsolved mystery behind. It may be suggested, as a possible explanation, that the compound of gambogic acid and gum is more hygroscopic than the pure acid.

PHYSIOLOGICAL ACTION.—The general physiological action of gamboge is that of a powerful, but at the same time rather uncertain, irritant of the alimentary canal. Taken in small doses, it increases the secretions of the alimentary canal and the frequency of the stools; and it is said by Abeille that if the small doses be gradually increased up to 20 grains daily, all purging ceases, and a copious diuresis takes its place. Given in single doses of from 3 to 5 grains, it causes nausea (sometimes bilious vomiting), colicky griping, tenesmus, and watery evacuations, and has also been said to increase the urine. But it is certain that a good deal of what was formerly stated about the amount of water discharged from the bowels and kidneys in consequence of the administration of gamboge was erroneous: in regard to the stool, it has been distinctly proved by Radziejewsky that there is no more water in stools produced by gamboge than in those that follow the use of croton oil. Taken in larger doses, gamboge produces actual inflammation of the intestines; the violent abdominal pain, vomiting, and purging are followed by collapse and death; such cases have been repeatedly recorded. Christison[1] quotes a fatal case (the dose was a drachm) from a German source: and mentions Morrison's quack pills, which contain (or then contained) gamboge, as a source of this danger. He cites several trials and inquests, in which death had resulted from inflammation produced by these pills, and in the latter it seemed probable that gamboge was the fatal ingredient.

At the same time, Christison mentions the very curious fact that certain constitutional states seem to fortify the body against the more violent effects of gamboge. Rasori and his followers, in Italy, used to give drachm doses in inflammatory diseases with no other result than brisk purging; and Linoli gave it in inflammatory dropsy in increasing doses up to the extraordinary amount of 850 grains in twelve days, and 1,044 in a month. Christison might also have quoted Rayer, who gave gamboge to the extent of nearly 40 grains daily, for six weeks together, in cases of dropsy,

---
[1] On Poisons, p. 603.

without injury. It would be exceedingly rash, however, to presume upon any such facts, to which we do not at present possess the clue, for the general administration of gamboge: and, in fact, very moderate doses have been known to produce unexpectedly violent effects. As already stated, these effects consisted in violent inflammation of the bowels; this was evinced, after death, by redness, ulceration, and even gangrene of the mucous membrane.

Therapeutic Action.—The position of gamboge in medicine is unsatisfactory. On the one hand, its known irritant properties make it a dangerous and improper medicine in any case where there is likelihood of the existence of intestinal inflammation or congestion, or any tendency to uterine hæmorrhage. And, on the other hand, it has now been proved that although gamboge is more irritant than jalap, scammony, or colocynth, it is by no means so hydragogue as was formerly supposed. On the whole, considering the great superiority of elaterium as a hydragogue cathartic, and of digitalis and bitartrate of potash as diuretics, there hardly seems any place left for gamboge in the treatment of dropsical affections.

In Dysentery.—It would appear, however, that there is one very different application of gamboge which is really of much use. Malgaigne and Betz found the use of very small doses (about ¼-grain in twenty-four hours) to be exceedingly valuable for dysentery, especially in young persons;—an apparently paradoxical fact, but established on good evidence, and, after all, not more strange than the completely opposite action of small and large doses of strychnia, and of many other drugs.

Preparations and Dose.—Gambogia, gr. j.—iv. (.06—.25); Pilulæ Cath. Comp. (No. 1—3.)

EUPHORBIACEÆ.

CASTOR OIL (Ricinus communis).

Active Ingredients.—Castor oil contains a peculiar glyceride, called ricinic acid, $C^{18}H^{34}O^3$, which was formerly supposed to be the purgative principle; but more accurate researches have discountenanced this idea. Buchheim showed the probability, that (as with croton oil) a saponification process, by mixture with the alkaline intestinal fluids, produces the substance which really acts as a purgative. Its action can be imitated by recinate of soda.

Physiological Action.—Truly poisonous symptoms are hardly to be produced by the administration of any practicable dose of the *oil;* but the seeds, eaten by mistake, have produced violent drastic effects, and even death. Remote effects are scarcely to be proved, except such as are the mere result of depression from violent local action.

Therapeutic Action.—Castor oil is particularly valuable in constipation arising from indurated fæces, or from such as arises from swallowing acrid substances, or from the accumulation of acrid secretions.

It is likewise employed with great advantage in diseases attended by irritation or inflammation of the bowels, diarrhœa, dysentery, and enteritis.

As a laxative, it operates so speedily and mildly, that it is constantly resorted to where similar operation by medicines of a more powerful nature would cause injury. Hence its value in hæmorrhoids; in inflammatory or spasmodic diseases of the urio-genital organs; in inflammation of the kidneys or bladder; in calculous affections, gonorrhœa, and stricture. The advantage of the use of castor oil as a purgative in these disorders cannot be over-estimated.

The same remark applies to affections of the rectum, especially in stricture.

Pregnant women employ it with considerable advantage.

Castor oil is a highly valuable evacuant also for children, for whom it is used, not only in Europe, but in India; and for infants it is the safest of cathartics. A larger relative dose may be given to infants than to adults, probably from their digesting a larger quantity of what is swallowed. [The leaves of the ricinus communis are decidedly galactagogue, and may be given in decoction or fluid extract when such action is desired.]

PREPARATION AND DOSE.—Oleum Ricini, ℥ ss.—j. (15.—30.)

CROTON OIL (CROTON TIGLIUM).

ACTIVE INGREDIENTS.—Croton oil, as expressed from the seeds of the *croton tiglium*, and employed in medicine, is a clear, darkish-brown, rather thick fluid, and contains, according to Schlippe,[1] 4 per cent. of a substance which he calls crotonöl, and on which the irritant action of the drug on the skin appears to depend. It is a colorless or pale-yellow turpentine-like substance, with a faint but characteristic odor; it is destroyed by heat; and the action of caustic potash changes it to a brownish resinoid mass. This, however, is not the *purgative* principle of croton oil, for Schlippe could not produce either diarrhœa or intestinal inflammation when he administered as much as $1\frac{1}{2}$ grain to rabbits. The oil also contains fixed fatty acids, and two volatile ones; angelicic acid, $C^5H^8O^1$, and crotonic acid, $C^4H^4O^2$. Neither of these acids appears to be purgative, and the latest researches render it probable that the drastic principle is only secondarily developed from a saponification process within the alimentary canal, which produces a resinoid body.[2]

PHYSIOLOGICAL ACTION.—Croton oil possesses both a local and a remote action; the former is well displayed in its effects upon the skin, which, as already remarked, are due to the crotonöl. The result of croton-oil frictions is the speedy production of redness, and a copious pustular rash, attended by great irritation and pain. Given internally, it produces the drastic purgation which is valued in medical practice, and, in large doses, causes also violent vomiting, collapse, and death; the general symptoms much resembling those of Asiatic cholera. If the patient survives a large dose, he suffers from gastro-enteritis.

These poisonous effects have been produced, not merely by the oil,

---

[1] Ann. Chem. Pharm., cv. 1.
[2] Buchheim.

but also by the accidental eating of the seeds. The inhalation of the dust by those who handle large quantities of the seeds has likewise caused epigastric pain, soreness of the eyes, and swelling of the face; and in one recorded case has even proved fatal.

THERAPEUTIC ACTION.—Croton oil is valuable only for a limited class of cases, in which it is desired to produce a speedy drastic purgation, combined with considerable vascular depression. In general, the action is speedy, and, compared with those of other powerful drastics, the effects are not so distressing to the patient. At the same time, it must be remarked that there is some uncertainty in the operation, for occasionally even 6, 8, or 10 drops fail to produce the desired effect.

PREPARATIONS AND DOSE.—Croton oil is administered simply in pill (with conserve of roses, or gum and sugar), or it is mixed up with powdered sugar, and placed upon the tongue, as when the patient is insensible, in apoplexy, &c. Dose, $\frac{1}{2}$ to 1 minim of the oil, repeated as required. For external use there is the liniment [B. Ph.], made with one part of croton to seven parts of olive oil.

CASCARILLA BARK (CROTON ELEUTERIA).

ACTIVE INGREDIENTS.—The active elements of cascarilla bark are probably the neutral crystalline principle Cascarilline, and, in less degree, two resinoid extracts. Cascarilline forms white, six-sided needles; odorless, and of bitter taste; very insoluble in water, but soluble in alcohol and ether. It gives a blood-red color with concentrated sulphuric acid, and a violet-red with concentrated nitric acid.

PHYSIOLOGICAL ACTION.—No exact experiments have yet been made upon the physiological action of cascarilla or its crystalline principle. It is known, however, that a very strong infusion of the bark in considerable dose will cause nausea, vomiting, and even diarrhœa. It has even been said to produce narcotic effects in very susceptible subjects; but this statement appears open to doubt.

THERAPEUTIC ACTION.—Cascarilla appears to have been first introduced into practice by J. N. Stisser, M.D., professor in the University of Juliers. This was in 1693; and the original employment of the drug was in arthritic and scorbutic cases. In 1719, an epidemic dysentery which raged in France was successfully treated with cascarilla; and soon after this it was extensively employed in every part of Germany.

The bitter-tonic qualities of this drug have recommended it as a substitute for cinchona; and although inferior to that medicine in tonic and febrifugal operation, it has the advantage of sitting easily upon the stomach, being not merely bitter, but aromatic. In irritable conditions of the alimentary canal, cinchona is apt to produce vomiting or purging; but neither of these results ensues upon the exhibition of cascarilla in anything like a moderate dose.

In England, cascarilla is chiefly employed in such forms of dyspepsia as require an aromatic stimulant and tonic.

It is also used in cases of debility generally; and in chronic bronchial affections, with a view to checking excessive secretion of mucus.

In Germany, where it is a favorite medicine, cascarilla has been used in many other affections, such as low nervous fevers, intermittents, dysentery, and the later stages of diarrhœa. Of late, however, it would seem that it has fallen much into disrepute.

PREPARATIONS AND DOSE.—Of the powdered bark : Pulvis Cascarillæ, gr. x.—xxx. (.65—2.); Infusum cascarillæ, ℥ ss.—iss. (15.—45.).

KAMALA (Rottlera tinctoria).

ACTIVE INGREDIENTS.—Of the three characteristic bodies detected by Anderson[1] in kamala,—Rottlerine, Rottlerared,[2] and a peculiar flocky substance,—it is probable that the first named is the true medicinal principle. This one—rottlerine, $C^{11}H^{10}O^3$, crystallizes out of ether in yellow silky crystals, which melt with heat, and are decomposed if the temperature be raised high enough. They are soluble in water, not readily so in alcohol, but easily in ether. Aqueous solutions of alkalies dissolve rottlerine with a deep-red color.

THERAPEUTIC ACTION.—The chief medicinal use of kamala is as a vermifuge; it is especially efficacious against tapeworm. Its value in this disorder was first brought prominently into notice by Dr. Mackinnon,[3] and various other observers have spoken very strongly in its favor. On the whole, it might perhaps be described as the best remedy for tænia *after* Filix-mas. An incidental advantage that it possesses over the latter is, that the patient does not need to take any other medicine, either before or after the dose of kamala.

It appears from the statements of Mackinnon, that the Hindoos also employ kamala for the destruction of external parasites: since they employ it successfully as a local application in itch and in *herpes circinatus*.

PREPARATIONS AND DOSE.—Pulvis Rottleræ, ʒ i.— ʒ ij. (4.—8.) as vermifuge.

CUCURBITACEÆ.

ELATERIUM (Ecbalium agreste).

ACTIVE INGREDIENTS.—According to the earlier of the modern writers upon elaterium, every portion of the plant is cathartic; and it is expressly stated by some, that the root may be employed in substance, and that an infusion of it is equally efficacious. It appears, however, from the experiments of Dr. Clutterbuck, that the properties of the herbaceous portions are very trifling, and that the pericarp of the fruit is itself of scarcely more importance, the only part really valuable as a medi-

---

[1] Edin. New Phil. Journal (new series), i. 300.
[2] A beautiful dye for silk.
[3] Indian Ann. Med. Sciences, i. p. 286.

cine being the juicy pulp which invests the seeds. This juice, as it issues from the natural orifice in the fruit, without pressure from the fingers, is perfectly limpid and colorless. After standing for a short time, it becomes turbid; and in the course of a few hours deposits a sediment, which, being dried without much exposure to light, becomes a yellowish-white powder, slightly tinged with green, very light and pulverulent, and constituting the pure and genuine "elaterium" of the Materia Medica. Much of the elaterium of commerce is not pure, being prepared by strong pressure of the fruit, and consequently contains the deposit of the ordinary juices, as well as the absolute principle. This compound elaterium is dark green, approaching black, and is in substance compact and heavy, breaking with a resinous and shining fracture. The best British elaterium contains 26 per cent. of the active principle; the worst about 15 per cent. The French elaterium is considerably inferior, the proportion being only 5 or 6 per cent.

The active principle is Elaterine ($C^{20}H^{28}O^{5}$), a crystalline substance, which forms colorless, shining, hexagonal tables; it has a pungent, bitter taste, and a neutral reaction; it melts at $200°$ C. into a yellow fluid, which on cooling forms amorphous yellow masses. It is neither soluble in water nor in glycerine, but is easily dissolved by boiling (not by cold) alcohol, and by chloroform. Concentrated sulphuric acid dissolves elaterine dark red: the addition of water gives a brown precipitate.

PHYSIOLOGICAL ACTION.—The active properties of elaterium are fully represented by elaterine, which has been made the subject of numerous experiments, the earliest being those undertaken by Morris (the discoverer) and Golding Bird. Given even in very small doses, either to animals or man, it produces the violent drastic purging characteristics of the action of the drug; and in larger quantity proves speedily fatal. It is remarkable that the drastic action only takes place when the drug comes in direct contact with the bile. Köhler experimented on animals by placing powdered elaterine in a portion of intestine, which was included between two ligatures (the contents having first been squeezed out of this part of the bowel), and found no effects, whether local or remote. He also tied the ductus choledochus, and then found that the aqueous solution of elaterium administered by the stomach, produced no purging, but only some remote poisonous effects.

Investigation of its action upon man has been made by Schroff:[1] he observed the effects of very considerable doses upon two of his pupils, as much as $\frac{3}{4}$ grain, an amount somewhat dangerous, being given. In one of these young men the earliest phenomenon was marked salivation; followed (45 minutes after administration) by nausea, retching, and vomiting, which was repeated four times in the course of two hours, and was eventually quite bilious: the salivation ceased when the vomiting began. Simultaneously with the sickness, the patient suffered from flatulent belchings, griping, abdominal pains, irritation in the throat, and cerebral torpor. Six hours after swallowing the elaterium, there occurred a very copious watery evacuation, soon followed by two similar ones, leaving the patient in a state of anorexia and depression. In the second case, a similar dose was given, and this quickly produced nausea and retching, which was a good deal relieved by a fit of *sneezing:* a stool occurred $6\frac{1}{2}$ hours after the administration, and a second two hours later. Eleven hours

---

[1] Lehrb. d. Pharmakologie.

after the administration there occurred *epistaxis*, and a single violent vomiting, which last did not recur, although there was constant nausea and a smell (subjective ?) of rotten eggs. On the second day there were nine fluid stools, and on the third day there were three; the nausea and retching, with great feeling of weakness and languor, continued till the fifth day.

It is certain that over and above the action upon the alimentary canal, elaterine can produce a train of poisonous effects upon the nervous system; and, according to the way in which it is given, the irritative or the nervous phenomena may respectively predominate. Marked toxic effects upon the nervous system have been produced in cats, by direct injection into the veins: here, again, a marked symptom was salivation, followed in one instance by restlessness, stertorous breathing, and death in twenty-two minutes. In a second experiment (half the former dose being employed) there were salivation, dyspnœa, and death from passive hyperæmia and œdema of the lungs in an hour and a half. In rabbits to which elaterine was given by subcutaneous injection, Köhler observed salivation, œdema of the lungs, coma, tetanus, and death.

THERAPEUTIC ACTION.—Elaterium, though only occasionally available, is one of the most powerful and valuable remedies in existence when employed in appropriate cases.

In Passive Dropsy, the remedial effect of elaterium consists in the abstraction of a large quantity of fluid by the channel of the intestines; and hence it is applied to the removal of passive dropsical accumulations which by their mechanical pressure are causing mischief that requires to be promptly stopped. It was much employed by Sydenham and his contemporaries; but, for a time, owing to its violent effects, it fell into disrepute. Dr. Ferriar, of Manchester, restored its credit by employing it in cases of hydrothorax; and Dr. Hope showed it to possess equal efficacy in other forms of cardiac dropsy. Since then elaterium has held its ground, at least in these respects, though many practitioners are very much afraid of it.

In Obstinate Constipation, elaterium has been proved, by Golding Bird and others, to act very effectively; but it is certain that there are numerous cases in which a happy result could not be looked for, and some in which such a drug as elaterium would do great harm: while, so far as I know, there are not any sufficiently accurate indications to allow of its being employed with confidence for obstruction of the bowels.

Gout.—A combination of elaterium and opium has been employed with good results in the treatment of gout.

In Apoplexy.—In those cases of sudden brain-mischief where it is thought that the action of a decided purgative may produce beneficial revulsion, elaterium has been strongly recommended. It is important to remember, however, that if any good is to be produced in this way, the action must be speedy: it is therefore not desirable to give elaterium for this purpose in ordinary doses by the mouth, as its action would probably be far slower than that of calomel and croton oil placed on the tongue with a little sugar. Elaterium should be given in suppository, a large dose (2 grains) being rubbed up with hard soap; or, still better, in injection, a solution of elaterine being thrown up well into the large bowel.

As an Errhine.—Elaterium was formerly valued as a member of the now almost disused class of remedies intended to provoke sneezing; but it is clearly very inferior to many familiar agents.

Preparations and Dose.—Usually given in the form of pills, upon a basis of extract of gentian.

Dr. Hope, to whom we owe very much of our accurate knowledge of the efficacy of elaterium in cardiac dropsy, was in the habit of combining elaterium with capsicum in these cases; and this is, very probably, a useful plan, as tending to avoid the extreme depression which is sometimes produced by the powerful action of the drug. Each dose might be accompanied by a draught containing 20 minims of the tinct. capsici.

Another combination of elaterium is that with hyoscyamus, which is said by G. Harley to be especially necessary when the drug is resorted to in the treatment of renal dropsy. Without this safeguard it is apt to set up a persistent diarrhœa, seriously dangerous to the patient, especially if uræmic symptoms have already appeared.

It is unquestionably very desirable that elaterine should be employed rather than a drug which is subject to such serious fluctuations as elaterium. Even in this case, however, it will be necessary to exercise care in choosing a manufacturer, and to make sure that the drug corresponds to the physical and chemical characteristics given above; for whatever may be the state of things now, it is certain that formerly there was great difference in the purity of different specimens: this is obvious from the varying results, as to intensity of effect, recorded by different experimenters.

As regards dose, it must be reckoned that elaterine is four to five times as strong as good elaterium; on this basis, the proper dose of the former, for drastic purposes, will be $\frac{1}{32}$ to $\frac{1}{8}$ gr. (according to the age and strength of the patient), repeated every four hours till decided action is produced, which will usually occur after the second dose. It is in some cases necessary to maintain the action for some time; in that case the elaterium should be combined with 5-grain doses of extract of gentian, and the doses be repeated every two or three days; and then an interval allowed, after Dr. Darwell's plan.

Smaller quantities ($\frac{1}{64}$ gr. elaterine, $\frac{1}{16}$th elaterium) are proper to be given in such maladies as gout, unless in the exceptional instances where it is desirable to relieve not only the joint affection, but a coincident dropsy.

COLOCYNTH (Citrullus Colocynthis).

Active Ingredients.—The active properties of colocynth depend upon the presence of the glucoside colocynthine, $C^{56}H^{84}O^{23}$, discovered by Herberger, and further investigated by Walz and by Bastick.

Colocynthine is usually seen in amorphous yellow masses, but by very careful evaporation of the alcoholic solution it is possible to obtain small bundles of yellowish-white crystals. It has an excessively bitter taste, and is very soluble in cold and hot water, and in alcohol, but insoluble in ether. Its character is at once displayed if, besides applying the above tests, the solution is boiled: colocynthine is then immediately changed into sugar, and a resin (colocyntheine, $C^{44}H^{64}O^{13}$), which is precipitated.

There is no reason to suppose that another substance, colocynthitine, a tasteless crystalline body, which Walz procured from colocynth by treatment with ether, has any share in the action of colocynth.

Physiological Action.—The experiments of Schroff show that colocynthine is a very active irritant: 5 to 7 grains killed rabbits in four hours with repeated purging, and the development of an extensive gastro-enteritis.

Colocyntheine, also (the resin which is artificially precipitated by boiling a solution of colocynthine), is stated by Solokowski to be capable of producing colic and diarrhœa, even when given in small doses.

THERAPEUTIC ACTION.—It unfortunately happens, that, when speaking of the medicinal effects of colocynth, the action of the pure drug is never referred to, but that of colocynth mixed with one or several other agents. In modern practice it has been found that colocynth is too sharply irritant to be conveniently given, except with the addition of carminative and soothing agents, and often in combination with other aperients. It is never given alone; and neither colocynthine nor colocyntheine have come into medical use. The ancient Greek physicians were accustomed to employ colocynth as a drastic purgative in dropsy, and in lethargic and maniacal cases. They were also well acquainted with its violent effects when injudiciously administered.

In Habitual Constipation, colocynth operates mildly, certainly, and effectively; but care must be taken that it be given at regular, and not too short, intervals.

In Alvine Obstruction, where the bowels have ceased to act for some days, but there is no reason to suspect the existence of any mechanical impediment, colocynth often proves invaluable, and may be conveniently given in enema, rubbed down with soap and water.

In Torpor of the Abdominal and Pelvic Nerves and Vessels generally, colocynth often acts as a most useful corrective stimulant. One of the best instances of this kind of action is its effect in certain cases of chlorotic amenorrhœa; its rousing influence upon the rectum seems to affect the uterus and ovaries by sympathy.

In Apoplexy, Cerebral Congestion (including Congestive Headache, in Cerebral Paralysis, and other Brain Diseases, colocynth sometimes appears to act favorably upon the principle of revulsion or counter-irritation.

In Dropsy, colocynth may prove useful as a hydragogue, though it is greatly inferior to elaterium, since the latter drug evacuates much more fluid in proportion to the degree of intestinal irritation which it causes.

As a Diuretic, colocynth only acts indirectly ; but it sometimes exerts this action very beneficially, over and above its purgative effects, when employed for the treatment of dropsy. It is only after a certain persistence in its use that this effect is likely to be attained, a delay which seems to be due to the cumulative action of colocynthine upon the kidneys when it has circulated for some time in the blood. In its general character as a purgative, colocynth, probably, most nearly resembles aloes, and, after this, gamboge. But it appears to act more uniformly upon the whole intestinal tract than aloes; and over gamboge it has a very decided advantage—namely, that while the latter seems to act almost exclusively by its topical acrid influence, colocynth exerts a double power—that is to say, over both the vermicular action and the mucous secretion of the bowels; it is, therefore, at once a more manageable and a more effective remedy. From aloes, again, it differs in being apparently devoid of the *tonic* qualities which sometimes prove inconvenient in the use of that drug.

Preparations and Dose.—Extr. Colocynthidis, gr. ss.—ij. (.03—.12); Extr. Colocynth. Co., gr. ij.—x. (.12—.65); Pil. Cath. Co., No. 1—3. Colocynthine is one of the principal ingredients of Laville's gout remedy.

BRYONIA (Bryonia dioica).

Active Ingredients.—The active principle of the *Bryonia dioica* is probably the neutral body bryonine, which appears to be a glucoside. It exists, when separated, in white amorphous masses, of very bitter taste; is readily soluble in water and alcohol; insoluble in ether. On boiling the solution sugar is formed, and a resin (bryoretine) is precipitated.

Physiological Action.—The poisonous effects of bryonia have long been known, the medicine having once been a favorite with the French physicians, who noticed, however, that in over-doses it often produced violent sickness, griping, watery evacuations, and fainting; some of the cases ending fatally. Orfila's experiments upon animals proved that the root is a violent irritant : introduced into the stomach, it caused intense and fatal gastritis; into the pleura, it caused fatal pleurisy, with fibrinous effusion.

The researches of Collard de Martigny showed that bryonine possesses an exactly similar power of inflaming the stomach when swallowed; also any wound or raw surface to which it may be applied. Of course it acts more energetically than bryonia. Further researches are needed, however, to discover whether there may not be other physiological properties in bryonine.

Administered in moderate doses, bryonia is purgative, and may be employed with advantage, provided that care be taken to suspend the use, should the irritant effects become developed. In this case opiates or cordials must be resorted to. To insure uniformity in the action, when employed as a medicine, the roots should be dug up in the spring, then cut into thin slices, and dried either in the sun or in a warm room, by adopting which means the acrid matter becomes partly dissipated. The whole of the acrid principle may be expelled by repeated washing with water ; the fecula which remains is not unlike that of the potato, and is said to be a nutritious food. [Bryonine, in doses of three to four grains, is a violent poison; in doses of one-sixth to one-third grain is a drastic purgative (Cazin).]

Therapeutic Action.—The medicinal reputation of bryonia is very ancient.

In Epilepsy, the juice was administered in the time of Dioscorides; the upper part of the root was laid bare, and a hole scooped in it, which in a few hours became filled. This method has been followed for the same purpose in our own day.

In Hysteria, it was recommended by Matthiolus in the sixteenth century.

In Mania, it was formerly much relied upon—*i. e.*, in days when this malady was habitually treated by free purgation. Among other authorities in its favor is found Sydenham. For all these purposes, however, it is probable that bryonia has been justly superseded by other remedies.

As a Drastic Purgative in Dropsy, it has been recommended, and probably might still be well employed. Dr. Pearson's opinion was that "it would very well supply the place of jalap in our hospitals." The infusion is the best form of administration: half an ounce of the dried root should be placed in a pint of boiling water, to which should. be added an ounce of spirit of juniper; of this preparation a wineglassful should be taken every four hours until copious watery motions are induced.

As a Diuretic.—In dropsical cases the infusion, given as above recommended, not merely purges, but also produces a vigorous action of the kidneys—a circumstance that appears to point to bryonia as a specially valuable drug in cases where it is desirable to get rid very rapidly of large accumulations of fluid. A good deal of care, however, is required, the known tendency of bryonia being to depress the action of the heart.

In Pleurisy, and other Serous Inflammations, bryonia is an exceedingly valuable drug; it is usually in the second stage, in which general pyrexia has diminished or disappeared, but exudation continues, that the best effects of this remedy are seen. It is just in these cases, in which aconite is so effectively employed in the earlier feverish stage, that bryonia afterwards proves most useful; it limits the extent of serous effusion, and actively helps its removal by absorption.

Bryonia is more especially effective in pericarditis and in pleurisy; in these maladies it fully equals any remedy that exists.

In Pleuropneumonia, bryonia is often of great service; here, as in simple pleurisy, it both limits effusion and assists absorption.

In Rheumatism, of various forms, it has also proved useful; but, although it will sometimes relieve joints that are swollen, it is more especially the merely painful and stiff rheumatic joints that are benefited by this drug.

In Liver Affections of various kinds, and also in the ordinary bilious headache with vomiting, I have myself found bryonia worthy of commendation.

[*Bryonia Alba* is a favorite remedy with the homœopaths, and possesses, so far as known, the same properties as the *B. dioica*.]

PREPARATIONS AND DOSE.—Non-officinal. In trade are found a tincture made from the dried roots of the *B. dioica* and *B. alba*, indiscriminately, and a tincture made from equal parts of the expressed juice of the fresh root of the *B. alba* and alcohol. Dose of the latter ♏ j.—iij. (.06—.18); of the former the dose will be indefinitely larger, depending on the variations in quality.

SPIGELIACEÆ.

NUX-VOMICA (Strychnos Nux-Vomica).

Active Ingredients.—The alkaloids strychnia and brucia [are said to] represent all the powers of nux-vomica.

Strychnia is colorless, and crystallizes out of an alcoholic solution in four-sided, right-rhombic prisms, with four-sided pyramidal apices. It is very insoluble in water; more soluble in alcohol of sp. gr. .863, especially when this is heated to, or nearly to, the boiling-point. The solution is intensely bitter, and leaves a metallic after-taste, which is very characteristic. Strychnia is one of the alkaloids which dissolve without change of color in pure concentrated sulphuric acid : but if a single crystal be thus dissolved on a white plate, and a drop of solution of some deoxidizing substance (*e. g.*, bichromate of potassium), be made to mingle at its edge with the dissolved strychnia, there is presented a perfectly unique play of colors, including blue, purple, crimson, and red-brown; all this passes in a few seconds, and then the red-brown gradually fades into a light red, which is persistent for some hours.[1] The formula of strychnia now generally accepted is $C^{21}H^{22}N^2O^2$.

Strychnia is a powerful base, completely neutralizing the strongest acids, and precipitating many metallic oxides from their solutions as salts; not unfrequently double salts are formed. Strychnia salts are mostly crystalline, and always intensely bitter.

Brucia ($C^{23}H^{26}N^2O^4$) forms exceedingly transparent four-sided prisms, or, if the solution be more hastily evaporated, pearly scales that resemble boracic acid, or which form a cauliflower-like mass; the taste is strongly and persistently bitter; it is soluble in 500 parts of boiling, or 850 of cold water. Brucia and its salts are both turned scarlet or blood-red by strong sulphuric acid.

[In addition to strychnia and brucia, Desnoix, in 1853, discovered a third alkaloid, to which the name of igasuria has been given. It is soluble in one hundred parts of hot water, from which a portion of it is deposited on cooling. It is very soluble in alcohol, chloroform, and essential oils. Its ultimate analysis has not been satisfactorily determined. If brucia be heated with iodide of methyl, a combination, called iodide of methyl-brucia, is formed.]

PHYSIOLOGICAL ACTION OF STRYCHNIA.—The action of this alkaloid very nearly represents the whole activity of nux-vomica, the effects of brucia being a weak though perfect copy of those of strychnia. [Igasuria stands midway between strychnia and brucia.]

Given in anything like massive doses, strychnia produces powerful and characteristic tetanic convulsions. The first symptoms are developed in from a few minutes to an hour after the administration; the patient feels a sudden sense of suffocation and dyspnœa; the head and limbs begin to jerk in a shuddering manner; the limbs are then suddenly stretched out rigidly; the hands are clenched; the head is bent backwards; and at last the whole body is stiffly arched, so as to rest upon the head and the heels. The soles of the feet are arched; the belly is hard and tense as a board; the chest is fixed, and breathing is nearly arrested. In the height of the spasm, the face looks dusky and congested, and the eyeballs stand out strongly. The jaw-muscles are also affected with spasm, and the throat is dry, with a sense of choking. Meanwhile the intellect remains quite unclouded, and the patient often expresses a sense of impending death. After the paroxysm has lasted for a minute or two, there is usually a relaxation; in this interval the patient suffers only from soreness of the muscles, but before long he experiences sensations which warn him that the fit is returning, and cries out for some one to hold him or rub his limbs. It is worthy of notice that the jaw-muscles are never so soon or so

---

[1] It has been erroneously stated that other alkaloids answer to this test, but Dr. Guy has very clearly proved that nothing but strychnia produces the particular sequence of color-changes above described.

powerfully affected as in tetanus from ordinary causes. After one or two or a succession of paroxysms, respiration stops in the middle of a fit, and the heart soon after ceases to beat. The smallest dose which has produced a fatal result is half a grain; but several persons have recovered after swallowing even larger quantities, the paroxysms gradually becoming less violent, and the intervals longer. The only remarkable fact to be noted after death is the rapid or almost immediate occurrence of rigor mortis, which, however, is not universal.

It is certain that the chief action of strychnia is exerted upon the spinal cord; the brain, in ordinary cases, does not appear to be noticeably affected. The action upon the cord produces an extremely exalted reflex irritability, through which it is, and not by direct action on the motor nerves, that the convulsions are produced. The sensory nerves are but seldom influenced, and then only in a partial and indirect manner.

Such are the phenomena of strychnia-poisoning by very large doses; sundry minor effects produced by smaller but still poisonous doses remain to be noticed. In the early days of the medicinal use of strychnia, when unwisely large doses were constantly administered, the patients were frequently thrown into a state of heightened reflex irritability, in which, though there were no absolute paroxysms, frequent twitchings of the muscles occurred, causing great discomfort and restlessness, and often rendering continuous sleep impossible. And although strychnia does not ordinarily appear to affect the brain directly, singular cerebral phenomena are occasionally induced by too large medicinal doses. The more frequent of these are greatly heightened irritability of the optic and the auditory nerves, so that brilliant light and loud sounds produce a painful impression; but in a few cases there is a true cerebral intoxication, resembling slight drunkenness. In short, though the more violent and characteristically poisonous effects of strychnia are due to the special action upon the cord above mentioned, it can hardly be doubted that strychnia exerts a minor influence upon every part of the nervous centres, and particularly upon the vaso-motor centres.

An important, but still undecided question is, whether the action of strychnia takes place directly upon the nervous tissues, or whether the nervous phenomena are only secondary to a general blood-poisoning. Dr. George Harley, some years ago, published an opinion, based upon experimental researches, to the effect that the nervous phenomena are due to the influence of the blood, the oxidation of which was believed by him to be suddenly arrested by the action of strychnia, as well as by that of several other poisons. So far as I am aware, this opinion has not gained ground in recent years. That strychnia, however, is in part chemically changed within the blood can hardly be doubted; for although it has been detected by various observers in the urine of animals poisoned by it, the quantity so found is not sufficient to account for the possibility of those occasional recoveries after large doses which have been observed. Some of these were, no doubt, merely examples of non-absorption, a large part of the strychnia having passed away with the stools, without ever having entered the blood. But this will by no means account for all the facts; and on the whole it seems probable that a portion of the alkaloid is destroyed by oxidation within the blood. It is much more difficult to accept Harley's opinion that the deoxidation of the blood accounts for the nervous symptoms; for upon that theory it seems inexplicable that the spinal cord should be so exclusively affected with the graver symptoms, and the brain be not disturbed in any high degree.

Strychnia has been proved (by Binz and others) to possess in a humbler degree the influence which quinine so powerfully exerts as a poison to the leucocytes, and also to various lowly-organized animal and vegetable organisms—a fact which it will be advisable to bear in mind in future attempts to extend the therapeutic applications of this alkaloid.

Lastly, it must be remembered that strychnia in concentrated solution is capable of acting as a direct irritant. It is possible that a part even of the nervous phenomena of ordinary strychnia-poisoning are reflex effects of stomach-irritation; but, at any rate, the singular facts brought to light by Mr. Barwell's plan of injecting the alkaloid into the substance of paralyzed muscles (to be presently noticed) is conclusive as to its irritant qualities.

[PHYSIOLOGICAL ACTION OF BRUCIA.—Like strychnia, brucia produces a condition of exalted reflex irritability with convulsions, the only difference so far discovered being that larger doses of brucia are required to produce the same effects. The iodide of methyl-brucia, discovered by Stahlschmidt and the sulphate of methyl-brucia appear to produce an opposite condition, namely, diminution of reflex irritability and paralysis of motor nerves, similar to the condition induced by Curara.]

THERAPEUTIC ACTION.—There are few medicines which have more remarkable therapeutic power than nux-vomica and strychnia, and few that have been more abused and misapplied, considering the short time during which they have been known to medical men.

The actions of nux-vomica and of strychnia so essentially correspond, that it is only necessary to say a few words about the former. Its employment in any case, in preference to strychnia, is purely a matter of convenience: the extract of nux-vomica is a convenient form, for example, for making into pills, while the tincture can be readily subdivided into doses representing extremely small quantities of strychnia. The latter is an especially favorite form for administration in certain complaints of children presently to be mentioned.

In Paralysis.—The earliest applications of strychnia in medicine were naturally directed to the cure of paralytic affections, and during some years it was employed very indiscriminately, and often in far too large doses. Those who prescribed it entertained the idea that the use of the drug should be pushed to a point at which actual symptoms of poisoning commenced: acting upon this impression, doses as high as the $\frac{1}{8}$ and $\frac{1}{4}$ grain, twice or three times a day, were often reached, with the almost invariable result of throwing the patient's nervous system into great disorder, manifested by twitchings of the limbs, hyperæsthesia of the retina and the auditory nerve, and a state of perpetual restlessness. It was discovered after a time that such effects were always injurious rather than useful: but it has further been found that violent strychnia-action is specially hurtful, and that the drug in any dose is valueless in particular forms and stages of paralysis. In paralysis of cerebral origin it is seldom of any value, and, on the other hand, its too early use in these affections, especially in such as proceed from hæmorrhage, has often proved very mischievous. It would be natural to expect better results from its employment in spinal paralysis, yet even here there is frequent disappointment, and in the early stages of organic lesions it often does much harm, especially if given in large doses. In the so-called "reflex" paralysis, it was supposed (by Brown-Séquard especially) to promise great results, but experience has hardly supported these expectations—a result, perhaps, not surprising, since the tendency of recent pathology is to render it probable that in these cases there is actual myelitis, produced by irritation transmitted to the cord through the nerves of the kidney, bladder, uterus, or whatever organ it may be in which the original mischief existed. In functional paralysis, on the contrary (*e. g.*, hysterical), strychnia is often of use, and decidedly promotes the recovery of power by muscles which have been paralyzed by the influence of lead. Brown-Séquard more correctly states that strychnia ought to be avoided as a most dangerous poison in those cases of paraplegia in which there are signs of congestion or inflammation of the spinal cord or of its membranes. For my own part, I think it is fully established that strychnia should not be employed except in cases where the symptoms are chronic, and thoroughly free from acute congestion.

In incontinence or in retention of urine, such as occurs in old people, strychnia is sometimes employed with much benefit.

There is one form of paralysis—that, namely, which is limited to one or two groups of muscles—in which a remarkable and very effective employment of strychnia has been devised by Mr. Barwell. He adopted it chiefly in cases of infantile paralysis of old standing, when the atrophic process had gone so far as to much impair the electric sensibility, even to a constant current. He employed a 2 per cent. solution of hydrochlorate of strychnia, and in some instances injected as much as $\frac{1}{4}$ grain. When Mr. Barwell first brought forward these cases, it was suspected by most persons that there must be some mistake, since very much smaller quantities, subcutaneously injected, had several times been known to induce alarming symptoms. But the solution, being examined, was found to be exactly of the represented strength; and as Mr. Barwell made a point of injecting quite into the substance of the paralyzed muscles, it was certain there was no waste of the fluid. After Mr. Barwell had brought forward a number of such cases in which great good had been effected by the strychnia injections, other persons ventured to try it; and the success has proved remarkable, while not a single accident has occurred. It is now conceded, by all who have studied the matter, that the high concentration of Mr. Barwell's solution is what localizes its action; such a solution as the one described is highly irritant, and, when injected into the belly of a muscle, it at once sets up inflammation around, and becomes enclosed, instead of mingling with the general circulation. A weaker solution would be far more dangerous to life. The local injection has been practised in France with much success for the treatment of prolapsus ani, only here the solution has been much weaker (1 in 1,000). Of such a solution 10 or 15 drops were used by Dolfear and Foucher. This operation would probably be far safer if performed with the concentrated fluid used by Barwell.

In **Anæsthesia** of functional character, the good effects of strychnia are often very marked—a circumstance which illustrates the difference between the poisonous and the therapeutic actions of the drug, since among the former we scarcely find any direct and evident effects upon common sensation.

In **Tremors and Atactic Movements** of many kinds (though not in true locomotor ataxy, which is an organic disease), strychnia has proved very useful. It is of much value, for instance, in the tremor of chronic alcoholism; here it is probable that a part of the beneficial action is due to its influence in removing the catarrhal condition of the stomach, an organ upon which we shall see that strychnia has a powerful influence. In chorea, particularly in those cases where fright, or the disturbing effect of commencing puberty, rather than rheumatism, is the principal cause, strychnia in minute doses ($\frac{1}{80}$ grain *ter die* for a child of ten years, $\frac{1}{40}$ to $\frac{1}{16}$ grain after puberty) has often been found of much use. In some cases of hysteric tremor, where the hysteria is merely the product of great bodily weakness, induced by illness or exhausting fatigue, it is similarly beneficial.

In **Dyspepsia** of the simple atonic form, strychnia, or tincture of nux-vomica, is often of the highest possible value; it is best alone, but an excellent combination is tincture of nux-vomica and diluted nitric acid, 5 or 10 minim doses of the first and 15 of the latter. I have already mentioned its good influence upon the irritated stomachs of drunkards; it may be added that heartburn, hiccough, regurgitation, and even pyrosis, when the origin is chiefly due to an atonic condition of the muscular walls of the stomach and functional languor, may frequently be cured by a short course of this medicine. It is often very useful in the morning-sickness of pregnant women. Abdominal cramps and spasm are quickly overcome by it.

In **Atonic Conditions of the Bowels**, also, strychnia is of great service, especially in that almost hopeless-looking form of constipation in which the large bowel almost ceases to contract, and becomes passively loaded with more and more fæces day after day. In prolapsus ani, arising from such a state of the rectum and sphincter, Dr. Schwarz strongly recommends nux-vomica; and I agree with him, and can speak with equal emphasis of its benefit in hæmorrhoidal tumors of the anus. Five to ten drops of the tincture of nux-vomica, taken in a tumbler of cold water before breakfast and dinner, act as a laxative, and often overcome most obstinate constipation.

In **Epidemic Dysentery and Diarrhœa**, nux-vomica is often of much service, especially when given in conjunction with mineral acids.

In **Coldness of the Feet**, from languid capillary circulation, Dr. Anstie has seen much benefit induced by minute doses of strychnia.

In **Neuralgias of various kinds**, strychnia has often proved most useful. Probably its utility is greatest in cases where the pain is visceral rather than superficial; in hepatalgia, for instance, and in the milder forms of angina pectoris, and in gastralgia, it has been found of much benefit. In the latter disease the use of tincture of nux-vomica has long been a favorite and very successful remedy, though there are intractable cases, of course, where this, like every other medicine, will fail. In all forms of neuralgia the dose of nux-vomica or of strychnia should be very small; there is seldom occasion to use more than $\frac{1}{100}$ or at most $\frac{1}{32}$ of a grain twice or three times a day.

In **Cases of the Impairment of the Nervous Apparatus of Sight and Hearing**, strychnia is believed by some authors to be very useful. As regards vision, no one, of course, expects it to do good in those diseases in which the nutrition of the retina is destroyed by complicated inflammatory or degenerative processes; but it has been rightly expected that cases of simple atrophy might be really improved, and there is reason to think that this sometimes takes place. The supposed improvement of nervous deafness by strychnia is more than doubtful.

In **Spasmodic Asthma and in Dry Cold in the Head**, nux-vomica and strychnia have often been used with great effect.

In **Heart Weakness**, strychnia may be spoken of as an excellent tonic; it is sometimes the first remedy which begins to do good in cases of fatty heart, when everything else that could be thought of has been tried. One caution must be added, however—namely, that any undue pushing of the remedy will produce, even more seriously than in other subjects, a state of nervous worry and restlessness, in which sleep is broken or even destroyed by a tendency to perpetual muscular movement. To the subjects of fatty heart this condition is not merely fatiguing and annoying, but positively dangerous; and we are therefore bound to inquire for the first symptoms of its appearance when administering strychnia to such persons.

In Intermittent Fevers, it has been proposed to substitute the use of strychnia for that of quinine; but so far as concerns the acute stages, there is no satisfactory evidence of its efficiency. In the stage of convalescence, however, the combination of strychnia with quinine and iron, in the shape of Easton's syrup, is spoken of with much approval by Dr. Maclean and other good authorities. [Strychnia has also been found of decided service in tetanus, chorea, and epilepsy.

"In 1847 Dr. Fell, of New York, published seven cases of tetanus, six of which were certainly of the traumatic variety, and which all recovered under its use. . . . In chorea there is strong evidence of its antispasmodic virtues. Attention was first attracted to its power in this affection by Rougier and Trousseau. The former published ten cases of boys, between the ages of six and sixteen, who had suffered from the disease during various periods of from one month to four years. The duration of treatment varied between one and eight weeks. Only one of the cases relapsed, and that was cured by the same means. . . . In epilepsy nux vomica was asserted even at the time of Murray to be a valuable remedy; and in 1836 Chrestien, of Montpelier, alleged that out of thirty cases treated by it, eight were entirely cured, and the remainder benefited. Since then nearly all who have written concerning the disease have either made no mention of strychnia, or have only done so to condemn it. But in 1867 Mr. W. Tyrrell, apparently unacquainted with the precedents to which we have referred, was led by the fact that conia, which controls the convulsions of strychnia, increased those of epilepsy, to conjecture that strychnia might be antidotal to epilepsy. Having treated sixty-nine cases with the medicine thus suggested to his mind, he found that in every case it exhibited a marked power in controlling and altering the convulsive attacks." (Stillé.)]

PREPARATIONS AND DOSE.—Tinct. Nucis Vomicæ, ℥ v.—xx. (.30—1.20); Extr. Nucis Vom., gr. $\frac{1}{6}$—$\frac{1}{2}$ (.01—.03); Strychniæ Sulphas, gr. $\frac{1}{100}$—$\frac{1}{20}$ (.0006—.012).

IGNATIA BEAN (Strychnos Ignatia).

Active Ingredients.—Analysis proves the constituents of ignatia seeds to be similar to those of nux-vomica, though in different proportions.

Physiological Action.—Given in small doses, gradually increased to a poisonous quantity, the symptoms which ignatia produces are as follows: Increase of the salivary secretion, nausea, heaviness, giddiness, headache, pain in the stomach, flatulence, a feeling in the limbs as if they had gone to sleep, accompanied by great weakness and twitchings throughout the body; constipation at first, and diarrhœa later on; constriction about the throat, numbness, torpor, mental depression, twitching of muscles, tetanic spasms; a feeling of intense anguish at the pit of the stomach; convulsions; cold sweats; and finally death by dyspnœa or asphyxia.

THERAPEUTIC ACTION.—St. Ignatius' bean contains a much larger amount of strychnia than the strychnos nux-vomica, yet its effects are not identical with, though somewhat similar to, those of the latter.

[There is a very decided difference between the finer effects of ignatia and nux-vomica that is not explainable by the results of chemical analysis; comparative clinical experience, however, will quickly demonstrate this to the careful observer. The two drugs are by no means interchangeable.]

Ignatia is useful in many cases of hysteria, where the patient suffers from a feeling of suffocation, and a sensation as of a ball rising to the throat, whether or not attended by the usual symptoms of an ordinary hysterical paroxysm. It not only relieves the convulsive attacks, but in most instances prolongs the interval between them, and frequently prevents their return. Its effects are considerable in controlling, and even permanently removing, the convulsive bursts of crying or laughing, as well as the hiccough, the flatulent distention, and the general hyperæsthesia (or the morbid increase of sensibility of the tissues). The intercostal neuralgia, so common in hysteria, is quickly removed by the agency of this drug; also clavus hystericus, the acute pain in the head, as if a nail were being driven into it. Where there is great mental excitement or depression, ignatia has a soothing effect. In hysterical women with whom aphonia frequently occurs, with few or no traces of catamenia, and often accompanied by profuse leucorrhœa, the symptoms named all disappear after a steady course of this medicine. It also corrects diseased appetite. I have often, with the happiest effect, used ignatia in the convulsions of children arising from intestinal irritation, such as worms or undigested food, but unattended with cerebral congestion. [In sleeplessness resulting from nervous erethism, ignatia in small doses will often afford relief more promptly and satisfactorily than morphia, and without the disagreeable after-effects of the latter. We have found it exert a marked control over the nervous exaltation that sometimes follows the prolonged use of quinine. Of all remedies with which we are acquainted it is *par excellence* the controller of functional phenomena of the cerebro-spinal axis. In large doses it excites, in small doses diminishes, the irritability of this organ.]

The constituents of the *lignum colubrinum* of Timor, the produce of the *Strychnos ligustrina*, correspond with those of ignatia; and as this *lignum colubrinum* is held by the natives in great estimation as a cure for neuralgia and for paralysis of the lower extremities, I do not see why ignatia should not be tried in similar cases, on a more extended scale, and perhaps be proved to be a successful agent.

Ignatia is useful also in the treatment of dyspepsia, hypochondriasis, and various nervous affections. Of the two preparations of ignatia in use I prefer the tincture.

PREPARATIONS AND DOSE.—The only officinial preparation is the Extractum Ignatiæ, the dose of which is somewhat less than that of nux-vomica, and should be large or small according to the kind of effect we desire to produce. The extract commonly dispensed is not, in our experience, a trustworthy preparation. We agree with the author in preferring a tincture.

CAROLINA PINK-ROOT (Spigelia Marilandica).

Active Ingredients.—The root and the leaves contain some ingredient which at any rate proves fatal to the life of intestinal worms. Whether this effect and the other not very well-defined symptoms recorded by Eberle depend on the peculiar bitter substance, or on the volatile oil, which Feneulle's analysis discovered in the root, we have as yet no knowledge.

Physiological Action.—The experience of Thompson is conclusive as to the fact that *some* narcotic effects can be produced by spigelia. He took large doses, and found that it produced quickness of pulse, flushed face, drowsiness, and stiffness of the eyelids. In another case, three-eighths of an infusion of three drachms of the root, in three gills of boiling water, were given to a child: after the third dose, the skin became hot and dry; pulse 110 and irregular; face (especially about the lids) much swollen; pupils widely dilated. There was strabismus of the right eye, and a wild, staring expression of countenance. The intellect seemed unaffected, but the child was seized with general tremor on trying to assume the erect position; the tongue also was very tremulous. Next morning every symptom had disappeared, except the swelling of the eyelids. It appears, however, that these symptoms are but rarely provoked in actual practice; and the probability is that the cathartic element prevents the influence of the narcotic.

Therapeutic Action.—In the United States, spigelia ranks as probably the best of the anthelmintics, especially for round worm; and it is in general and constant use as a vermifuge. There are, also, some cases of intestinal derangement in children which simulate those produced by worms, in many respects, and are accompanied by much fever and nervous irritability, which are often cured by the use of spigelia. Both in these and also in the case of worms, it is often advisable to unite a purgative with it—for instance, senna.

Preparations and Dose.—Spigelia, ℨ ss.—ij. (2.—8.); Extr. Spigeliæ Fluid., ℨ ss.—ij. (2.—8.); Extr. Spigel. et Sennæ Fluid., ℨ ij.—iv. (8.—16.); Infus. Spigeliæ, ℨ j.—ij. (30.—60.).

DEMERARA PINK-ROOT, or WORMGRASS (Spigelia Anthelmia).

Physiological Action.—Spigelia Anthelmia produces vomiting, giddiness, stupor, dilated pupils, diarrhœa, subsultus tendinum, irregular pulsations throughout the body, palpitations, dyspnœa, convulsions, and death. When cattle eat it, they perish in violent agony.

Therapeutic Action.—S. Anthelmia is useful in rheumatic pericarditis and endocarditis, in rheumatic ophthalmia and facial neuralgia. In rheumatic fever, with little inflammatory swelling of the joints, but where the pain quickly shifts from one joint to another, I have found this medicine of great service. It appears to have the power of centering the inflammation in certain joints and steadying it there. In palpitations arising from mitral and aortic disease, attended with considerable dyspnœa, the symptoms are much relieved by tincture of the same. It is also employed as a vermifuge.

PREPARATION AND DOSE.—None officinal. A tincture (1—8) may be given in doses of from five to twenty minims.

CURARA (SOUTH AMERICAN ARROW-POISON).

DESCRIPTION.—Curara, as seen in England, is a blackish-brown substance, of consistency varying from that of an extract to a resin. The taste is very bitter. It is readily soluble in water, and when heated gives out an odor resembling that of chocolate. The physiological action is highly characteristic, and serves at once for the identification of this curious drug. If the smallest possible drop of the solution be injected beneath the skin of a frog, the animal immediately becomes paralyzed in all its voluntary muscles.

ACTIVE INGREDIENTS.—The whole activity of curara has been satisfactorily proved to depend upon the presence of the alkaloid curarine, discovered by Roulin and Boussingault in 1830, and procured in well-defined crystals by Preyer in 1865. These crystals are four-sided, colorless prisms, of very bitter taste, and having a weak alkaline reaction. They readily attract moisture, are very soluble in water and in alcohol, but not at all soluble in pure ether, sulphide of carbon, benzol, or turpentine. No chemical formula can at present be safely given.

PHYSIOLOGICAL ACTION.—Curara induces a very marked and peculiar effect upon the system, and exactly the same follows the use of its alkaloid curarine, only that the intensity of the action of the alkaloid is far greater. The course of the symptoms is the same in mammalia as in frogs; the limbs are first paralyzed, the respiratory muscles then rapidly lose power, and the breathing becomes slower and slower, until at last it ceases entirely. The circulation is at first quickened, the heart acts no longer; and as death comes on, there occurs a flow of saliva and of tears. Slightly convulsive movements are likewise perceived in the first stage of certain cases. In mammalia this remarkable action of curara can be set aside by prompt resort to artificial respiration. In fact, it is now a constant practice with experimenters to curarize animals, and to establish artificial respiration, as a preliminary to physiological experiments which require absolute stillness. Under the influence of the restored respiration (if the dose be not too large) the heart resumes and maintains its action, although the paralysis of the limbs persists. If a rather considerable dose be injected into the jugular veins, the blood-pressure in the systemic arteries becomes greatly lowered, and the pulse-frequency is much increased. Later on, both of these conditions disappear. The intestinal movements are greatly accelerated, and there is also a much heightened sensibility of the intestines to stimuli—circumstances which present a marked contrast to the absolutely paralyzed condition of the voluntary muscular system. A singular feature is the occurrence of decided diabetes, which usually sets in very soon after the injection. The mode of administration exerts a marked influence upon the activity of curara: injection into the jugular vein produces the most powerful effects of all; injection into the subcutaneous tissues comes next; after that comes injection into the lungs; while the stomach absorbs curara so very slowly, that the poison has no time to accumulate in the system, being eliminated with equal rapidity by the kidneys; and, consequently, none of the usual effects are produced on the nervous system. Curarine, in fact, is perhaps more rapidly and completely eliminated than any other alkaloid. Poisoning, it is said, can be induced even by gastric administration of curara if the kidney-elimination be stopped: this was effected in the experiments made by Hermann upon rabbits; his method being to ligature the renal arteries, and then inject curara into the stomach. The effects, however, were anomalous: some of the animals died in convulsions—a very uncommon mode of action for curara; while in others the convulsions were quickly stopped by artifical respiration: they appear to have been due to suffocation from sudden paralysis of the respiratory nerves.

To give a complete account of the physiological action of curara is not possible; but certain facts are clear. It operates in no degree upon the spinal cord; it affects the brain only in an indirect manner. The weight of its operation evidently falls upon the motor nerves: the action commencing towards the periphery, and extending a greater or less distance along their trunks. The muscular substance itself remains unaffected: thus, if the motor-nerve trunks of a curarized animal be Faradized, there is no movement whatever, but direct electrization of the muscles makes them contract quite actively. The vaso-motor nerves appear to be paralyzed: in all probability the increased flux of secretions, and also the curara-diabetes, are due to this paralysis. At a later stage, the pneumogastric branches to the heart and to the stomach are disabled, as are also the sympathetic branches to the iris and the vesical nerves; last of all (often after a considerable interval), the musculo-motor cardiac nerves are affected. As regards the sensory nerves, the general opinion has been that curara does not influence them in the least; but the contrary view has been adopted by v. Bezold, and again of late by Lange, who thinks that the sensory nerves are powerfully affected, and that the reflex function of the cord is also impaired.

It has already been mentioned that curara is absorbed only by very slow degrees when introduced into the stomach. It appears from the researches of Claude Bernard and Preyer that curarine shares this peculiarity; its activity when given by the stomach is trifling in comparison with that which it shows when subcutaneously injected, especially when thrown into a vein. The fact of its elimination by the urine was most distinctly shown by Voisin and Lionville, who injected the urine of curarized frogs into healthy frogs, producing the characteristic curara effects; and the matter was carried still further by Bidder, who paralyzed a third frog with urine taken from one that had itself been poisoned by the urine of the animal to which the original dose of curarine had been given. There is no reason, however, to think that the whole of the curarine is eliminated without change: for curarine is known to be exceedingly sensitive to ozone, which destroys its poisonous properties; and it can hardly be doubted that a certain amount of it always gets oxidized and destroyed in the body.

The only other symptom that requires mention is the protrusion of the eyeballs, as observed by Voisin and Lionville. On the whole, the position of curara among narcotic poisons must be considered as very peculiar; and it is certain that only a few closely resemble it in action,—conia, for instance, and some of the methyl-bases, of which Drs. Fraser and Crum Brown have made such interesting experimental trials.

THERAPEUTIC ACTION.—The medicinal properties of curara have as yet scarcely been investigated.

In Tetanus, as might be expected, it has received considerable attention, and the results obtained have undoubtedly been encouraging. Four cases under Italian physicians are mentioned in the *Union Médicale*, 1869. The first was one of idiopathic tetanus, treated by Dr. Capozzi. Here the curara was injected subcutaneously in doses of $\frac{1}{4}$ grain, increased to $\frac{1}{2}$ grain, twice daily; about five grains in all completed the cure. A case under Dr. Merra was treated by injection with $\frac{1}{4}$-grain doses, and the patient recovered in a fortnight. A third patient was under Dr. Nobis, in whose hands about 12 grains, administered in small doses, proved successful. A fourth was a traumatic case, under Dr. Gherion. Sixty-four injections, varying from $\frac{1}{4}$ grain to $\frac{1}{2}$ grain of curara, quite cured the patient. An important case occurred also under Chassaignac.[1] A young man had a gunshot wound of the foot; at the end of a fortnight trismus and opisthotonos set in. Chassaignac gave 1$\frac{1}{4}$ grains of curara by the mouth, and ordered the wound to be syringed with an aqueous solution of curara, 1 part in 1,000. This produced a diminution of the spasms; the strength of the solution was then increased by 50 per cent. On the sixth day of tetanus there was a relapse; the internal doses were increased, and the curara lotion was raised to double its original strength. On the following day there was much improvement, and on the sixteenth day the tetanus had altogether disappeared. This case has been criticised as being probably an example of chronic tetanus spontaneously subsiding; but for this to be true, the connection between remedy and effect seems too close.

One of the most important papers on the use of curara in tetanus is that of Professor Busch, of Bonn; it is drawn from Busch's experience of the war in Bohemia in 1866. Busch had 21 cases of tetanus in the field hospitals, 14 of which were fatal. Out of these 21, he treated 11 with curara, the patients recovering in 6 of them; though one of these last seems to have owed his escape rather to the repeated use of hypodermic morphia than to the curara with which the treatment was commenced. Busch administered the curara hypodermically in doses of from $\frac{1}{10}$ to $\frac{1}{4}$ grain. His own conclusion was, that in the very severest cases of all, curara was inferior in power to hypodermic morphia.

The testimony of Demme is also very striking: he obtained his experience in the Italian war of 1859, in which there were 86 cases of tetanus on the Austrian side, and 140 on the Italian. He treated 22 cases with curara, and out of these obtained 8 cures. In order to appreciate the value of such facts as those of Busch and Demme, it is well to call to mind what is the usual mortality of traumatic tetanus in war: this, it is needless to say, is enormous—greater than that which occurred under curara-treatment in the hands quoted. Thus Demme, out of the whole number of cases which he treated, only saved 7 per cent.

There is no doubt that the subcutaneous plan is far preferable to any

---

[1] Journal de Méd. et de Chir. pratiques, 2ᵉ série, ¶ 5722.

other mode of administration, and that curara by the mouth is unnecessary and untrustworthy. On the whole there is a very encouraging amount of evidence in favor of curara in tetanus, if we look at the matter with sober and reasonable expectations. Certain cases of tetanus will, it may be feared, always remain among the list of necessarily fatal maladies; the personal or inherited nervous constitution of the patients rendering them peculiarly easy victims to the kind of influence which sets up fatal tetanus. It ought to be no serious discouragement to the use of curara that it has failed in such exceptionally severe and rapid cases as those recorded by Vulpian.[1] Some other instances that have been cited against curara in tetanus are equally inconclusive: thus, a case (from Professor Schuh's Clinic[2]) of traumatic tetanus was fatal; but it was one of extreme severity, as attested not only by the rapid course, but by the unusual amount of positive material lesion found after death in the cord; yet even here the curara produced very striking temporary amendment.

On the whole, it would appear very desirable that curarine should be exclusively employed in order to insure uniformity of effects; but it appears that at present there is much difficulty in obtaining this alkaloid, which is not easily prepared. In the meantime, curara can now be obtained of one or two of our best pharmaceutical chemists, of a considerably more uniform strength than formerly.

In Strychnia-poisoning, curara has also been warmly recommended; but its claims to confidence are very doubtful. It would appear that against the violent spasms of the respiratory muscles which strychnia produces, the doses of curara, to be effective, would have to be so large that there would be very great probability of inducing complete curarism, during the maintenance of which it would be necessary to keep up constant artificial respiration. Considering the apparently much greater benefits which are offered by nicotine (without these disadvantages) in strychnia-poisoning, it can scarcely be worth while to use curara for this urpose.

In Chorea, curara has been tried with very different results in different hands; and, on the whole, nothing can be said to be definitely established in its favor.

In Epilepsy, also, curara was loudly vaunted at first; but the comparatively recent experiments of Beigel give little encouragement to the idea that it is really effective. He employed it in both large and small doses.

In Facial Spasm, Gualla found curara effective when other remedies had failed.

PREPARATIONS AND DOSE.—Non-officinal. A solution should be given subcutaneously in the commencing dose of $\frac{1}{10}$ grain, repeated as may be necessary. As different samples of curara vary in quality, the first few doses should be administered with caution.

Of curarine it is unfortunately impossible at present to state the proper dose with any precision. Preyer considered that the alkaloid has twenty times the power of curara, but Beigel's researches with Preyer's own curarine make it very doubtful whether the proportion is so high as this. Subcutaneous injection is, of course, the only mode of administration to be thought of in the case of curarine. The dose can at present be only

---

[1] Union Méd., No. 7, 1857.
[2] Schmidt's Jahrbuch, 118, p. 292.

stated with reserve, as probably ranging from the $\frac{1}{60}$ to $\frac{1}{30}$ grain. Marquart of Bonn is the best maker.

AKAZGA (West African Ordeal Poison).

ACTIVE INGREDIENTS.—From an alcoholic extract of the bark, Fraser obtained by Stas's process an alkaloid which he named akazgia, and which he proved to represent all the physiological powers of the bark. Akazgia is a white, amorphous substance, but may be crystallized with some difficulty from an alcoholic solution, in small prisms. It is very insoluble in water; very soluble in alcohol of 85 per cent.; less so in absolute alcohol; very soluble in ordinary, less so in absolute ether; and has an alkaline reaction. It is probably the only alkaloid in existence which presents the same remarkable series, of reactions (color-tests) to sulphuric acid, with bichromate of potash, or peroxide of lead, as those of strychnia. Akazgia, and the neutral salts which it readily forms with acids, are bitter tasting, but the bitterness is far less intense and lasting than that of strychnia and its salts.

PHYSIOLOGICAL ACTION.—There is a strong general resemblance between the action of akazgia and that of strychnia, but not so strong as its likeness to that of brucia and igasuria. Fraser found that for a rabbit of 3 lb. weight the minimum fatal dose was $\frac{6}{10}$ grain, subcutaneously injected. With this quantity the reflex movements were evidently exalted at the end of nine minutes; tetanus occurred a minute later, and death at eleven minutes after the administration. Smaller doses than this produced at first serious effects, but the animals slowly recovered. Pécholier and Saint Pierre concluded that akazgia acts in the same manner as strychnia, *i. e.*, its chief effects are on the sensory nervous system, producing, first, exaggerated sensibility, then tetanic convulsions, and at last paralysis and death, and that it only acts secondarily upon the motor-muscular system. Fraser remarks on the interesting fact that a certain quantity of the poison is able to produce a condition which as nearly as possible approaches death, and that an extremely small augmentation of this dose is capable of quickly causing violent symptoms and a rapidly fatal termination.

THERAPEUTIC ACTION.—Nothing satisfactory is known upon this subject at present; but the above description of the physiological action of the plant and its alkaloid gives considerable room to hope that akazgia may have special therapeutic virtues of its own. It appears probable, for instance, that in many cases where strychnia in small doses is employed for a considerable time, but with the inconvenience of producing a nervous erythism, which from time to time obliges us to suspend its use, the milder akazgia would prove more practically manageable.

PREPARATIONS AND DOSE.—Of course nothing can be precisely stated. It remains to be seen how far akazgia can be obtained commercially in sufficient quantities; and we cannot tell how far the production of the alkaloid, which seems a difficult process, would allow of its being sold at anything like a moderate price.

RUTACEÆ.

RUE (Ruta Graveolens).

Active Ingredients.—Rue contains a bitter, neutral, crystalline principle called rutine, $C^{25}H^{28}O^{15}$, not known to possess any physiological activity, and an ethereal oil, which appears to be the source of the virtues of the plant. This ethereal oil is of a light-yellow color (turning brown on keeping), has a bitter acrid taste, and a very disagreeable odor: sp. gr. 0.911. It consists of two substances, of which one has for some time past been recognized as a hydro-carbon of the turpentine-camphor type; and the other is an oxygen-containing oil, which until lately was not understood. The researches of Gorup-Besanez and Grimm, and of Giesecke, which have proved that it is methylcaprinol, or methylpelargonylketon, $C^{10}H^{19}(CH^3)O$; which substance, which boils at about 440° F., is colorless if looked at from the side, but shows a violet-blue fluorescence when looked down upon; it has a specific gravity of .8268, and crystallizes at 42.8° F.[1]

Physiological Action.—Rue is a powerful irrritant narcotic. Orfila injected 15 minims into the venous system of a dog; in two minutes there followed vomiting, vertigo, and semi-paralysis of the hind-limbs; the animal, however, recovered. With human beings, the effect of an overdose on the general nervous system is to produce great prostration, combined with restlessness and excitement, confusion of ideas, and dimness of vision; besides which, there is violent vomiting or retching, with hiccough. Rue has also, from very ancient times, been reputed to possess the power of exciting the uterus; on this account it was formerly much used as an abortifacient by rustics and by irregular practitioners. In England the common people and the "herb-doctors" now more frequently use savin for this purpose, but in Hindostan the women still put faith in rue.

Another phenomenon which has been noted as the result of an overdose of rue is weakness and slowness of the pulse. Probably the alleged weakness referred merely to the small size of the pulse, for the true condition is most likely that of high arterial tension.

This view is justified by the experiments of Dr. Helie,[2] which show that rue acts specifically on the uterine walls, and simultaneously slows the pulse very considerably. It seems to me that this slowing is due to increased arterial resistance from stimulation of the muscular coats of the small vessels.

Topically, rue exerts a powerful irritant action; if the bruised and moistened leaves be applied to the skin, they speedily cause redness and inflammation; if chewed, they excoriate the mouth.

Therapeutic Action.—Rue has somewhat unaccountably fallen into disuse and discredit with the majority of the regular profession in

---

[1] Giesecke (N. Jahrbuch der Pharm., xxxiv. 366, quoted in Husemann, Pflanzenstoffe).

[2] Med.-Chir. Review, lviii., p. 604.

this country; and, in Germany, Nothnagel speaks of it as a quite superfluous drug, to be classed with thyme, origanum, and some half-dozen other remedies of the like efficacy. This is certainly very unjust; probably even the exaggerated ancient ideas of the value of rue were nearer the truth. The essential oil, which is undoubtedly the active element, has long been reputed as antispasmodic, stimulant, emmenagogue, and anthelmintic. In the last of these capacities it need not be mentioned here, since there are plenty of much more trustworthy anthelmintics.

In **Flatulent Colic**, more especially when occurring in children, oil of rue is an exceedingly useful remedy; it may be either given by the mouth, or administered as an enema, which Dr. A. T. Thompson recommends for infants.

Infantile Convulsions, of the kind which depend on flatulence of some part of the alimentary canal, can be most beneficially treated with oil of rue, in enema.

Hysteria, especially in the form which is associated with amenorrhœa, is sometimes apparently much benefited by rue; but of course in this disease all medicinal treatment must be combined with moral discipline.

As an Emmenagogue, rue for centuries enjoyed a very high reputation, though at present the tendency is to treat it as worthless, or not to be relied upon. It is, of course, always difficult to know whether a remedy which has power to act on the uterine muscular system—as rue confessedly has—is also a true stimulant of the ovarian secretion, if we may so call the discharge of ova from the Graafian vesicles. No doubt, in a considerable proportion of cases, where the menstrual flow is restored under the influence of remedies, this is only effected in a much more indirect manner. But among the class of true emmenagogues, those which tend to produce the menstrual flow without reference to improvement of the general health, the quality of the blood, etc., rue will probably be found to occupy a very genuine position. M. Beau says that it holds the same position towards savin that ipecacuanha holds towards tartar emetic. M. Alibert regards it as useful in hysterical dysmenorrhœa; and it seems likely that the cases in which it is most useful are those examples of amenorrhœa and dysmenorrhœa, in which muriate of ammonia has been found to do good service.

In **Epilepsy**, rue has been vaunted as a specific, its reputation having been originally founded by Galen, who directed rue-leaves to be sprinkled in the beds of those who suffer from priapism and erotic dreams. Murray and others have claimed for it the power of curing epilepsy; but the utmost that can be allowed as even probable is, that in cases where the malady is wholly or partly dependent on seminal weakness, small doses of rue, by their action on the sexual organs, may limit the amount of nocturnal discharge, and thus mitigate the nervous prostration so favorable to a continuance of the fits.

The Eyesight.—M. Elgâjaki says that the excessive use of rue produces dimness of vision; but he also alleges that by taking it in moderate doses the eyesight becomes improved. I have myself certainly seen good effects follow the continued use of minim doses, night and morning, in dimness of vision dependent, apparently, upon a functional amaurotic condition.

The bruised leaves of rue laid upon the forehead arrest epistaxis by their revulsive action. The oil, mixed with honey, and placed on warts, is said to have the power of destroying them.

PREPARATIONS AND DOSE.—Ruta, gr. x.—xxx. (.65—2.); Oleum Rutæ, ♏ij.—vj. (.12—.35).

ANGUSTURA, or CUSPARIA BARK (Galipea Cusparia).

ACTIVE INGREDIENTS.—Angustura has a strong, peculiar, and slightly offensive odor, with an abiding, bitter, and aromatic taste which is said to produce salivation. It yields its properties to water and alcohol; these depending upon the presence of a bitter, crystallizable principle called angusturine, or cusparine; a hard and bitter resin; a soft and balsamic resin; volatile oil; gum and woody fibre. The cusparine exists in tetrahedral crystals; is slightly soluble in water; more so in alcohol, acids, and alkalies; insoluble in volatile oils and ether. Its composition is unknown; and from the experiments of Saladin it would appear to have little activity.

The hard resin is brown, soluble in potash, alcohol, and acetic ether, insoluble in oil of turpentine and sulphuric ether. The soft resin is slightly acrid, greenish-yellow, soluble in alcohol, ether, oil of turpentine, and oil of almonds; insoluble in solution of potash.

The volatile oil is obtained by the distillation of the bark with water. It retains the peculiar odor of the bark, and in taste is acrid; it is lighter than water, and is of a yellowish-white color.

PHYSIOLOGICAL ACTION.—In its operation, angustura is said to resemble cascarilla; but the affinity is nearer to that of the pure bitters, and the stimulating effects are less decided. Administered in large doses, it induces nausea, followed by flatulence and purging. Observations are still wanting, however, both as to its absolute properties and its relative ones; and meantime it is safest, perhaps, to regard this drug as an aromatic tonic and stomachic, adapted for substitution when there are objections to cascarilla, cinchona, or calumba, and especially for use in the tropics.

THERAPEUTIC ACTION.—Dr. Hancock, who had excellent opportunities for watching the effects of angustura, declares it adapted to the worst forms of bilious fevers. All fevers of an adynamic type, especially when accompanied by severe stomach disorders, are considered amenable to it; as are likewise most of the ordinary kinds of muscular debility, and such forms of dyspepsia and anorexia as are referable to similar causes.

In Diarrhœa and Dysentery, when chronic, angustura may likewise be employed with great benefit to the patient; and in the later stages of these disorders, when there is often a considerable amount of mucous discharge, though not chronic, its efficacy again seems unquestionable.

PREPARATIONS AND DOSE.—Angustura, gr. x.—xxx. (.65—2.); Infus. Angusturæ, ℥ ss.—j. (15.—30.).

BUCHU (Barosma Crenulata).

ACTIVE INGREDIENTS.—Buchu leaves contain 7 per cent. of a volatile oil, which gives them a powerful and penetrating odor. This oil is yellowish-brown and lighter than water. They contain, also, a peculiar bitter extractive, called barosmine or diosmine, which is brownish-yellow, bitter, and slightly pungent; soluble in water, but insoluble in alcohol and ether. Resin, gum, and lignine are likewise present.

PHYSIOLOGICAL ACTION.—Taken in large doses, buchu produces a burning sensation in the stomach, with vomiting, purging, and strangury. Smaller doses induce in the same part a sense of heat, which soon diffuses itself over the body. It quickens the pulse, relieves nausea and flatulence, increases the appetite, and causes slight moisture of the skin. At the same time the flow of urine is augmented, the color of the fluid becomes darker, and a brown sediment is deposited, accompanied by a strong aromatic smell.

THERAPEUTIC ACTION.—Appearing to exercise a specific influence upon the genito-urinary mucous membrane, buchu is eminently useful in chronic maladies of the organs themselves, especially those arising from chronic inflammation of the mucous membrane of either the whole or a portion of this tract, and when accompanied by copious secretion, independent of real organic change—such, for instance, as occurs in cases of catarrh of the bladder following gonorrhœa, or the improper use of injections. When the catarrh implicates the ureters, and even the kidneys themselves, producing considerable mucopurulent discharge, and often incontinence or retention of urine, buchu renders more help in combating and curing these affections than any other drug known; but it cannot be relied on in every constitution, any more than many other drugs. It is also given in irritable conditions of the urethra, such as spasmodic stricture and gleet. Lithiasis, attended by rapidly increasing secretion of uric acid, has again been beneficially treated by buchu, the formation of the acid being checked. In such cases the medicine should be administered in combination with an alkali, such as solution of potash. Diseases of the prostate are likewise relieved by buchu. It is recommended, also, in atonic dyspepsia depending upon urinary disease, in chronic rheumatism, for cutaneous affections, and for dropsy. The use of buchu appears to be unknown, or quite neglected, in Germany. Buchheim, Husemann, and Nothnagel do not even mention it.

PREPARATIONS AND DOSE.—Extractum Buchu Fluidum, ℨ ss.—ℨ j. (2.—4.); Infus. Buchu, ℥ j.—ij. (30.—60.).

[JABORANDI (Pilocarpus Pinnatifolius).

ACTIVE INGREDIENTS.—In 1875, MM. Byasson and Hardy both obtained from this plant an alkaloid which is now known as pilocarpia. It is probable that jaborandi also contains another alkaloid and a peculiar acid. Of pilocarpia several salts, especially the hydrochlorate and nitrate, have been prepared. The writer was the first, he believes, to prepare an iodohydrargyrate. This has since been prepared in much larger quantity by Chas. Rice, Ph.D., chemist to Bellevue Hospital, New York, who furnishes $C^{11}H^{16}N^4O^4HI.HgI^2$ as its probable formula.

PHYSIOLOGICAL ACTION.—Jaborandi produces two most marked physiological effects—sweating and salivation, with a promptness and to a degree unequalled by any other known drug. If an infusion made of from one to two drachms of jaborandi leaves be administered to an adult, in from five to ten minutes the face will flush, and in a few moments more become covered with droplets of sweat. Soon the entire body becomes bathed in perspiration. Coincident with the sweating, free salivation occurs. These phenomena gradually disappear at the end of three to six hours. In some cases nausea and vomiting may occur. The face, at first flushed, afterwards becomes pale, and, according to Ringer, the temperature of the body falls from 4° to 1.4° F. The physiological effects resulting from continued use of the drug have not as yet been sufficiently studied; and it is by no means impossible that the violent perturbation of the system, which results in the profuse sweating and salivation, may seriously impair the functional and even organic integrity of important organs. The sweating is apparently not due to increased afflux of blood to the surface, but rather to a paralysis of the inhibitory influence of that portion of the nervous system that presides over the sudatory apparatus. The remarkable experiments of Adamkiewicz, in which he produced sweating at will by the faradization of certain nerves, should be considered in this connection.[1]

THERAPEUTIC ACTION.—Like all new drugs possessing marked and decided action, jaborandi has been employed in almost every disease in which it seemed desirable for any reason to stimulate the perspiratory function. In this respect it is a marked example of the methods too much employed in so-called "physiological therapeutics." In a given case it seems desirable to fulfil a certain indication, and a drug capable of effecting this result when given in immense doses is brought into requisition, utterly regardless of the ulterior or even immediate effect which it may have on organs other than those which it is desired to influence. Instances of this are too numerous and too well known to need citation. With respect to jaborandi, the notes of warning are already being heard. In his seventh edition Ringer says: "I have used it in many cases" (of Bright's disease) "not only without benefit, but, I must confess, not without doing harm, for it has always much depressed the patient, has generally excited sickness, and the effects have been so disagreeable that in almost every case the patient has begged the discontinuance of the treatment." In a case of syphilis complicated with renal anasarca, under the writer's care, the administration of a teaspoonful of fluid extract of jaborandi was followed by diminution of the urine and rapid and marked increase of the œdema without sweating. More recently Dr. Fordyce Barker[2] has called attention to certain dangers accompanying its use in puerperal albuminuria. When first introduced, it was, so to speak, "warranted safe;" but its inconveniences are now beginning to come to the surface. In the spring of 1877 the writer employed the hydrochlorates of pilocarpia in a case of severe ichthyosis. A daily dose for two weeks entirely removed the scales, but the patient preferred the disease to the remedy, and discontinued the treatment. It has been employed by others in different cutaneous affections, but so far, as we are aware, with but exceptional benefit.

Ringer makes the statement that to many may seem extraordinary that "pilocarpine, in doses of $\frac{1}{10}$ grain, given thrice daily, will check pro-

---

[1] Die Secretion des Schweisses, Berlin, 1878.
[2] Medical Record, 1878.

fuse perspiration." In the spring of 1878 a case of mild mercurial salivation occurred in one of the writer's wards at the Charity Hospital; a few weeks later a case of severe salivation occurred in connection with his dispensary service. In the first case pilocarpia, and in the other jaborandi in small doses, was employed. In both immediate improvement and rapid recovery ensued. With the exception of the class of cases just referred to, we are unable at present to specify any diseases in which jaborandi can unreservedly be recommended, and are forced to conclude, with Ringer, that "though of great physiological interest, jaborandi has not yet proved very serviceable in the treatment of disease." The iodohydrargyrate of pilocarpia, already alluded to, suggested itself to the writer as a theoretical substitute for Zittmann's decoction in the treatment of syphilis. We have not, however, as yet had sufficient experience with it to hazard any conclusions concerning its practical utility.

PREPARATIONS AND DOSE.—Non-officinal. To procure sudation, an infusion or a decoction of one to two drachms of the leaves, or an equivalent amount of the fluid extract, may be employed. The corresponding dose of hydrochlorate of pilocarpia is about gr. $\frac{1}{4}$—$\frac{1}{2}$.]

GRANATACEÆ.

POMEGRANATE (PUNICA GRANATUM).

ACTIVE INGREDIENTS.—The rind of the fruit contains about 20 per cent. of tannin, with extractive and mucilaginous matters. The root-bark contains about the same proportion of tannic acid, some gallic acid, resin, mannite, and other less important substances. An acrid principle called punicine has also been detected in the fresh bark. The chemistry of pomegranate bark is still, however, very uncertain. It is doubtful whether the "granatine" of Landerer is an independent body, or merely some form of mannite; and punicine is a very imperfectly identified substance.

PHYSIOLOGICAL ACTION.—Decoction of pomegranate, taken in full doses, causes nausea, flatulence, vomiting, occasional cramps in the legs, and purging, the stools being generally of a light-yellow color; sometimes giddiness, faintness, dimness of vision, and a general numbness in the limbs; while the urine is often increased in quantity. The fresh bark of the root is believed to be the only part of the plant which produces these effects, though the flowers have a certain measure of reputation.

THERAPEUTIC ACTION.—The bark of the root has obtained great celebrity as a specific in tænia. Two ounces of the fresh bark of the root are boiled in a pint and a half of water, till the quantity of fluid is reduced by one-half; and of this, when cold, about a third is taken at intervals of thirty minutes until finished. It occasionally produces a little nausea, but rarely fails to expel the worm.

Before the decoction is administered, vegetable broth and spare diet should be prescribed. The evening before the patient commences using it, he should take an ounce and a half or two ounces of castor-oil, combined with an equal quantity of syrup of lemons. The decoction itself should be taken as above described; and in an hour or two the worm will come away entire, wound into a ball, and in many places strongly knotted.

Sometimes the first and second doses are rejected from the stomach, owing to the nausea above mentioned. Should this be the case, the remainder of the medicine must nevertheless be taken; and should the worm not come away, similar doses must be given on the following day.

The bark should be dry, but fresh and *bruised;* the two ounces being macerated in water for twenty-four hours without heat, and then boiled gently, and reduced as before mentioned, when it only needs straining.

M. Bourgeoise, who has published his observations and experience in the *Bibliothèque Médicale*, never found occasion to use a smaller quantity, but considered it desirable rather to increase the strength.

From my own experience, there is some uncertainty in the action of this drug; and it is said by some " to produce serious consequences."

The astringent properties are readily communicated to water, so that pomegranate bark is strongly recommended for employment as the basis of a gargle for relaxed gums and throat. It is also employed with advantage in chronic diarrhœa; also in menorrhagia, and in prolapsus uteri, prolapsus ani, etc.

In phthisis pulmonalis, again, during the profuse perspirations, and in the colliquative diarrhœa which is often so distressing an accompaniment towards the close, pomegranate rind is prescribed with considerable benefit.

To patients suffering from ardent fever, and its attendant thirst, the juice of the pomegranate, especially if tempered with sugar or honey, is particularly refreshing; many consider it preferable to that of the orange.

PREPARATION AND DOSE.—Decoctum granati radicis corticis; dose as above specified.

SCROPHULARIACEÆ.

FOXGLOVE (DIGITALIS PURPUREA).

ACTIVE INGREDIENTS.—Until quite lately, the chemistry of digitalis was a mass of confusion, owing to the want of concert between the different observers, and the careless use of particular names, which were employed in distinct and even in quite opposite senses. This confusion was at length substantially cleared up by the researches of Nativelle, the results of which are shortly summed up as follows in the treatise of the Husemans: Digitalis contains three peculiar substances, the active, crystalline, bitter-tasting digitaline;[1] the likewise active amorphous, bitter-tasting digitaleine; and an inert, crystalline, and tasteless substance. Of these three substances, digitaline is found only in the leaves, which also con-

---

[1] We here employ the nomenclature of Wiggers. Nativelle names the digitaline here referred to digitaleine, and, conversely, the digitaleine of our description he calls digitaline.

tain the other two bodies, while the seeds contain only digitaleine and the inert substance. The wild plant is, according to Nativelle, much richer in the active ingredients than the cultivated; the maximum richness is obtained in May, before the development of the flowers. The respective percentages of the above-named substances require further accurate researches for their determination. Nativelle found only $\frac{1}{10}$ per cent. of crystallized digitaline in the leaves; of digitaleine, mixed, however, with the inert substance, he obtained 1 per cent. from the leaves and 2 per cent. from the seeds.

In the present article, the active crystalline body will be called digitaline, in accordance with the terminology which I have elsewhere adopted; and the active amorphous substance will be named digitaleine.

Digitaline forms small, white, silky needles, aggregated in masses, and having an exactly neutral reaction, an intense and lasting bitter taste—which, from the great insolubility of the crystals, only slowly develops itself—and no smell. Water—even boiling—ether, and benzol are almost powerless to dissolve it; but it is readily soluble in alcohol of 90 per cent.; still more so in chloroform. Concentrated sulphuric acid dissolves it with a green color, which, on the application of bromine vapor, turns to red, but is restored to green on the addition of water. Nitric acid at first gives no color, but afterwards the solution becomes yellow. Muriatic acid turns it to a yellowish-green, which afterwards changes to emerald-green; water precipitates it from this solution as a resinoid mass. The chemical formula is not yet accurately established.

Digitaleine is a colorless, amorphous body; soluble in water in all proportions; and, like the first described substance, digitaline, is destitute of nitrogen.

The above is all that need be borne in mind by physicians in regard to the chemical and physical nature of the active ingredients of digitalis.

Concerning the commercial forms in which the active principles can be obtained, it will be sufficient to mention the so-called digitaline of Homolle and Quevenne, which is very insoluble in water, and evidently consists mainly of digitaline, as above described. This is the preparation best known in England, and perhaps the most to be relied on; but it would appear that there is no great difference between it and the preparations supplied by Merck, Marquart, and other German manufacturers of repute. They all contain both digitaline and digitaleine, besides other matters that are practically of no account, the first being very much the most active in its effects on the body, as has been proved by the experiments of Nativelle and of Schroff. Fothergill[1] throws doubt on the superior activity of digitaline to digitalis itself, because of his finding a tincture of the leaves act more powerfully upon frogs. But this probably arose from the alcohol having dissolved little or nothing but the digitaline, while Homolle's preparation contained also digitaleine and digitalose. [There are four main substances which can be isolated in a pure state from the leaves, and which have poisonous or else physiological properties, viz., digitanin, digitalin, digitaleïn, and digitoxin (Binz). Commercial digitalin is a varying mixture of these, and the eclectic "digitalin" of still more uncertain composition.]

PHYSIOLOGICAL ACTION.—Taking the digitaline of Homolle and Quevenne as the standard, the following may be said to be the true

---

[1] Prize Essay on Digitalis.

physiological actions of digitalis, so far as authorities are agreed about them:

(1.) It is admitted on all hands that digitalis is a cardiac poison; given in large doses, it brings the heart to a standstill.

(2.) In doses which just fall short of a fatal effect, digitalis produces faintness, diarrhœa, nausea, and vomiting, with irregularity of the heart's action.

(3.) In still smaller doses, the heart's pulsations are much reduced in frequency, and the arterial blood-pressure is remarkably raised.

(4.) Doses large enough to slow the heart's action usually reduce the temperature.

Concerning the manner in which these effects are produced, there is, however, diversity of opinion. Traube put forward the theory that the action of digitalis on the heart takes place through the medium of the vagus; this is at first stimulated, and exercises increased inhibitory action on the heart, reducing the number of its beats; after a time, however (with large or repeated doses), there is vagus-exhaustion, the result of which is fluttering, rapid, and irregular cardiac action. It is evident, nevertheless, that the slowing action of digitalis is not exerted through the vagus-centre, since it is manifested in animals in which the trunks of the vagi have previously been divided; and on this ground, chiefly, certain authors have denied its action on the vagus altogether. Yet it can hardly be disputed that digitalis acts on the vagus, though only on its peripheral cardiac branches; for when atropine (which has the power to paralyze the vagus down to its very terminal twigs) has been previously administered to an animal, it is found that digitalis fails to slow the heart. Such is the argument of Traube, though disputed by various writers. The theory of direct stimulation of the cardiac ganglia giving rise to increased propulsive action of the heart, and overcoming the resistance of the inhibitory vagus branches, was put forward by Dybkowsky and Pelikan, supported by Handfield Jones and Fuller, and accepted by Fothergill; and the experiments of Eulenberg and Ehrenhaus, who plunged the separated, but still pulsating, hearts of frogs into a solution of digitaline, and observed strengthening of the contractions and reduction of the frequency of the beats, are considered by Ringer[1] to prove that the "effect of digitalis is not due to any action on the pneumogastric nerve." One of the most recent writers on digitalis, Professor Ackermann, supports the theory of Traube; and, indeed, it is difficult to see how the evidence afforded by the control experiment of v. Bezold and Bloebaum with atropine can be set aside. Granting, however, that there is an action on the vagus, and that the primary excitement and subsequent exhaustion of the vagus-terminals account, in part, for the slowing and subsequent irregularity of the heart's action, there is no particular reason for denying the possibility of a simultaneous stimulation of the cardiac ganglia. There is, moreover, a considerable body of evidence tending to show that the muscular walls of the ventricles pass gradually into a tetanized condition, and are found so after death.[2]

One of the most indisputable facts respecting the action of digitalis is the remarkable increase of blood-pressure which it occasions in the arteries. Of this fact, again, different explanations are possible. Nearly all experimenters admit that digitalis produces contraction of the capil-

---

[1] Handbook of Therapeutics, 3d ed., p. 390.
[2] H. Fagge and Stevenson; Nunneley; Fothergill.

laries of the smaller arterioles, when taken into the general circulation;[1] and this in itself would undoubtedly heighten the arterial pressure. But, on the other hand, the blood-pressure may be heightened by increased force of the cardiac contractions; and the observation of Meyer, that there is a continuous rise in the pressure up to a point almost immediately before death, when it suddenly falls to zero, has been made by Dybkowsky one of the chief arguments for the direct stimulation of the cardiac ganglia.

As for the contraction of the small arteries, this is believed by Traube and Böhm to depend on excitation of the vaso-motor centre in the medulla oblongata; but Ackermann discredits this idea, and holds that the contraction is due to direct action on the peripheral ends of the vaso-motor nerves, or even on the muscular coats of the arteries themselves.

The next point of importance is, whether digitalis is a necessary physiological provoker of diuresis, or whether the diuretic effects undoubtedly obtained in disease are due to the accidental morbid conditions which are present in the latter case. Over this question there hangs some doubt. The diuretic effects do not appear to have been very striking in any case, and in many instances severe digitalis-poisoning has been accompanied by partial suppression of urine.[2] Ackermann states that digitalis acts diuretically only by increasing the fluidity of the blood, by facilitating the resorption of exudates, and suggests that by this means a part of the increase of blood-pressure may be caused.

Another interesting question concerns the cause of the reduction of temperature which is so marked and constant an effect of digitalis. This is probably due to an increased rapidity of circulation in the peripheral blood-vessels; for it must not be supposed that the slow pulsations of the heart indicate a diminished rapidity of the blood-current; on the contrary, their increased vigor causes the blood to circulate with abnormal swiftness: this causes increased transpiration, and a large loss of heat from the skin.

The more remote phenomena of digitalis-poisoning, especially those connected with the brain (giddiness, delirium, etc.), seem to be due entirely to changes in the circulation; there appears to be no direct action of the drug upon either brain or spinal cord.

THERAPEUTIC ACTION.—When and by whom this now celebrated drug was first employed as a medicine is not known. The date of the earliest employment was certainly prior to 1597, since in that year a treatise on Foxglove was published by Gerarde. Parkinson shortly afterwards recommended it for external application in diseases of a scrofulous character; he also administered it internally "against the falling sickness." In 1721 a place was given to it in the London Pharmacopœia, but in the ensuing edition (1746) it was omitted. In the Pharmacopœia of the Edinburgh College, digitalis has experienced corresponding alternations of repute and disfavor.

In approaching the subject of the therapeutics of so important a medicine, it may be well to observe, as a preliminary, that it is particularly adapted to persons of a sanguine or indolent temperament, with soft and lax muscles, and light hair. Dr. Withering long ago expressed his opinion that "digitalis seldom succeeds with men of great natural strength,

---

[1] Nunnelly, however, denies this, except as a very limited secondary effect.
[2] Christison on Poisons. Brunton on Digitalis.

of tense fibre, warm skin, and florid complexion, nor yet with people of tight and cordy pulse."

The employment of digitalis in disease includes some of the most interesting and valuable applications of remedies that are to be found in the whole range of the medical art. The following are the principal points, though they do not exhaust the list of the uses of digitalis.

As a Tonic to the Heart, digitalis justly enjoys very high favor in a variety of morbid conditions. It is a singular fact that digitalis should have been regarded for so long a period solely as a cardiac sedative, seeing that its virtues as a stimulant tonic had been recognized by more than one old writer. The tide has now fully turned, and there is no better recognized fact in modern medical practice than the power of digitalis to sustain and strengthen cardiac action in a variety of morbid conditions.

To Traube we owe the credit of pointing out that the value of digitalis in correcting the disturbances in the circulation, caused by organic disease of the heart, is constantly referable to mechanical conditions; and that this is especially true in regard to the general venous hyperæmia which commonly accompanies such disease. The venous hyperæmia is dependent on general anæmia of the arteries, and originates when hypertrophy causes insufficient compensation, and when the heart is no longer competent to drive a sufficient supply of blood into the aorta. Two physiological effects of digitalis then become remedial—namely, the increase which the drug produces in the force and regularity of the heart's contractions, and the increase which it causes in the contractions of the smaller arteries. The action of digitalis is therefore most effective in cases of insufficiency of the mitral valves. In disease of the aortic orifice, compensation by hypertrophy continues to take place for so long a period that the use of digitalis may be dispensed with. Niemeyer[1] remarks that "in digitalis we possess a very powerful means of moderating, not only hyperæmia of the lungs, but also engorgement of the aortic venous system which arises in mitral disease." "If we can succeed," he continues, "in retarding the action of the heart by means of digitalis, we afford time for the auricle to drive its contents into the ventricle through the contracted passage."

Digitalis plays an important part in the treatment of the state originally pointed out by Stokes as "weak heart," a state which is referred by German physicians to fatty and granular conditions of its muscular tissue.

The above principles, though expressing much of the truth, are perhaps too absolutely laid down; for there is not so great a difference between the relations of digitalis to mitral and to aortic disease as is here indicated. Sydney Ringer[2] is right in saying that it is better to consider "all the symptoms than to confine the attention simply to the nature of the valvular affection;" and in the following picture represents the class of cases in which digitalis is most strikingly useful. He says: "There is dropsy, which may be extensive. The breathing is much distressed in the earlier stages of this condition only periodically, and especially at night; but when this reaches its worst stage, the breathing is continuously bad, although it becomes paroxysmally worse. The patient cannot lie down in bed, and is perhaps obliged to sit in a chair, with the head either thrown back, or more rarely leaning forward on the bed or some other support. The jugular veins are distended, and the face is dusky and

---

[1] Text-book of Practical Medicine, i. 357.
[2] Handbook of Therapeutics, 3d ed. p. 395.

livid; the pulse is very frequent, feeble, fluttering, and irregular. The urine is very scanty, high-colored, and deposits copiously on cooling. The heart is seen and felt to beat over a too extensive area; and the chief impulse is sometimes at one spot of the chest, and sometimes at another. The impulse is undulating, and the beating very irregular and intermittent. The physical examination betrays great dilatation of the left ventricle, with often a not inconsiderable amount of hypertrophy. A murmur is ordinarily heard, having the characters of one produced by mitral regurgitant disease; and there may be also disease of the aortic valves."

The expression above quoted, that "in disease of the aortic orifice, compensation by hypertrophy continues to take place for so long a period, that the use of digitalis may be dispensed with," is certainly not universally true. In the case of aortic insufficiency, it is, indeed, approximately correct; this affection is usually attended with great hypertrophy, and the patients, for the most part, do not well tolerate digitalis. Theoretically, also, there is the danger of exciting a tetanoid contraction, from which the heart might not recover. Even here, however, it is not always the case that hypertrophy is developed in an excessive, or even in a sufficiently compensating, degree; and in instances such as I have occasionally witnessed, in which the inadequate hypertrophy made the overburdened left ventricle liable to sudden failure, with serious or fatal results, digitalis has proved of real service; and, as for the case of aortic obstruction, any one who has given digitalis practical trial will certainly recognize it as a remedy capable of exerting most beneficial influence in suitable cases of this affection. Dr. Fothergill has pointed out that in some such patients, especially when the obstruction has suddenly arisen, it may even be right to disregard the occurrence, during the administration of the drug, of intermission of the heart's pulsations, except as a reason for increasing the dose, upon which we shall find relief given to the cardiac symptoms. Here the failure of the heart has not been really due to the drug, but to the disease, hypertrophy not having been sufficiently established to do the work of compensation without powerful artificial assistance.

It may be well to say a few words respecting the old notion of a cumulative poisonous action after repeated doses of digitalis. This notion, as every one remembers, prevailed with most physicians some twenty to thirty years ago; but the dread seems to have been quite devoid of foundation. It must be recollected that in those days the perfectly false notion prevailed that digitalis was very useful in hypertrophy, but very dangerous in dilatation and in muscular weakness of the heart.[1] Recent experience has proved that the exact reverse of this is the truth; and there is much reason, therefore, to distrust the whole of the observations upon which the idea of cumulative poisoning rested; or, rather, there is good reason to think that the cases of dilated or degenerated heart, with weak and fluttering pulse, which were supposed to have ended fatally from the cumulative effects of digitalis, really sank from natural causes. And, indeed, it is certain, that were digitalis so dangerous, when continuously given in the very small doses to which it was formerly restricted, numerous persons must have fallen victims to the comparatively reckless administration of it which has distinguished many of the recent experiments, and especially to the German employment of digitalis in febrile affections.

As a general rule, it may be laid down that those cases of heart-dis-

---

[1] See, for example, Dr. Walshe (Diseases of the Heart, p. 664) for a clear enunciation of this principle.

ease will receive most benefit from digitalis in which there is weakness of heart-pulsation, irregularity (with or without dyspnœa), venous engorgement (with or without dropsy, but especially when combined with the latter), and a scanty secretion of urine. In such cases the drug induces effects as striking as are ever produced by any drug in any disease. The form which is decidedly to be preferred is the fresh infusion, made from good leaves; powdered digitalis, which was the form chiefly used by the German physicians in their researches on febrile diseases, is apt to vary greatly in strength. Of the infusion, from one to two drachms should be given every night and morning, and increased, if necessary, to a dose every three or four hours until a decided diuretic effect is produced. It is remarkable how unfailing is the occurrence of this beneficial action; for, while in health it is difficult to procure decided diuresis at all, either with small doses or large ones, in the morbid conditions described, digitalis produces a flow of urine with a certainty almost magical.

As a Diuretic, digitalis would hold very high rank, even were its power of relieving cardiac embarrassment unknown. In nearly all varieties of dropsy, except those where there is aortic regurgitation and very great cardiac hypertrophy, this drug is of the greatest service; it is wholly free, also, from the tendency to irritate the kidneys by which many so-called diuretics defeat their own immediate end, besides inflicting further mischief in cases where the kidney is either already diseased or mechanically compressed by ascitic fluid. The diuretic action of digitalis is also especially valuable in purely dropsical hydrothorax, and in those more passive kinds of pleuritic effusion which most nearly approach the character of the dropsical exudate. In all such cases there is a very good prospect of effecting a speedy resorption of the fluid by the use of digitalis in the manner just now described.

The Sudden Suppression of Urine, which occurs in young persons after scarlatina, or from simple exposure to cold and damp, is doubtless to be looked upon as part of a catarrhal inflammation of the uriniferous tubes, which requires little more than a few days' rest in bed, with almost a certainty that the urinary secretion will reappear. In this simple form the catarrhal inflammation scarcely needs the use of a diuretic; but should danger threaten, digitalis is still far preferable to any other drug. It is a different matter when, in the course of a chronic Bright's disease, in which many of the other organs have probably become impaired, a sudden suppression, induced by cold or any other cause, takes place. Here it is clearly advisable to resort to direct measures for the restoration of the urinary secretion, and digitalis is again one of the best agents for the purpose.

As the diuretic effect of digitalis in disease is very different from any effect which it is capable of producing in health, it may be well to consider the probable nature of the action. The idea that this consists merely in the heightened arterial pressure and the swifter blood-current (which, as already explained, exist simultaneously with slackened heart-pulsation) will not bear investigation; for, on that supposition, a constant effect of digitalis (in certain doses) on healthy persons would be diuresis, but such is not the effect. It appears evident that the presence of venous hyperæmia, with or without actual dropsy, is necessary to the production of anything like marked diuresis by digitalis; and the opinion of Legroux and others, that digitalis necessarily compels increased urination by virtue of the heightened arterial pressure which it causes, must be set aside.

In Febrile Diseases, digitalis has been largely employed of late years, especially in Germany. It is not easy to say who deserves the credit of first observing the power of digitalis to lower the human temperature;[1] but the first considerable steps in the use of the drug for this purpose were taken by Traube (in 1850), and from that time forward Germany has contributed a constant succession of observations on this subject.

The original paper by Traube announced a number of conclusions, physiological as well as therapeutical; the former of which there is no need to repeat, as I have already given the drift of his opinions. The therapeutical bearing of Traube's researches amounts to this: That, in the first place, digitalis as a stimulant to the regulating cardiac nervous system, diminishes the lateral pressure in the blood-vessels, and also the rapidity of the blood-current, and simultaneously lowers temperature. And, secondly, that by virtue of these actions, it limits the inflammatory process. It has already been shown in my discussion of the physiological action of the drug, that, though Traube was probably right in ascribing a portion of the action of digitalis to its effects on the vagus, it is impossible to account for the whole of its effects in this manner; and, in two important particulars, Traube was certainly wrong. The arterial pressure is not lowered, but heightened; and, as stated above, this heightening of arterial pressure is probably of great service in a therapeutic point of view. Secondly, although the pulse is undoubtedly slowed by moderate (therapeutic) doses of digitalis, there is great reason to think that the blood-current is rendered more rapid—a phenomenon which, since the re-researches of Edward Weber and of Ludwig, is easily intelligible. Traube was right, however, as to the influence of digitalis in lowering temperature, and diminishing the intensity of inflammatory and febrile processes generally—at any rate for a time. The treatment of pyrexial diseases of all kinds with digitalis has become very widely spread on the Continent, and has found a few followers in this country.

Leaving the subject of acute inflammations aside for the moment, we may refer to the remarkable manner in which digitalis has been pushed, especially by German physicians, in the treatment of typhoid fever, typhus, erysipelas, pyæmia, rheumatic fever, and other general constitutional fevers. The plan of reducing high temperature by cold bath has, it is true, to some extent, overridden the use of drugs for this purpose; nevertheless, some of the most enthusiastic supporters of the bath-treatment insist on the great value of digitalis as an auxiliary—as, for instance, Liebermeister, in the remarkable treatise founded on his experience of typhoid fever in Basle. The doses there given have been very large. Traube commonly employed as much as two grammes (31 grains) of the powdered digitalis leaves in twenty-four hours. It seems clear that doses that were unnecessarily large have been employed, and very probably the mortality has been rendered needlessly high. E. Hankel reports that in an epidemic of true exanthematic typhus observed in the Leipzig Klinik, eighty cases of the disease were treated, the ordinary dose of digitalis being 47 to 62 grains per diem, but as much as 100 grains being occasionally given. There were thirty-five deaths; and although Hankel explains this immense mortality by the statement that the type of the disease was uniformly severe, and that twenty-three cases were admitted in a moribund condition, there is nothing very remarkably successful in a mortality of 15 per cent.; and it is difficult not to suspect that some of the patients may have been

---

[1] Probably the earliest mention may have been that by Currie, in his "Report on the Effects of Water, Cold, and Warm," etc.

fatally affected by the large quantities of digitalis that were administered. It is not easy to draw definite conclusions from Hankel's report; but it may be mentioned that, as regards the remote influences of digitalis, such as its power to check delirium, only the milder cases seemed to benefit at all permanently. It appears, also, that the patients treated with digitalis emaciated greatly, and made slow convalescence. Wunderlich[1] especially recommends digitalis in the second week of typhoid fever, when the temperature is high, and the pulse rapid; two or three days of the treatment often produce a fall of 2° or 3° (Fahr.) of the temperature, and thirty to forty beats of the pulse.

In **Erysipelas**, digitalis was particularly recommended by Traube; and certainly there is not only much practical evidence in favor of its use, but a strong antecedent probability that it may do good in this complaint from its tendency to control the smaller arteries. But it is greatly inferior in this respect to belladonna, which has proved itself to be a powerful agent against erysipelas; and, on the other hand, the experience of Ferber declares it to be a dangerous remedy in this disease, because, when pushed beyond a certain point, it tends to bring out a skin exanthem, peculiar to itself, which much complicates matters. Probably, however, many of the objections to digitalis in erysipelas have been based on the use of improperly large doses.

In **Rheumatic Fever**, digitalis has been employed by various authorities; but the most complete account of its use in this way has been given by M. Oulmont, who was induced to try it, in consequence of the enthusiastic encomiums of Traube, and especially of Hirtz, of Strasburg,[1] upon its action in fevers generally. M. Oulmont took special care to employ only the same preparation of digitalis-powder which had proved itself very efficacious in the hands of Hirtz; with this he made an infusion of 15.6 grains of digitalis to about four oz. of eau-sucrée, which, in divided doses, was the allowance for twenty-four hours. The quantity was thus very considerably smaller than that employed by most of the German authorities. Twenty-four patients were taken quite indiscriminately, being the whole number that presented themselves during a certain period of time; and the results seem to have been clear and decisive. No effect was ever produced on pulse and temperature till after about thirty-six hours, but at this time, in simple and non-complicated cases,[2] a sensible fall of the pulse took place, and soon afterwards a decline of the temperature; this went on very gently till the third or fourth day, when nausea and vomiting ensued; the next day the pulse invariably fell from twenty to forty beats, and the temperature from $1\tfrac{3}{4}°$ to $2\tfrac{1}{4}°$ (Fahr.) The use of the digitalis was then suspended; nevertheless, the lowering of the pulse and of the temperature persisted for several days, and the morbid symptoms disappeared, sometimes gradually, and sometimes with surprising rapidity. A few cases were tardy in recovering, and a few had relapses; but, on the whole, the cases of pure febrile rheumatism seem to have done remarkably well. Several patients were cured in five or six days, and left the hospital at the end of ten days. Oulmont, however, states very clearly that digitalis was only of use in the primary period of simple joint-fever. When relapses took place, the drug seemed to effect no real good, although the characteristic stomach symptoms were evoked just as

---

[1] Medical Times and Gazette, 1862.
[2] Gaz. Méd. de Strasbourg, 1862.
[3] By this is meant, not complicated with visceral inflammations.

before. And in cases which were already complicated with visceral inflammations, it only produced a temporary and unimportant lowering of pulse and temperature. On the other hand, Oulmont seems decidedly to think that it tends (used early enough) to avert cardiac complications. When old cardiac disease existed prior to the commencement of the rheumatic attack, the influence of digitalis seems usually to have been highly favorable in rendering the pulse regular and strong; but in one instance of extensive valvular mischief, where the febrile excitement recurred several times, at the moment of the fall of the pulse, the patient was seized with an attack of suffocation, with rapid pulsations of the heart, and precordial anxiety, and seemed in great danger; the paroxysm, however, soon passed off. Oulmont speaks of this as a reason for caution in administering digitalis in instances of old and advanced heart-disease; but he does not distinguish between different valvular affections, nor does he say whether this particular patient was the subject of excessive hypertrophy or not: a matter which we now know to be very important in reference to the effects of digitalis.

The general conclusion to which Oulmont comes, as to the action of digitalis in febrile rheumatism, is, that it benefits solely in so far as it relieves the febrile state. It does not touch (he says) the cases where the malady is more deeply rooted, and is either complicated with serious internal inflammation, or tends to repeated relapses of the fever and the joint-affection. On the other hand, in first attacks, and generally in cases which are of a simple type, it greatly shortens the febrile period, averts cardiac and cerebral complications, and hastens the convalescence. These conclusions are of great value, though, of course, they only apply to the use of digitalis in a certain dose and manner. Whether as good, or even better, effects might not be obtained with smaller doses, is a further question for careful consideration.

In Acute Inflammations, digitalis is reckoned by many authorities to be all but the most powerful direct remedy in existence. It is necessary, however, quite to disregard certain observations, which, though professing to refer to digitalis treatment, really have nothing to do with it. This remark applies to all cases where the alcoholic tincture has been used in very large doses; for example, Mr. King, of Saxmundham[1] announced that by the use of single very large doses of the tincture—half-ounce to an ounce—with twenty-four hours' interval, it was quite possible to cut short various acute inflammations, provided the remedy were administered before the organs involved became disorganized. The influence of Traube, Thomas, Hirtz, and other physicians has, on the whole, maintained a high reputation for this drug in such affections, though individuals have not been wanting by whom its efficacy has been doubted and even denied.

In Pericarditis, Niemeyer[2] strongly recommends digitalis in cases in which the heart's action is very rapid and feeble, and where there are cyanotic and dropsical symptoms.

In Pleurisy, with high fever, Niemeyer recommends digitalis in infusion (gr. x. ad. ℥ j.). In more chronic and less febrile cases he gives the powder in one-grain doses, with equal parts of quinine. In the first stage of pericarditis and of pleurisy, my own experience leads me to prefer aconite, as it reduces the pulse and temperature more quickly than digitalis.

---

[1] Quoted by Ringer, "Therapeutics." 3d ed., p. 411.
[2] The Practitioner, 1868, vol. i., p. 179.

In Pneumonia, Bleuler made extensive trials of digitalis, employing usually about half a drachm (in the form of infusion) daily; this was continued until a decided impression was produced on the pulse; vomiting also set in at the same time. The influence on the mortality seems to have been unfavorable, amounting to 21 per cent. under the digitalis, as against 14.5 per cent. with merely expectant treatment; but Bleuler hesitates to draw a positive conclusion from this fact. In the patients who recovered he was unable to trace any distinct influence of digitalis in cutting short the inflammatory process; the defervescence occurred as early as from the fifth to the seventh day, and both Bleuler, and Niemeyer, who comments on Bleuler's cases, regard this as being unusually early. But, according to English authorities, this is by no means an uncommon date for defervescence, especially in one-sided pneumonia; everything, therefore, depends on the question whether Bleuler's cases were of a high average severity. He records rather frequent toxic phenomena occurring after the fall of the temperature; but these disappeared as soon as the medicine was discontinued, and convalescence was never delayed. Niemeyer employs digitalis in pneumonia when the pulse ranges from 100 to 120, in combination with nitrates of potash and soda. He states that it reduces the fever, but does not alter the plastic processes.

On taking the balance of the statements which have been made by various authors as to the action of digitalis in pneumonia, it seems clear that a good case is made out for continued trials of it. The approval of so cautious an observer as Ziemssen is of much importance. He gives of an infusion (gr. v. ad. ℥ j. water) a teaspoonful every two hours, or about the equivalent of 7½ grains daily (for children); and though he admits that this occasionally produces intermittence, he does not think that symptom of itself dangerous. Certainly, there is no kind of justification for the statement of Stillé, that the favorable opinions respecting digitalis in inflammations are totally unsupported by direct evidence. To quote one more affirmative authority, I may mention the strong testimony of M. Rory-Sancerotte,[1] who, in thirty-five carefully observed cases in pneumonia, found that digitalis reduced the fever and kept down the graver symptoms, while intolerance of the drug was very rare. He remarks that the antiphlogistic operation is less rapid than that of leeches, but more durable, and nearly the same with all patients.

Digitalis has been applied, in some instances, with remarkably good effect, as a remedy for local inflammations. In orchitis, it was employed with very good results by Dr. Besnier,[2] who was led to its use through having seen M. Debout apply it with success in cases of hydrocele. Besnier kept the patients at rest, and in a recumbent posture, with the scrotum conveniently raised, and constantly enveloped within compresses soaked in a very strong infusion of digitalis leaves, either warm or cold, according as the one or the other was found most comfortable. A covering of oil-silk was kept over all, to maintain continual moisture.

In concluding my remarks on digitalis in inflammation, it may be observed that one of the great questions still requiring to be answered is, whether the beneficial effects of the drug can be produced without nauseating and depressing the patient. Alison[3] asserted that the latter result could not be avoided, if the drug were given in sufficient doses to

---

[1] Practitioner, 1869, vol. ii., p. 180.
[2] Bull. de Thérap., Feb., 1870.
[3] Outlines of Pathology, p. 243.

produce an impression with the rapidity which is required in cases of acute disease. Most of the recent observers insist, also, on the use of large doses; and, in their writings, nausea and vomiting are treated almost as matters of course: but, at any rate, Alison was wrong in his further opinion as to the danger implied by this state of things; for, whatever may be the worth in other respects of the German observations, they have conclusively proved that even marked intermission of the pulse is seldom followed by serious consequences, and that doses quite tenfold larger than any which Alison could have been referring to were given with impunity.

In **Hæmorrhage** of various kinds, digitalis has, during the last few years, assumed an important place. In uterine hæmorrhage its efficacy was, I believe, first distinctly proved by Dr. W. H. Dickinson,[1] who made a remarkable communication on the subject. In doses of 1 to 1½ fl. oz. of the infusion, he found that hæmorrhage of this description, when unconnected with organic disease, was speedily arrested by it. Dickinson for a long time treated all his menorrhagic patients in this way, and never failed to arrest the bleeding by the second to the fourth day, according to the quantity of the drug that had been given; thus demonstrating the greatly superior action of large doses to the small quantities which had previously been held in doubtful repute for the same purpose. These statements were afterwards fully confirmed by Dr. Barclay and Dr. Tilt,[2] and also by the illustrious Trousseau. It has even been shown that in organic disease of the uterus, digitalis has some power to arrest the bleedings; for M. Decaisne treated a case of hæmorrhage from fungoid growths with complete success by the use of digitalis in the daily dose of 6 granules.[3]

In **Hæmoptysis**, digitalis has received warm praise from many writers. Dr. Brunton[4] states that, as soon as the characteristic slowing of the pulse is produced, the bleeding from the lung is stopped; he even speaks of it as the most powerful remedy for hæmorrhage from cavities. Dr. Fuller recommends it (in conjunction with dry-cupping, and ice to the spine) in full doses, 1 to 2 drachms of tincture, or 6 to 8 grains of the powder, daily. It may be said, perhaps, that these are scarcely what we should call "full doses" at the present day; but I may remark that, from experience with other remedies, as well as with digitalis, it would appear that smaller doses of hæmostatic medicines are required to check hæmorrhage from the lungs, than of those which are needed to arrest uterine bleeding. It may be observed that, though most of the authors who have written on the subject mention, that the bleeding stops when the peculiar slowing of the pulse has been produced, there is much doubt whether the mere reduction of the frequency of pulsation is the curative agent; it is much more probable that the cessation of hæmorrhage is chiefly due to the contraction of the smaller arteries, and the consequent prevention or diminution of venous stagnation.

In **Epistaxis**, digitalis, especially in the form of infusion, appears to be quite as efficacious as in pulmonary or uterine hæmorrhage. This is markedly the case in the bleeding from the nose which sometimes complicates acute rheumatism, and is occasionally so severe, if unchecked, as

---

[1] Medical Times and Gazette, Dec. 15, 1855.
[2] Handbook of Uterine Therapeutics, 3d ed., p. 225.
[3] The granules of Homolle and Quevenne contain .0156 gr. each of digitaline.
[4] Brunton on Digitalis, p. 4.

to greatly confuse the patient and delay his convalescence. Indeed, in all cases of epistaxis, except those which depend upon a general hæmorrhagic tendency, such as is developed in scurvy, in purpura, and in the rarer diathetic disease which is now called hæmophilia, digitalis appears to be a very prompt and powerful remedy. It should be given in one or two large doses, ½ oz. of the infusion, repeated twice, if necessary, at half-hour intervals.

In **Nervous Diseases**, digitalis has probably very useful functions to fulfil; but it must be admitted that the subject has not yet been accurately inquired into. Speaking generally, it would appear that the effects, physiological and therapeutical, of digitalis upon the nervous system are, for the most part, strictly dependent on its action on the circulation. Yet there appears to be something more than this, as indicated by certain curious and, apparently, well-established instances of its acting in a localized manner, not to be easily explained as a circulation-effect. For instance, there is a considerable amount of evidence as to its favorable action on local nerve-pain. Fuller[1] recommends it on this ground for sciatica of pure neuralgic type, and Lehmann[2] speaks of it as an effectual local remedy in earache. There are many facts, also, which show its sedative action on the nerves of the sexual apparatus. Brugmans[3] found that it was equally applicable for this purpose to both sexes, and in a great variety of complaints. It is possible that these effects are wholly or partly due to the reduction of the quantity of blood in the organs, an idea to which some slight probability is given in Dr. E. Mackay's[4] observations on the great relief to the congestion and pain of hæmorrhoids which is produced by digitalis; for there is, undoubtedly, much sympathy between the disturbances of circulation in the lower bowel and in the genitalia. But there is reason to think that digitalis also acts upon the sensation of different parts through the spinal cord, possibly through the vaso-motor centre in the medulla oblongata, primarily. Its generalized action is seen in the instance of the moderately successful treatment of intermittents, which I do not include under the section of digitalis in fever, properly so called, as it stands on a very different footing from the positive effects which have been obtained in typhoid, etc. Of the localized sedative action, perhaps the most useful form is the employment of digitalis in spermatorrhœa, as originally recommended by Corvisart,[5] and approved by Ringer,[6] who says few remedies are more successful in arresting spermatorrhœa than digitalis; he recommends 1 or 2 drachms of the infusion twice or thrice daily. In this view these writers are supported by several others.

The principal therapeutic effects of digitalis upon cerebral affections are produced, there can be little doubt, through modifications in the cerebral circulation.

In **Delirium Tremens**, digitalis first attracted attention as a remedy from the bold experiments of the late Mr. Jones, of Jersey.[7] This gentleman was in the habit of giving ½-oz. doses of the tincture at first, and 2-drachm doses subsequently, till calm and sleep were produced.

---

[1] "Rheumatism, Rheumatic Gout, and Sciatica," 3d ed., p. 426.
[2] Amer. Journ. Med. Sciences, v., p. 34.
[3] Journ. de Méd. de Bruxelles, Nov., 1853.
[4] Brit. Med. Journ.
[5] Bull. de Thérap., xi. iv. 18.
[6] Handbook, 3d ed., p. 412.
[7] Med. Times and Gaz., 1860, vol. ii.

Mr. Jones's facts were probably correct, for they have been substantially confirmed by the careful observations of Dr. Peacock;[1] though the latter considers digitalis chiefly applicable to young and robust persons. Very obviously, however, the treatment has no right to be called a digitalis-treatment; for, as Dr. Anstie[2] observes, the ½-oz. doses of proof spirit were probably of much more consequence than the drug itself. It is otherwise, perhaps, with the tincture in 20 or 30 minim doses, more frequently repeated, as advised by some. Nothing can be reckoned satisfactory except the evidence from the use of the infusion of the powder or of digitaline, and this must at present be pronounced too scanty to admit of positive conclusions.

Acute Mania.—In various forms of this disease digitalis has been strongly recommended by Homolle and Quevenne,[3] Lockhart, Robertson.[4] Maudsley,[5] Blandford,[6] Van der Kolk,[7] and others; and in a considerable proportion of cases bids fair to supersede the old and dangerous routine administration of opium. It is tolerably certain that the chief proximate cause of the excitement and the sleeplessness in affections of this kind is the existence of a very irregular and ill-balanced state of brain circulation; and for this digitalis can supply a remedial influence which opium cannot afford. It would, also, be a great mistake to suppose, as was thought formerly, that digitalis is a remedy too depressing to be employed in any cases of acute mania which occur in persons of generally feeble health. This is far from being the fact; some remarkable instances have been observed, in which the maniacal patients were the subjects of certainly weak and probably fatty heart, and yet the use of digitalis in tolerably full doses was followed by a simultaneous strengthening of the pulse, which became regular, and by a subsidence of the maniacal symptoms. In fact, one cannot but remark, that with digitalis, just as with opium, the old theoretical notions of the action of the two drugs led to conclusions directly opposed to fact, and practically very mischievous. This is particularly the case in regard to the views which prevailed during the first half of the present century, as to their employment or non-employment in the treatment of acute insanity.

Neuralgia.—As a direct remedy in this disorder, notwithstanding the evidence of Fuller, above quoted, respecting sciatica, and the more dubious observations of Debout and Serre, respecting migraine, digitalis is probably not at all effective or to be relied on. According to my own experience, it succeeds only casually and rarely—if, indeed, we can suppose it to succeed at all; and certainly it is not worth naming in comparison with the powerful agents with which we are now in the habit of controlling the disease. In those forms of migraine in which there is much reason to suppose that there is great vaso-motor disturbance at the time of the attacks, and especially in those cases where there is a very evident connection between the attacks and a difficulty at the commencement of each menstrual flow, it would probably be worth while, nevertheless, to make careful and systematic trial of digitalis, pushing the doses to a rather high mark. Experience has shown that this form of migraine

---

[1] Ibid, 1861, vol ii.
[2] Art. "Alcoholism," Reynold's Syst., vol. ii.
[3] Gaz. des Hôp., 1850. No. 53. Union Méd., 1851, Nos. 69, 70.
[4] Journal Mental Science.
[5] Ibid., Jan., 1869.
[6] Practitioner, Feb., 1869.
[7] Pathol. and Therap. of Mental Diseases. English Translation.

is not without its serious dangers, one or two recorded cases having terminated in fatal cerebral hæmorrhage: and it would therefore be very desirable, upon the appearance of any threatening symptoms, to try the effects of tolerably full doses of digitalis, as a means of regulating the brain circulation.

PREPARATIONS AND DOSE.—Digitalis, gr. j—v. (.06—.30); Extr. Digitalis, gr. $\frac{1}{4}$—j. (.01—.06); Extr. Digitalis Fluidum, ℞ j.—x. (.06—.60); Tinct. Digitalis, ℞ v.—xxx. (.30—2.); Infusum Digitalis, ℥ ss.— ℥ j. (15.—30.); Digitalinum, gr. $\frac{1}{16}$—$\frac{1}{12}$ (.001—.005). There is little doubt that the watery and alcoholic preparations of digitalis differ greatly in their physiological and therapeutic effects. Since it has become the fashion to use the infusion in preference to the tincture, in order to obtain a diuretic effect, or to exert a "tonic" influence on the heart, the drug is frequently given for several days or weeks in succession. The ordinary infusion is inconvenient for this purpose, as it must be freshly made every day or two. To obtain from digitalis the active principles that are soluble in water, and to preserve them unimpaired for an indefinite period, is a problem that awaits a satisfactory solution. Dr. E. M. Hale, of Chicago, suggests that an infusion be made with boiling water, and when cold strain, and to twelve ounces add two ounces each of alcohol and glycerine. Some pharmacists have the reprehensible habit of simply dispensing the fluid extract diluted with water when a prescription calls for the infusion.

[EYE-BRIGHT (EUPHRASIA OFFICINALIS).

ACTIVE INGREDIENTS.—The plant has not been subjected to very thorough chemical analysis. Enz, however, has obtained a tannin-like body, euphrastic acid (*Euphrastannsäure*) of the composition, $C^{14}H^{10}O^{17}$.

THERAPEUTIC ACTION.—When the curious doctrine of signatures reigned supreme in medicine, this plant enjoyed an extended reputation in various diseases of the eyes, and led to its popular appellation (in German, *Augentrost*). As a mild astringent in catarrhal conjunctivitis it is undoubtedly of service, and Paullus[1] speaks of it as a drying remedy ("quia oculorum fluxiones sistet"). Its moderate virtues in this respect are more than equalled by its decided utility in acute nasal catarrh (cold in the head). A few drops of the tincture, taken at the beginning of the affection, and repeated every two or three hours, will often abort it. This we know by personal experience. During the past year a number of friends have, at our suggestion, used it in *catarrhus æstivus*, and report very distinct mitigation of their sufferings.

PREPARATIONS AND DOSE.—There are no officinal preparations. We have used with satisfaction the imported tincture from the fresh plant, dose ℞ j.—v., in a wineglass of water; as a collyrium, ℞ v.—xx. to an ounce of distilled or of rose water. As the plant is found in the White Mountains (N. H.), about Lake Superior and northward (Gray), extended trials of it could without difficulty be made.]

---

[1] Quadripartium Botanicum, Frankfort, 1708.

[YELLOW JASMINE (Gelsemium Sempervirens).

Active Ingredients.—The active ingredients of gelsemium were first investigated by Wormley,[1] who succeeded in isolating an alkaloid gelsemia which existed in combination with an acid termed by him gelseminic acid. He was led to these researches by a case of fatal poisoning from gelsemium that came under his notice.

Physiological Action.—This has been investigated by Bartholow and others. According to Ott, gelsemia produces in man double vision, ptosis, want of co-ordination, disagreeable feeling in the head, great muscular relaxation, drooping of lower jaw, tongue stiff, sensation blunted, pupils dilated, respiration slow, irregular, pulse slow, surface cold and congested, unconsciousness, and death by asphyxia. Ringer and Murrel state that during the diplopia the images in the upper part of the field of vision appear at different heights, although actually in the same plane. Locally the drug contracts the pupil, while internally it contracts and dilates it, paralyzing the third pair. The drug affects the sixth nerve before the third, as the external rectus is the first muscle weakened. Taylor states that it increases the urine and reduces the pulse, respiration, and temperature. Tweedy states that it impairs the power of accommodation of the eye for near objects.

Therapeutic Action.—The precise sphere of the therapeutic action of gelsemium has not as yet been clearly defined. Its first and perhaps most frequent use has been the treatment of the bilious remittent fevers of the Southern States, its value in this connection having been accidentally discovered. The prostration produced by excessive doses has suggested its use in cases of cerebral excitement and spinal congestions, and led to its successful employment in delirium tremens, chorea, epilepsy, tetanus, and other disorders. It is also serviceable in the acute stage of an acute gonorrhœa, and in trigeminal neuralgia.

Preparations and Dose.—Extract. Gelsemii Fluidum, ♏j.—xv. (.06.—1). Fluid extracts of gelsemium vary in quality to a degree that makes it almost impossible to fix the dose in a given case with any accuracy, unless the extract employed has been carefully tested in advance. In the majority of cases it is best to commence with a small dose frequently repeated until the early physiological effects (diplopia, etc.) are developed. In neuralgia its effects appear to be specific, and in suitable cases small doses answer as well if not better than large ones.]

---

[1] Amer. Jour. of Phar., Jan., 1870.

ANACARDIACEÆ.

RHUS (Rhus Toxicodendron).

ACTIVE INGREDIENTS.—When a branch or leaf of this plant is broken off, there exudes a yellowish milky juice, which in a little while turns black. So poisonous is this juice, that even the atmosphere surrounding the shrub is said to be tainted with emanations from it; the poisonous matter appearing to be volatile, and thus capable of diffusion, even without breakage of the stem or leaf. The researches of Maisch have proved that the cause of the acridity of the juice is the presence of a volatile acid—Toxicodendric acid—which, when isolated, is found to produce effects upon the skin and mucous membranes exactly analogous to those which will be mentioned under the physiological action of rhus.

[According to Cazin, the exhalations of rhus, collected in full daylight, are a nitrogenous gas with a little water, neither of which are irritating; but, on the contrary, the gas collected after sundown is found to be a carburetted hydrogen, mixed with a peculiar acrid principle. The dried leaves do not furnish noxious emanations.]

PHYSIOLOGICAL ACTION.—The effects produced by rhus, whether it be taken internally or absorbed by the skin (either from exhalations from the plant, or otherwise), are redness and swelling of the affected parts; and, if referable to exhalations, most particularly in the face and eyes, in which last there is burning, with inflammation of the lids, and agglutination of these organs in the morning. Subsequently there is swelling, with pain, and often a considerable increase of temperature, and the inflamed surface is generally studded with vesicles. Combined with these symptoms, there is an almost unbearable amount of itching, which is not confined to the patches of inflammation, but diffuses itself, more or less, over the entire surface of the body, the hairy portions appearing to be very specially affected. The condition induced thus appears to be of an erythematous or erysipelatous type. It is superficial but spreads rapidly over the surface, and speedily involves large areas of the body; eventually extending to the mucous membranes, as indicated by redness and swelling of the throat and mouth, with, ordinarily, great thirst, irritable cough, nausea, vomiting, vertigo, dulness and stupefaction of head, and colicky pains throughout the abdomen. These last are chiefly experienced during the night, and are aggravated by eating or drinking. Diarrhœa frequently ensues, accompanied by tenesmus, and the stools are often bloody. There is often retention of the urine, or else diuresis, and the water is frequently accompanied by blood.

Rhus also induces pains, apparently of a rheumatic kind, and which are felt not only in the limbs, but in the body, though most especially about the joints. Pain and stiffness in the lumbar regions are often induced, and to these affections is often added a sense of numbness in the lower extremities. The structures most powerfully affected appear to be the fibrous ones. The pains in question are accompanied by a very slight amount of swelling; and, singular to say, they become intensified by rest and warmth. Sleep is greatly disturbed, the patient becoming restless, constantly turning about, and often suffering from great nervous depression.

The fever which sometimes accompanies the effects of rhus, though by no means an universal symptom, usually occurs, when present, in the later stages, and generally partakes of a typhoid character. It is often attended by delirium; the lips are apt to become dry and parched, and to be covered with a brownish crust. Sometimes it assumes an intermittent character, and is then usually marked by profuse perspiration.

The above-described effects of rhus, though so distressing to whoever may have to endure them, appear, however, to be very seldom fatal; and it is remarkable that a certain constitutional predisposition appears requisite to their occurrence, so that it is only individuals who are in danger. Were it otherwise, a plant so common in its native country as the present would be a perpetual source of trouble to the persons dwelling near. I have myself witnessed several instances of its poisonous influence, and can personally vouch for the manifestation of nearly all the phenomena that have been indicated.

THERAPEUTIC ACTION.—The properties of the *Rhus Toxicodendron* were first brought into notice about the year 1798, by Dufresnoy, a physician at Valenciennes. Alderson, in England, likewise made some interesting observations with regard to them. Dufresnoy's attention was attracted to the plant by the circumstance of a young man, who had suffered from a six years' eruption upon his wrist, being cured by accidental subjection to its influence; and shortly afterwards he employed it successfully in various cases of obstinate herpetic eruption. Herpes zoster, pemphigus, and eczema, especially when accompanied by burning or itching sensations, represent the class of eruptions which are very readily subdued by the external and internal exhibition of rhus; and in erythema and eryipelas, especially when accompanied by vesicles and bullæ, it is without question a very useful remedy.

[As Dufresnoy was the first to study the plant from a therapeutic stand-point, it may not be out of place to introduce the following anecdote as related in his original work:[1] One day when lecturing on Rhus at the botanical garden of Valenciennes, a mischievous student said to a young florist who was present, that the professor's account of the noxious properties of rhus was incorrect, as the plant growing in France was perfectly innocent. To convince him of this, he plucked some leaves and rubbed them freely on his hands and wrists, as he knew by previous experience he could do with impunity. The florist thus persuaded, followed his example, but in a short time had occasion to repent his imprudence. The next day, finding himself in trouble, he consulted the student, who gravely assured him that he had caught the itch somewhere, and advised him to rub into his hands half an ounce of citrine ointment, and to purge himself freely with mercurial pills. This did not mend matters, and finally Dufresnoy was made acquainted with the state of affairs. In about ten days the young man recovered from the effects of the rhus, and to his great surprise found that a chronic eczema of six years' standing, for which he had vainly sought relief, had disappeared at the same time.

---

[1] Des Charactères, du Traitement et de la Cure des Dartres, et de la Paralysie des extrémités inferieures . . . . par l'usage du Rhus Radicans, etc. Paris: An VII. de la République.

This led Dufresnoy to experiment freely and successfully with rhus in this class of affections.]

Rheumatic Affections.—Rhus is a very powerful therapeutic agent in various subacute and chronic rheumatic affections of the fibrous tissues generally. The synovial membranes are less amenable, however, than the fibrous tissues outside of them, such as the tendons, the ligaments, and the fascia. When the muscles seem to be affected and relieved by rhus, it is through the extension of the rheumatic affection from the joints and not from rheumatism primarily in the muscular substance.

In the after stage of acute rheumatic fever, when aconite may have been employed, and when the temperature has fallen to 100°, or below it, and where the patient still suffers from wearing stiffness, and aching, of a subacute character, in the neighborhood of the joints, rhus is positively invaluable. It should be applied externally, in the form of lotion, with compresses, and be given internally, in small doses, every two to four hours.

In cases of scarlatina, accompanied or followed by rheumatism of a type more or less acute, rhus is often of the greatest service, especially if typhoid symptoms are present.

[Our own experience fully coincides with that of the author as regards the effect of rhus in subacute rheumatism. For several years we have been subject to subacute muscular or tendinous rheumatism, the pain usually occupying the right shoulder, and most severe in the neighborhood of the insertion of the right deltoid, shifting thence to other parts, sometimes apparently to the fascia of the outer and posterior aspects of the leg. The pain was always worse at night, and during the winter of 1877–8 became so continuous and severe as to interfere with sleep. After four months of serious discomfort, complete relief followed the use of a few doses of rhus. At the Charity Hospital it not unfrequently happens that patients complaining of rheumatism, "worse at night," are sent to the venereal wards on the hasty diagnosis of syphilitic rheumatism. In a number of these cases in which we have been unable to verify the diagnosis of syphilis, we have obtained most decided and prompt relief from rhus. In private practice many cases of chronic eczema have been encountered in which this form of rheumatism has been present as a complicating or alternating symptom. Some of them had been dosed, *ad nauseum*, with colchicum, alkalies, iodide of potassium, and salicylic acid without much, if any, benefit. In these cases prompt and decided relief was obtained from rhus.]

Paralysis.—Dufresnoy administered the dried leaves, in doses of half a grain or a grain, twice a day, in cases of paralysis. The patients recovered, so it is said, to a certain extent; the first symptom of improvement consisting in an unpleasant sensation of pricking or twitching in the limbs, analogous, apparently, to that induced by strychnine. Subsequent experience had not at that time defined the exact class of paralytic affections amenable to rhus. It is now certain, however, that this medicine is efficacious in cases 'that depend upon a rheumatic condition.'[1] It is a curious fact, pointed out by Dufresnoy, that persons not constitutionally susceptible to the disorders induced by rhus, regarded as a poisonous agent, are not so likely to receive benefit from it if used as a medicine. This circumstance is important to be borne in mind by those who would try the therapeutic powers of this remarkable drug.

---

[1] [Eberle reports a case of paralysis cured by rhus. (*Western Jour. of the Med. and Phys. Sciences*, 1831.)]

[The statement of Dufresnoy and others that those who are not subject to rhus poisoning (by external application) are likewise not likely to receive benefit from its internal administration, we are not prepared to unreservedly accept. Personally we have handled the fresh leaves of both the R. *toxicodendron* and the R. *venenata* with impunity, and have frequently daubed the tinctures on our skin without the slightest result; tinctures that we had experimentally found active when applied locally to others. On the other hand, we have experienced decided physiological effects from the internal use of very small quantities. Similar experiments with other irritants (*cardol, succus euphorbiæ, cantharides,* etc.), lead us to suspect that the unsusceptibility may be local and not general, as asserted by most writers.]

When semi-poisonous effects have to be dealt with, the treatment should be lowering, and consist in rest, low diet, and laxatives. The extreme irritability and the burning sensations may be greatly alleviated by the use of opium and of strong black coffee; and externally by the use of ice or a solution of acetate of lead. Fomentations and lotions of limewater and linseed-oil are likewise useful. [In the editor's experience, the most serviceable local applications are a solution of bichloride of mercury, originally recommended by Barton (Thacher's Dispensatory, 1810), and the fluid extract of Grindelia robusta.

Botanists are pretty well agreed that the plant called Rhus *radicans* is the same as the one under consideration; its differing form and characters depending on the circumstances of its habitat. Besides these there is met with in this country a Rhus *venenata* (formerly called R. *vernix*), which is even more poisonous than the others. The writer possesses a tincture of R. *ven.* made by Dr. Manlius Smith, of Syracuse, in 1847. A drop of this applied to the arm of a medical friend, produced an active dermatitis. From California we have accounts of a Rhus *diversiloba*, the poisonous powers of which are said to exceed those of the Eastern varieties.]

PREPARATIONS AND DOSE. — Non-officinal — The dried leaves, Toxicodendri Folia, were included in the second edition of the U. S. Ph. (1830) — the dose usually assigned being from one to two grains. These leaves when freshly dried are doubtless reasonably active, but if long kept become comparatively inert and cannot be depended on. It is perhaps to this cause or to the fact that physicians of that day did not know how or when to use them, that the drug fell into disuse and was discarded from the Pharmacopœia. A tincture made by macerating one part of fresh leaves in two parts of alcohol is the only preparation that we have employed. Of this preparation we cannot definitely formulate the most useful dose. In general, however, it may be stated that in rheumatic and cutaneous affections a small fraction of a minim has proved most serviceable. In suitable cases of paralysis, however, the drug may be pushed to the limit of tolerance. The Tinct. Toxicodendri of the Ger. Ph. is made from five parts of the fresh leaves to six parts of alcohol. The Extract. Toxicodendri of the older pharmacopœias is an unreliable preparation on account of the variability of its strength. Fifty years ago the late Dr. Jacob Bigelow declared that :[1] "It is one of the most uncertain of medicines, sometimes having been carried to the extent of an ounce of the extract at a dose without effect; at other times having produced alarm-

---

[1] A Treatise on the Materia Medica, Boston, 1822, p. 378.

ing consequences in an almost imperceptible quantity." The latest pharmaceutical writer, Bernatzik,[1] dismisses the extract in a few words: "*In der Regel ganz unwirksam.*" A good tincture, however, will be found sufficiently active.]

MYRRH (BALSAMODENDRON MYRRHA).

ACTIVE INGREDIENTS.—Myrrh contains two per cent. of Myrrhol, $C^{10}H^{14}O$, a volatile oil, soluble in alcohol, and in ether; from 30 to 40 per cent. of a resin, called myrrhus, $C^{40}H^{30}O^{10}$, soluble in alcohol, ether, and acetic acid, and capable, by prolonged heating, of changing into myrrhic acid, $C^{14}H^{34}O^{8}$; and 40 to 50 per cent. of gum.

PHYSIOLOGICAL ACTION.—Like other resins, myrrh is a stimulant. Mucous surfaces are affected by it, and the secretions therefrom are augmented, those, in particular, which belong to the bronchial tubes. Taken in small or medicinal doses, it excites sensations of warmth, and quickens the desire for food, promoting the healthy action of the assimilating organs, and giving tone in general to the system. Large doses are somewhat hurtful, the sensations of warmth in the stomach being so great as to become unpleasant, and the pulse being increased in fulness and frequency. In its remote action this drug is somewhat tonic, but it exerts none of that influence over the nervous system which marks the action of gum-resins derived from the Umbelliferæ.

THERAPEUTIC ACTION.—Myrrh is found useful in certain forms of phthisis and bronchitis. It checks the excessive mucous discharge in pulmonary catarrh; and in phthisis it is said to assist in diminishing the puriform expectoration. Whether these ends are accomplished directly or indirectly, there can be no doubt that myrrh strengthens the system, when debilitated, by its energizing powers, perhaps over the heart and the vascular system generally. In dyspepsia arising from an atonic condition, myrrh is reputed highly serviceable. It is also much esteemed in disorders connected with the menstrual functions. Mucous discharges, such as leucorrhœa, particularly when accompanied by amenorrhœa, and when arising from an enfeebled state of the system, are likewise treated beneficially with myrrh, which is then best associated with iron or aloes. Removing the leucorrhœa, it often does away with the amenorrhœa.

EXTERNAL APPLICATIONS.—Myrrh is found very serviceable in cases of foul and indolent ulcer. Spongy and ulcerated gums are likewise well treated with the tincture; and the same, diluted with water, forms an excellent gargle in ulcerated throat.

As a dentifrice, usually in combination with other substances, myrrh has long enjoyed celebrity.

PREPARATIONS AND DOSE.—Myrrha, gr. x—xx. (.65—1.30); Tinct. Myrrhæ, ℨ ss.—j.; Pilulæ Aloës et Myrrhæ; Pil. Aloës et Mastiches; Pil. Ferri. Comp.; Pil. Galbani Comp.; Pil. Rhei Comp.; of either of these pills the usual dose is one to three.

---

[1] Specielle Arzneiverordungslehre, Wien, 1878.

THE CASHEW NUT (Anacardium occidentale).

The pericarp of the nut contains abundance of a thick and caustic oil, capable of blistering the skin, and which has been used as a caustic for warts, corns, ringworm, and obstinate ulcers. The vapor of the oil, while the nut is being roasted for eating, will often produce violent swelling and inflammation. [The oil referred to is called *cardol vesicans* to distinguish it from the *cardol pruriens* obtained from the anacardium orientale. Properly diluted with simple ointment, we have found it a useful stimulant in alopecia areata and in circumscribed patches of indolent chronic eczema.]

Oil of cashew has been much lauded as a remedy for leprosy. The discoverer, Dr. Beauperthuy, of Cumana, communicated his secret to Dr. Bakewell, at whose instigation Dr. Milroy was sent out to Cumana by the College of Physicians to examine into the value of the alleged remedy and the truth of the reports. The result is that, while admitting that the oil has a certain value in alleviating the disease, Dr. Milroy is of opinion that much of Dr. Beauperthuy's success is to be attributed to the employment of hygienic measures in connection with a liberal allowance of fresh meat.

COMPOSITÆ.

ELECAMPANE (Inula Helenium).

Active Ingredients.—The thick and substantial root, when dried and sliced, has an agreeable and camphoraceous odor, and an aromatic, rather bitter, and slightly pungent taste. It contains a neutral crystalline principle, insoluble in water, somewhat resembling camphor, and called helenine, $C^{12}H^{20}O^3$; also a peculiar kind of starch, called inuline, $C^6H^{10}O^5$; bitter extractive, soluble in water; an acrid resin, gum, lignin, albumen, and salts of potash, lime and magnesia.

Physiological Action.—Elecampane acts as a gentle stimulant to the organs of secretion, and is said to be expectorant, diaphoretic, and diuretic. Large doses cause nausea and vomiting.

Therapeutic Action.—In pulmonary affections, such as catarrh, when accompanied by profuse secretion, but without concomitant febrile disorder, elecampane is considered decidedly efficacious. It is also administered with benefit in dyspepsia, when attended by relaxation and debility; and is found useful in exanthematous cases, on account of its promoting the eruption.

Formerly, elecampane was supposed to possess emmenagogue properties, and with this no doubt is connected the origin of the classical name.

Preparations and Dose.—Non-officinal. It is usually given in the form of decoction containing one or two drachms of the root.

PYRETHRUM (Anacyclus Pyrethrum).

Active Ingredients.—The principal ingredients of pyrethrum are three in number; (1) an acrid fixed resin, pyrethrine, insoluble in caustic potash; (2) a second resin; (3) an acrid yellow oil, both of which latter are soluble in the alkali mentioned. The root contains likewise inuline, tannin, and various unimportant substances.

Physiological Action.—The root of this plant, when dry, is scentless. On chewing it, there is soon perceived on the lips and tongue a peculiar pricking sensation, accompanied by heat. Acridity and pungency are then detected, and an abundant flow of saliva and of buccal mucus soon ensues. "The heat," says Grew, "is joined with a kind of vibration, as when a flame is brandished: this heat is by no means painful." When applied to the skin, the root acts as a rubefacient; and in all cases the effects appear to be due to the pyrethrine.

Therapeutic Action.—The chief employment of this drug is as a masticatory and sialagogue; it is recommended also in rheumatic and neuralgic affections of the head and face, and in cases of palsy of the tongue. When resorted to for any of these purposes, the patient should chew the root.

Pyrethrum is again very useful as a stimulant, in the form of gargle or lotion, when there is partial or entire obstruction of the sublingual or sub-maxillary glands. The gargle prepared from it was formerly prescribed also for relaxed uvula, and for partial paralysis of the tongue and lips. The tincture is useful for toothache.

Preparations and Dose.—Tinctura Pyrethri (B. Ph.). Not used internally;—when required for toothache—a few drops should be applied on cotton-wool. When used as a gargle, two or three drachms of the tincture in a pint of water.

WORMWOOD (Artemisia Absinthium).

Active Ingredients.—These consist of a green volatile oil, having a strong wormwood odor; and a bitter extract, yielding absinthine, $C^{15}H^{20}O^4$ (Kromager), which last is the essential bitter principle of the plant, and presents itself in the form of a faintly crystalline powder, of bitter taste and disagreeable odor. It is soluble in alcohol, ether, and the alkalies; the reaction is neutral. Another ingredient is absinthic acid, an inert substance. The bitterness of the plant is brought out by water as well as by spirit, especially the latter.

The vegetable alkali kept in the shops under the name of "salt of wormwood" is an impure carbonate of potash, obtained by incineration of the plant, and possessing none of the intrinsic qualities of wormwood itself.

Physiological Action.—Taken into the system, wormwood operates in the manner usual with aromatic tonics. It increases the appetite, promotes digestion, slightly accelerates the circulation, and to some small extent augments the secretions. Large doses are at first ex citant, causing a pleasurable degree of warmth to permeate the whole body; subsequently, irritation of the system is induced, attended by considerable pain in the stomach, nausea, giddiness, headache, confusion of ideas, faintness, insensibility, and occasionally, contraction of the extremities, often followed by convulsions. At the same time there is in many cases fixedness of the jaws, and foaming at the mouth. Small doses, long persisted in, as in the shape of " bitters "—the " Absinthe " of the French—seriously injure the nervous system, over which wormwood exercises a specific influence. The infatuation of the Parisians more especially, for absinthe, is extraordinary, and the extent to which they indulge in it is frightful. The idlers of the *pavé*, and even the lower classes, drink *petits verres* of the liqueur, one after the other, till they absorb an amount of alcohol which in itself is very pernicious, but over and above this the wormwood has special effects, the dangerous character of which has recently been investigated by M. Magnan. Absinthe-drinkers are distinguished, for instance, by a particular tendency to epileptiform symptoms, referrable, no doubt, to the action of the oil;[1] for experiments on animals have shown that the oil causes muscular tremors and shock-like spasms in the neck and fore-limbs,[2] and, if given in very large doses, trismus and tetanus, alternating with clonic convulsions, foaming at the mouth, involuntary defecation and discharge of semen, and (apparently) delirious hallucination. After death, the membranes of the brain and cord were always found injected, especially in the region of the medulla oblongata, and there was ecchymosis in the pericardium and endocardium, with hyperæmia of the lungs; the brain was but slightly congested. Symptoms almost exactly similar occurred in a case reported by Mr. W. Smith,[3] where a man who swallowed about half-an-ounce of wormwood oil became completely insensible, and had epileptic fits, with foaming at the mouth, trismus, and retching. He recovered in 48 hours.

Wormwood itself, when taken in any way into the system, impregnates the whole body with the bitter crystalline principle, a fact which is shown by the taste of the flesh of animals poisoned with it. It is said that the milk of nursing mothers likewise becomes bitter if they take wormwood, and according to Borwick, the infant suffers.

THERAPEUTIC ACTION.—Wormwood has been much extolled as a stomachic and tonic. It is adapted to atonic dyspepsia, occurring in torpid and debilitated constitutions; and that it is capable of promoting the assimilation of food there can be no doubt. Before the introduction of the febrifuges now in use it was celebrated also as a cure for intermittents, but chiefly in domestic medicine. Wormwood is further said to be efficacious as a vermifuge, destroying (but only by long continued use) both the lumbrici and the ascarides.[4] Haller recommends it as suitable for warding off attacks of gout.

PREPARATIONS AND DOSE.—Absinthium, gr. xx.—xl. (1.30—

---

[1] Epilepsy is doubtless an occasional effect of prolonged excess in mere alcoholic drinks, but it is a much more common result of absinthe-tippling.

[2] Magnan, Comptes Rendus, tom. 48, p. 14.

[3] *Lancet*, vol. ii., 1862.

[4] " M. Cazin recommends a preparation made by digesting equal parts of wormwood and garlic in a bottle of white wine, of which the dose is from one to three ounces every morning."—Stillé, Therapeutics, p. 647.

2.60) is alone officinal. Infusum (℥ j—℥ xx. water), dose ℥ j.—ij (30.—60.); Oleum Absinthii essentialis, ♏ iv.—viij. (.25—.50) with a little ether or sweet spirits of nitre.

SANTONICA (ARTEMISIA SANTONICA).

ACTIVE INGREDIENTS.—These consist of a volatile oil, and a peculiar, crystalline, neutral principle called santonine, $C^{15}H^{18}O^3$, which, though procurable from other plants, is contained in quantity worth extracting only in the present one. It occurs in flat, four-sided, colorless prisms, which are inodorous and feebly bitter; in cold water it is scarcely soluble, requiring 10,000 parts, and in boiling water very little more so, requiring 5,000 parts: it is soluble in solutions of lime and alkalies, abundantly so in chloroform, and in boiling rectified spirit, but not at all in dilute mineral acids, though strong nitric acid is said to convert it into succinic acid. Santonine is fusible, and sublimes at a moderate temperature; the crystals, on continued exposure to light, become disintegrated, and assume a yellow tint.

PHYSIOLOGICAL ACTION.—Few medicinal substances have more curious or interesting physiological effects than santonine. Apart from its fatal action on intestinal worms, especially the round worm (*Ascaris lumbricoides*), to which it is the deadliest of all anthelmintics, it has remarkable influences on the animal organism generally. The most singular of these, though not the most important, are the color-changes. The urine has long been known to become deep orange-yellow during its use, and to Mr. Spencer Wells is due the observation (1848) that vision also becomes affected in such a manner that objects look as if they were seen through a yellow medium.[1] Since that time the phenomenon of colored vision has been closely studied by many investigators, but it is not definitely cleared up. The most complete researches are those of E. Rose.[2] This observer distinguishes two forms or grades of the colored vision; the first, or earliest, is a violet coloration which is by no means a constant phenomenon, but, when seen at all, is more intense than the yellow tint which follows, and is perceived most strongly in looking at dark objects, while the yellow tint, on the contrary, is best seen against white or fully lighted surfaces. All these phenomena of colored vision are of brief duration, much briefer, for example, than that of the coloring of the urine. Besides the peculiar appearance of surrounding objects, there is one very remarkable visual effect, which Rose declares to be constant; the spectrum, when looked at by the patient, is apparently shortened, especially at the violet end. It is extremely difficult to conjecture what the mechanism of these symptoms can be. The only change in the eyes observable from without is that the pupils are always dilated[3] (the dilatation is so considerable as to cause amblyopia); there is no staining of the conjunctivæ as with any diffused coloring-matter; and it may be noted here, by the way, that it is not possible to produce the colored vision by the local

---

[1] Mr. Wells's observation was independent; the coloration had, however, been noticed as an occasional effect of worm-seed by Hufeland as early as 1800, as an occasional effect of santonine by Callord (1843), and as a constant effect of santonine by Itzstein (1846).

[2] Virchow's Arch., 1868, p. 233.

[3] Even this is disputed, but I personally believe it is a constant symptom.

application of santonine to the eye. Neither the fluids nor the solids of the eye are at all colored by the action of santonine; there is no jaundice or other coloration of the skin; and although the urine, as already mentioned, is dark colored, the change does not depend on bile-pigment, but on some pigment formed probably in the kidneys. The latter conclusion is based on the fact that the coloring-matter is not to be found in the blood-serum, which is quite unchanged, nor in the lungs, the retina, the sweat, or in the amniotic water of pregnant animals. Rose himself considers that the yellow appearance depends on the retina becoming hyperæmic, and the violet appearance on a positive injury to the fibres of the optic nerves, independent of any central lesion. The phenomena are clearly different from those of the hallucinations of sight which occur in migraine and some other nervous affections, inasmuch as they are never observable when the eyelids are closed. There is much difference between the reports of different observers as to the possibility of a long lasting amaurosis or other serious damage to vision.

Symptoms of general severe poisoning with santonine, though very rare, have occurred from time to time. There have been several cases of poisoning by santonine lozenges, and two or three from taking excessive doses of worm-seed. The quantities of santonine that have proved poisonous in this accidental manner seem to have ranged from $1\frac{1}{2}$ up to 10 grains for children. Among these cases there is one in which a child of $5\frac{1}{4}$ years was actually killed in fourteen hours by two doses of about a grain each. Upon this last there rests, however, some degree of doubt, since far larger doses have often been taken almost or quite without bad effect. The symptoms of santonine-poisoning are partly concerned with the alimentary canal, and partly with the nervous system. The disturbance of the former is indicated by vomiting or retching, pain in the abdomen, anxiety, and restlessness; of the latter by twitchings of the muscles of the face and of the limbs, and by obscure epileptiform or tetaniform symptoms. The pulse seems to show a diminution both in heart-force and in arterial tone. The general phenomena disappear in twelve or fourteen hours if recovery takes place at all; but the coloration of the urine persists for some time longer.

Animals which have been poisoned by santonine exhibit marked hyperæmic changes in the cord and medulla oblongata, in the membranes of the brain, and in the lungs; the appearances in the nerve-centres are more considerable in dogs than in rabbits. The more powerful physiological effects of santonine are best produced by injecting a chloroformic solution beneath the skin; for when taken by the mouth the crystalline santonine in part passes through the alimentary canal unchanged; the remainder is taken up as an alkaline salt into the blood. (Santonate of soda, by the way, is an available form in which to give santonine for the physiological effects; many of the experiments of Rose and others were made in this manner.) As regards the ultimate chemical destiny of santonine in the body, it is not possible at present to speak with certainty; the coloring-matter which appears in the urine has not yet been completely identified in its chemical relations.

THERAPEUTIC ACTION.—Though it will not improbably be employed, at some future time, for other medicinal purposes, santonine is at present practically known only as an anthelmintic. It is tasteless and inodorous, and can thus be given with great convenience. To the round-worm it is absolutely fatal, and in somewhat less degree to the smaller ascarides. Its effects are most evident in cases where the presence of worms has given rise to intermittent or remittent febrile symptoms—the so-called "worm-fever." It should be given, like most other anthelmintics, on an empty stomach, and preferably at night and in the morning.

PREPARATIONS AND DOSE.—Santoninum: for adults, gr. ij.—vi. (.13—.40); Trochisci Santonini (adults) No. v—x.

CHAMOMILE (ANTHEMIS NOBILIS).

ACTIVE INGREDIENTS.—Chamomile contains a peculiar dark-blue or dark-green essential oil, the source of its activity, and which has lately assumed an unexpected importance, from the researches of Binz and Grisar. The composition of the oil has been a matter of considerable dispute. Until quite of late the most generally accepted opinion was that of Gerhardt,[1] who represented it as a compound of angelic aldehyde, valerianic acid, probably a small quantity of resinous matter, and a peculiar camphor-like body, $C^{10}H^{16}$, with a boiling point of 175° C., to which he gave the name of chamomilline.

More recently Demarcay[2] has examined the oil, and has come to the conclusion that Gerhardt was wrong in supposing oil of chamomile to contain an aldehyde; on the contrary, he says that it is a mixture of ethers, among which the angelates and valerianates of butyl predominate. The supposed hydrocarbon, chamomilline, appears really to possess the composition of valerianate of butyl.

PHYSIOLOGICAL ACTION.—By far the most important physiological effect of chamomile oil is its power to lower the reflex excitability. The important paper of Binz[3] first drew attention to this property of chamomile among several other essential oils, and Grisar[4] has since worked out the research with special care and very important results. His experiments were conducted on the principle introduced into practical physiology by Türck. This consists in suspending frogs with one limb immersed in dilute acid; the time which suffices so to irritate the limb as to cause it to be spasmodically withdrawn from the fluid, is carefully marked by a metronome, and forms the test of the degree of reflex excitability. The element of volition is got rid of by the preliminary adoption of Goltz's process—the division of the cerebral hemispheres by a knife passed through the skull in a line from one posterior orbital canthus to the other; this reduces the frog to the condition of a perfect machine for testing reflex irritability. Frogs so prepared are exposed to the acid, their degree of reflex irritability is tested by metronome beats, and then the chamomile, or other ethereal oil, is injected beneath the skin, after which successive observations are taken as the system becomes more and more impregnated with the drug. The result of experiments made in this way, and also with decapitated frogs, leaves no doubt at all that chamomile oil, even in doses that are not fatally poisonous, reduces the

---

[1] Ann. Chim. Phys. (3) xxiv. 96.
[2] Comptes Rendus, t. lxxvii. p. 360.
[3] Med. Centralbl., Feb. 8, 1873.
[4] Experimentelle Beiträge zur Pharmakodynik der ätherischen Oele. Von Vincens Valerius Grisar. Bonn, 1873 (pamphlet).

reflex excitability of frogs in a very marked degree. But the most important fact evolved by the researches of Grisar was that reflex excitability which has been artificially excited by strychnia or brucia can be calmed again by chamomile oil; or, rather, that an animal fortified with a dose of chamomile oil is not capable of being tetanized by a dose of strychnia which throws an unprotected frog of similar size into characteristic spasms.

THERAPEUTIC ACTION.—Our knowledge of the medicinal virtues of chamomile has been hitherto of an entirely empirical kind, but the experiments just spoken of seem to point the way to a better understanding of the matter.

In Poisoning with Strychnia it will henceforth be a consideration whether we should not employ oil of chamomile, or some of the other ethereal oils which will be mentioned in their proper places, as antidotes. The experiment is at least well worth trying. At present we have no knowledge of the doses that might be required, but it may be mentioned that five drops of chamomile oil, subcutaneously injected, neutralized the effect of the (for a frog) very powerful dose of six milligrams (.09 grains) of strychnia.

In Cough which mainly depends on heightened reflex irritability, particularly in the kind which afflicts hysterical women, chamomile oil may take rank with valerian as a remedy of unmistakable potency; but for this purpose it must be given in sufficient doses (4 to 8 minims).

In Pulmonary Catarrh with excessive secretion and difficulty of expectoration, chamomile oil in smaller doses (2 to 4 minims) is a very useful remedy, though it has been pushed out of the field by other substances which also depend for their effectiveness upon the presence of some ethereal oil.

In the Spasmodic and Pseudo-neuralgic affections of hysterical persons, chamomile oil in sufficient doses (4 to 6 minims) is a very excellent remedy; more especially in the pseudo-angina-pectoris, and the colicky attacks to which such patients are very liable; also in hysterical pain in the fifth nerve.

In Atonic Dyspepsia, small doses (2 minims) of the oil are exceedingly useful; also in the diarrhœa of children, especially that arising from worms.

In Spasmodic Asthma, and in Whooping-cough, chamomile oil has been found by German physicians very useful. Undoubtedly the therapeutics of chamomile will now attract increased attention; and as regards

PREPARATIONS AND DOSE: It is important that the infusion should be discarded from use, as also the extract; and that only the oil, of first-rate quality, and presenting its original green or blue tint (which fades to yellow after a time) should be employed; the doses ranging from two to eight minims under the varying circumstances already described. Sugar is the best vehicle for the oil. [Phillips.]

DANDELION (Taraxacum Dens Leonis).

ACTIVE INGREDIENTS.—These are Taraxacine and Taraxacerine, $C^{10}H^{10}O^8$. The former is dissolved out by hot water, the latter by alcohol from the insoluble residue.

PHYSIOLOGICAL ACTION.—Nothing whatever is known about this. It is supposed by some (on empirical evidence) that taraxacum acts on the liver; but there is no scientific proof of the fact. It has also been credited with diuretic powers, as the French name for it —"*pissenlit*"— indicates; but again there is no evidence to justify the notion.

THERAPEUTIC ACTION.—Although there is no exact physiological knowledge respecting the action of dandelion, there is no reason to doubt that it has some medicinal efficacy. The nature of that action, however, has in all probability been completely misunderstood by the majority of those who have spoken well of it. It has been supposed to act specifically on the liver; but the very magnitude and diversity of the statements respecting its action on that viscus, expose them all to grave suspicion. It has been supposed not merely to specifically increase the biliary secretion (as to which Dr. Hughes Bennett's experiments are flatly contradictory), but to cure chronic inflammation, indolent enlargements, and even to prove effective in commencing scirrhus of the liver. A great deal of the evidence on which such statements rest has been based on medication in which powerful drugs have been given in conjunction with it. For my own part, the only distinctly efficacious action of taraxacum appears to be as a mild stomachic, and possibly a duodenal tonic. It certainly does good in simple atonic dyspepsia, and even (temporarily) in failures of digestion which depend upon disease of the liver and other viscera. But there seems to me to be a complete absence of any proof that it is a specific remedy for biliary disorders. It appears to act solely as a tonic to the earlier digestive functions, and the extent to which it will relieve disorders of these depends wholly on the degree in which the original cause was severe, and continuous in operation. There is no pretence for believing that it modifies distinct organic changes in any viscus. [From Rutherford and Vignal's experiments it "appears that taraxacum is a very feeble hepatic stimulant."—(*Brit. Med. Jour.*, Nov. 13, 1875.)]

PREPARATIONS AND DOSE.—Extract. Taraxaci, gr. xx.— ℨj. (1.30 —4.); Extract. Tarax. Fluid., ℨj.—ij. (4.—8.); Infusum Tarax., ℥j.—iv. (30.—120.); Succus Tarax., ℨj.—iv. (4.—15.). No doubt the freshly prepared succus is the best form in which to give taraxacum; but there is much difference in opinion as to whether the thick albuminous juice obtained in winter, or the thinner and more acrid juice of early summer, is the more active. [Phillips.]

ARNICA (ARNICA MONTANA).

ACTIVE INGREDIENTS.—A very important question, since great mistakes have been occasioned by erroneous views. The ingredient long supposed to be of most consequence is *arnicine*, $C^{20}H^{30}O^4$, a bitter principle, which is insoluble in water, but freely soluble in alcohol and in ether, and forms amorphous masses of a golden yellow color; or else the ethereal oil, which is also insoluble in water. For a variety of reasons, hereafter to be mentioned, it is now probable that neither arnicine nor the oil, but "trimethylamine," is the really useful ingredient of arnica. Trimethylamine, C^3H^9N, is a clear, colorless fluid, which boils at a very low temperature, and then emits a fishy smell. It is quite freely soluble in water, in alcohol, and in ether, and its vapor is absorbed by water with great avidity. It has a strong alkaline reaction, and readily ignites, on the application of flame, even when diluted with an equal quantity of water.

PHYSIOLOGICAL ACTION.—The physiological action of trimethylamine, or of concentrated aqueous solutions of arnica, which contain trimethylamine, without arnicine, is as follows : Placed in simple contact with the skin, neither of these excites irritation; but if either of them be rubbed in for some time with flannel, the surface will become reddened. Like ammonia, they dissolve the little plugs of fat at the orifices of the sebaceous ducts. Applied to the mucous membrane, they act in a stimulant and caustic manner : pure trimethylamine is a decided caustic to mucous membranes. Taken internally, in large doses, it greatly reduces both the frequency and the force of the pulse, and causes a burning in the throat and stomach, but no sweating, no diuresis, no colic, and no diarrhœa. A drop of pure trimethylamine placed upon the lip, produces burning and a flow of saliva: the mucous membrane is first reddened, and then the epithelium is cast off, leaving a slight sore.[1]

The statements concerning the actions, both physiological and therapeutic, of trimethylamine, have been very various, as have been those respecting arnica itself. Buchheim, for example, regarded it as a substance of little power; but the recent experiments of Dujardin-Beaumetz—one of the highest living authorities upon the action of drugs—seem to render it clear that trimethylamine has a very definite physiological action, and that among other things, it diminishes the excretion of urea. And we shall see presently that he speaks most highly of it as a remedy. Trimethylamine, as employed by M. Beaumetz, was prepared either from herring-brine or from human urine, both of which fluids yield it to chemical processes. The external effect of arnica involves important questions, for while it is known that many persons have found it an excellent application for bruises and for wounds, other observers have complained that it produces either an actual erysipelas, or a peculiar violet-colored eruption, attended by great heat and pain. I venture to affirm that these are physiological consequences of the alcoholic, and not of the aqueous solution, which latter contains neither arnicine nor the oil. I have never seen inflammatory consequences follow the application of the purely aqueous lotion to wounds or bruises.

THERAPEUTIC ACTION.—Arnica, which has always been so favorite a medicament with the homœopathists, is a remedy much older than homœopathy, and some of the most valuable evidence in its favor has been given by non-homœopathic physicians. Among the most interesting of these testimonies is that of Schröder Van der Kolk, who employed it largely in the form of infusion of the flowers and of decoction of the root.

Mental Diseases were the field upon which Van der Kolk chiefly tested the powers of arnica. He employed the infusion of the flowers in the milder cases; and the decoction of the root when a more powerful remedy was required. He found arnica invaluable in that condition of idiopathic mania where the first excitement having diminished, the head nevertheless remains hot, and where a tendency to imbecility or to paralysis is shown. Exhausting diarrhœa and general cachexia are also checked by arnica with great certainty. Van der Kolk's results are the more interesting because obtained with aqueous preparations.

---

[1] Guibert.

In Paralytic Affections of various kinds, arnica has been found useful by numerous observers, among whom are Alibert and Meyer, who by means of it cured paralyzed bladder. Mannoir employed it with success in amaurosis, for which disorder it has long been a popular remedy in Germany.

In Typhoid and Typhus Fevers, arnica has been very highly extolled, though one of the latest writers, Nothnagel, speaks of it disparagingly. He does not, however, advance any good reasons for this, and as he allows that the general "picture" of the physiological actions of arnica gives every indication of the existence of a substance which has definite powers as a remedy, we may fairly put against his rather vague opinion, and against the prejudice which British physicians have widely felt (chiefly because of its repute with "homœopathists"), the very large body of German and French experience which exists as to the action both of arnica and, in more recent days, of trimethylamine.

In Rheumatism there has always been good evidence of the utility of arnica. Even in England, Dr. Fuller has spoken strongly of the value of the tincture and of the infusion in rheumatic gout. But the investigations concerning trimethylamine have given a new and important turn to the subject of arnica. As long ago as 1854, Awenarius employed trimethylamine in acute rheumatism. In the course of three years he treated 213 hospital patients, besides many in private practice. He frequently found the joint-pain and the fever arrested in a single day's treatment. Guibert confirmed the utility of trimethylamine in acute rheumatism. On the other hand, many observers denied (and still deny), its efficacy, and during several years very little was heard of the remedy. But the recent researches of M. Beaumetz have attracted general attention to the matter; the results attained by him being very remarkable. It should be mentioned that for a long time there was a confusion between the real trimethylamine and an isomeric body, "propylamine," which was conclusively proved, however, by Winckler and Mendius (1854—1857) to possess different physical and chemical qualities. Under the name of propylamine various commercial samples have been sold, of different composition in detail, but all containing alkalies, ammonia, and trimethylamine. On account of the impurity and variableness of these preparations, Beaumetz determined to employ a definite salt, the hydro-chlorate of trimethylamine. This substance crystallizes in long needles; is very deliquescent, and its solution, when very concentrated, acts as an irritant to the skin and the mucous membranes. Its solution does not possess the stinking-fish smell of trimethylamine unless it be heated or mixed with an alkali.

M. Beaumetz ascertained, by experiments upon himself, that this salt has the power to distinctly slow the pulse, and to diminish the bodily temperature. Its action as a sudorific, or as a diuretic, is very irregular, and corresponds pretty closely with what is known, in these respects, of ammoniacal salts. But its effects, as tested in rheumatic fever, are remarkably uniform and striking in the following respects: lowering of pulse and temperature, relief of the articular pain and swelling, and diminution of the excretion of urea. In the last particular, the effects of the drug are shown in an elaborate table of urine-analysis in a case where the urea-discharge (commencing at 40.74 grams per diem) was reduced in six days to 8.55 grams per diem. The doses employed by M. Beaumetz were from 7.8 to 15.6 grams in 24 hours, in solution. I believe the best vehicle is about a tablespoonful of water to each dose. These quantities of the drug cause no irritation of the throat or stomach.

In **Inflammations of the Serous Membranes**, Awenarius and others have obtained excellent results with trimethylamine; but it must be confessed that these require the confirmation of more extended experience.

The only preparation of arnica recognized by the British Pharmacopœia is the alcoholic tincture; and herein, for reasons already stated, it is evident that a great mistake has been committed. But the tincture has special virtues of its own. It is very probably more tonic and stimulating than other preparations, as containing not only trimethylamine, but also arnicine and the ethereal oil. Quite possibly, also, the latter ingredient, besides being stimulant to general nervous power, may be sedative, like chamomile oil, etc., to hyper-excited reflex irritability.

For **External Bruises and Cuts**, arnica is undoubtedly very useful; and, as already observed, the mischances that have attended its use, have probably resulted from the fact that the tincture containing arnicine and the volatile oil has been employed. The infusion or decoction alone should be used; and it would be better to give up employing all liniments and lotions in which the tincture is present.

For **Internal Bruises**, arnica is a most excellent remedy, neutralizing the ill effects of blows, falls, and other mechanical injuries. Ecchymosis and sanguineous effusions are rapidly dispersed by it, provided the medicine be administered shortly after the injury has been sustained. In cases of shake, concussion, and shock, resulting from railway accidents, it is also very serviceable. Under these circumstances I recommend that 5 to 10 minims be taken every 2 or 3 hours in a wineglassful of water. I believe there is no drug that can so well restore the contused muscular fibre to its healthy condition in a short space of time as arnica; and I consider it a great pity that it has not come into more general use in cases of this description. When used after amputations, arnica certainly has the power of uniting the surfaces very rapidly.

In **Hæmorrhages** arising from mechanical violence, bleeding from the nose, and hœmoptysis, arnica is also of great service; and the same may be said of pulmonary congestions arising from fractured ribs. In cases of concussion of the brain, induced by a fall, I cannot speak too highly of it.

As an **Electuary**, the dried and pounded flowers, mixed with honey or syrup, are sometimes used. A weak infusion may also be employed for this purpose, especially in chronic dysentery. When the motions are slimy and purulent, and attended by tormina and cutting pains in the bowels, the tincture may be given internally with good effect.

Preparations and Dose.—Extractum Arnicæ, gr. ij.—v. (.13—.32); Tinct. Arnicæ, ℳv.—xx. (.30—1.20); Emplastrum Arnicæ.

COLTSFOOT (Tussilago Farfara).

Physiological Action.—Coltsfoot is emollient, demulcent, and slightly tonic.

Therapeutic Action.—Coltsfoot is a popular remedy for pulmonary complaints, especially chronic coughs.

Dr. Cullen employed the expressed juice in scrofulous cases.

The leaves, as well as the capitula, may be employed, a handful being boiled in a couple of pints of water till the quantity of fluid is reduced by one-half.

The ancients were accustomed to smoke coltsfoot for the relief of obstinate cough, whence the expressive name Tussilago.

The root, prepared with honey, was anciently used for ulcerations of the lungs.

PREPARATIONS AND DOSE.—Non-officinal. May be taken almost *ad libitum*.

GENTIANACEÆ.

GENTIAN (Gentiana Lutea).

ACTIVE INGREDIENTS.—The active principle of gentian is not, as was formerly supposed, the gentianine or gentianic acid, $C^{14}H^{10}O^5$, a substance which crystallizes in yellow and silky needles, and is tasteless, and neutral in reaction; but gentio-picrine, $C^{20}H^{30}O^{12}$, a bitter glucoside, first obtained by Kromager, in 1862. Gentio-picrine forms colorless crystals, which on exposure to the atmosphere become dull white. This substance is readily dissolved in cold water and in cold proof spirit, while gentianine is scarcely soluble even in boiling water, and but little soluble in spirit. Potash and soda solutions and hot solutions of ammonia dissolve it yellow. Cold concentrated sulphuric acid dissolves it without color; but the solution, when gently heated, changes to a lively carmine red. On the addition of water, gray flocculi are deposited.

PHYSIOLOGICAL ACTION.—There can now be little doubt that gentio-picrine, the true active principle of gentian, is a bitter closely allied to quinine, alike in physiological and in therapeutic action. The conflict of evidence which occurred respecting its power as an antiperiodic, unquestionably arose from the confusion between the true bitter and the tasteless gentianine. The former has been proved to be undoubtedly efficacious in cases of intermittents, by Lange, who published a series of 34 cases in which the attacks were cut short or prevented by ½ drachm doses. The effects of an overdose of gentian itself are dulness, weight in the head, oppression of forehead, and slight giddiness; symptoms, in fact, much resembling those induced by cinchonine. The face becomes flushed, and the conjunctivæ are injected. The sweat and the urine acquire a bitter taste. The bowels are relaxed, and the stools have a bilious character. It is probable that, besides the gentio-picrine, there is some volatile ingredient in gentian which has a slightly inebriant action, since Planché states that water distilled over pure gentian possesses the latter quality.

THERAPEUTIC ACTION.—Gentian root has long been employed as a valuable tonic; and, before the discovery of the cinchonas, occupied the first place in medicine as a febrifuge. It is very properly regarded as a pure and simple bitter, that is to say, as a bitter without the accompaniment, or nearly so, of either astringency or aroma. It agrees best with patients of a torpid and phlegmatic habit, but should be avoided where the temperament is irritable. Given in small doses, gentian is found beneficial in dyspepsia (especially in gastric dyspepsia connected with a gouty diathesis), also in hysteria and jaundice, and generally in all those cases of debility for which tonics are exhibited, and which are unaccompanied by symptoms of inflammation. Gentian is likewise valued in scrofula, in intermittents, and as a vermifuge; and, in the form of infusion, becomes an excellent vehicle for the administration of chalybeates, mineral acid, and neutral salt, with which it is often necessary to combine it.

PREPARATIONS AND DOSE.—Extr. Gentianæ, gr. ij.—x (.13—.65); Extr. Gentian. Fl., ℳ x.—xx. (.60—1.20); Tinct. Gentian Co., ʒ j.—ij. (4.—8.); Infus. Gent. Co., ℨ j.—ij. (30.—60.).

CHIRETTA (Ophelia Chirata).

ACTIVE INGREDIENTS.—These are probably chirettine, $C^{28}H^{48}O^{15}$, a bitter, neutral, resinoid substance, discovered by Höhn (which does not reduce an alkaline solution of a copper salt), and ophelic acid, $C^{13}H^{20}O^{10}$. The latter forms a yellowish brown syrup, tasting, at first, sour, and then persistently bitter, having also a peculiar gentianaceous odor.

PHYSIOLOGICAL ACTION.—The physiological effects of chiretta are much the same as those of gentian.

THERAPEUTIC ACTION.—Chiretta is an excellent tonic, and is held in high estimation by the European practitioners in India, who employ it for the same purposes as cinchona when the last-named drug is not procurable. The effects, like those of gentian, appear to be relaxing to the bowels rather than constipating. It strengthens the stomach, obviates flatulency, and diminishes the tendency to acidity. As a stomachic, chiretta appears to be especially serviceable in the dyspepsia of gouty subjects.

PREPARATIONS AND DOSE.—Chiretta, gr. xv.—xxx. (1.—2.); Infus. Chiratæ (B. Ph.), ℨ j.—ij. (30.—60.); Tinct. Chiratæ (B. Ph.), ʒ ss.— ʒ ij. (2.—8.).

CONVOLVULACEÆ

SCAMMONY (Convolvulus Scammonia).

ACTIVE INGREDIENTS.—The medicinal powers of scammony depend principally upon the presence of jalapine (pararhodeorctine or scammonine), $C^{34}H^{56}O^{16}$, a glucoside,[1] in many respects corresponding very closely with convolvuline, the active principle of jalap.

Jalapine is a resinoid body, thin pieces of which are translucent. At 30° C. it melts, resolving into a colorless fluid, destitute of taste and smell, and with faintly acid reaction. In water it is nearly insoluble, but

---

[1] By the action of alkalies upon the glucoside is produced jalapinic acid, the operation of which corresponds with that of jalapine.

in alcohol, hot acetic acid, chloroform, and alkaline solutions it dissolves with facility. The purest virgin scammony contains about 80 per cent.

PHYSIOLOGICAL ACTION.—Scammony and jalapine, taken in full doses, are powerfully and drastically purgative. This effect appears to result chiefly from a local irritant action upon the intestinal mucous membrane, since the activity of the operation is found to depend very much upon the amount of mucous secretion that may be present in the bowels. If there be a thick lining of this secretion, the purgative action is comparatively mild and painless, whereas in opposite conditions there is griping. The extreme effects are seen in numerous bloody evacuations, accompanied by painful tenesmus, colic, and enteritis. The general action is therefore much the same as that of jalap, but more powerful; while as a medicine it has the advantage over the last named in being less nauseous both in smell and taste. Compared with gamboge, scammony is decidedly less irritant.

In addition to the irritant action, there are probably other and remoter operations, especially upon the nervous system. The nature of these, however, is imperfectly known. Buchheim and Hagentorn injected animals with 7 to 8 grains of jalapinate of soda, and produced convulsions, contraction of the pupils, and oscillation of the eyeballs, followed by death in half-an-hour, the pupils dilating immediately before the fatal termination. [According to Rutherford and Vignal, scammony is a cholagogue of feeble power.—(*Brit. Med. Jour.*, Nov. 13th, 1875.)]

THERAPEUTIC ACTION.—Scammony has for ages been employed in medicine. Hippocrates used it as a drastic purgative, endeavoring to modify the violence of its action by means of sulphur. The ancient Arabian physicians likewise resorted to it, but always with hesitation. Both the Greeks and the Arabians prescribed it for gout, rheumatism, and other chronic diseases; they were accustomed also to order an acetous decoction to be mixed with meal, when poultices were required for painful affections of the joints. In the materia medica of past empirics, scammony always held an important place. We read that Robert Dudley, Earl of Warwick, so strongly recommended a combination of this drug with antimony and cream of tartar, to Marcus Cornachinus, of Pisa, that the latter wrote a book in praise of it. The book passed through several editions, and conferred such renown upon its subject that, in France, scammony is known to this day by the name of *Poudre Cornachine*.

At the present day, scammony is found to be chiefly valuable as a smart purgative for children, on account of the smallness of the dose required to produce the desired result, also by reason of the energy, yet perfect safety of the operation, and, what is of importance in such cases, the slight *taste*. When administered to children, it is generally in combination with calomel. If the purgative required be only a mild one, rhubarb or sulphate of potash may be employed instead of calomel. Scammony may be used to open the bowels in cases of constipation; also to expel worms, especially when occurring in children; and again in dropsies, and in affections of the head: in the latter, acting as a hydragogue purgative, it brings relief on the principle of derivation. Scammony, in a word, is well adapted for all purposes which demand the employment of an active cathartic; and for torpid and languid conditions of the abdominal organs, accompanied by much slimy intestinal mucus. Hence, also, it is the appropriate purgative in mania and in hypochondriasis. When aloes produces unpleasant effects upon the hæmorrhoidal vessels, scammony may be substituted with advantage; while to modify, in turn, its own occasionally too severe operation, as above mentioned, sugar, gum, or almonds, should be combined with it; or, best of all, it should be associated with some other purgative, calomel, as above stated, or sulphate of potash.

PREPARATIONS AND DOSE.—Scammonium, gr. v.—x. (.30—.60); Resina Scammonii, gr. ij.—vj. (.12—.36.)

JALAP (IPOMŒA JALAPA).

ACTIVE INGREDIENTS.—Jalap contains about 15 per cent. of a resin upon which its cathartic properties depend; also about 20 per cent. of a watery extractive matter, with starch, sugar, and other substances of minor importance.

The pure resin (that of the true jalap plant) is of a grayish color, acrid, brittle, and opaque. It is soluble in alcohol; slightly soluble in ether; readily so in nitric and acetic acid, and in solutions of potash; but in water and in oil of turpentine it is not soluble. Sulphuric acid turns it crimson.

Jalap also contains a strongly purgative substance called convolvuline (rhodeoretine of the German chemists), colorless, transparent, insoluble in ether, and having for its formula, $C^{31}H^{80}O^{16}$. It has a slightly acid reaction, and exhibits the chemical characteristics of a glucoside.

Spurious, or fusiform, jalap, the product of the *Convolvulus Orizabensis*, has for its chief ingredient jalapine, already described as the active ingredient of scammony.

PHYSIOLOGICAL ACTION.—Jalap is a well-known and valuable cathartic, and one upon which great reliance can be placed. In full doses it produces nausea, vomiting, griping in the alimentary canal, often attended by colic and flatulent rumbling, with copious liquid and sour evacuations from the small intestines, and in less degree from the large ones. The action is not accompanied by any febrile symptoms; it is said, also, that constipation less frequently follows the employment of jalap than of most other purgatives. The watery extract prescribed by the Dublin Pharmacopœia is said to purge without griping, and therefore to be well adapted for administration to children. Compared with other medicines, the operation of jalap closely resembles that of scammony. It is more drastic than senna, and less irritant to the mucous membrane than gamboge.

The action of convolvuline upon the animal frame has been experimentally determined by various observers. The researches of Hagentorn appear to show that the pure glucoside is four times more powerful (as a drastic) than the soft resin. It rapidly produces pain in the belly, with liquid purging. Larger doses (3½ gr.) have been known to kill a guinea-pig in three hours, the purging being followed by gastro-enteritis; and administered in considerable quantity, there can be no doubt that it is an active irritant poison. The researches of Bernatzik seem to indicate that when given internally, or by the stomach, it acts only in a local manner, affecting the alimentary canal. Hagentorn and Buchheim observed, on the other hand, that the subcutaneous injection of the convolvulinate of soda,[1] in quantities exceeding 1 gram (15 to 16 grs.), produced, like jalapine, the symptoms of violent and fatal narcotic poisoning. It is remarkable that convolvuline appears to exert little or no irritant action upon the skin, the conjunctivæ, or the nasal mucous membrane.

As a purgative, convolvuline is much more active than the substances derived from its partial decomposition. These are convolvulinic acid and convolvulin oil, which, together with sugar, are the characteristic products of the action of the mineral acids upon the glucoside. Neither convolvuline nor the substances just mentioned appear to be eliminated in the urine, nor (unless administered in enormous doses) to make their appearance in the fæces. Hence they are supposed to be absorbed, and then to be destroyed by combustion.

["1. Jalap is a hepatic stimulant of considerable power. It renders the bile more watery, but at the same time increases the secretion of biliary matter. 2. Its effect on the liver is, however, far less notable than its effects on the intestinal glands. Its hydragogue cathartic effects on the latter were fully manifested in these experiments."—R. & V. (*Brit. Med. Jour.*, June 9, 1877.)]

THERAPEUTIC ACTION.—Jalap was first known in England, as a medicinal substance, about the year 1610, though it is only of late that the source of the truest and best description of the drug has been accurately determined. Since the discovery of it by Europeans, it has probably come to be more generally employed as a cathartic than any other of vegetable origin.

In Dropsies.—Because of its hydragogue properties, jalap is especially useful in dropsies, over which it exerts a marked power, evacuating the effusions, especially when combined with bitartrate of potash or with calomel.

As a Vermifuge, jalap acts very well in cases of tapeworm and of lumbricoid worms. It is said to operate upon them as a poison, and certainly, in many instances, the worms have been found dead when expelled. When used for ascarides, this medicine should be given in combination with calomel.

Habitual Constipation, arising from dryness or scanty secretion, by the intestines, is overcome by jalap administered in a moderate dose, and also by the resin, the medicine being taken before rising in the morning, and followed, an hour afterwards, by a tumbler of cold water.

In inflammatory affections of the brain, or its membranes, when a purge is required, jalap often proves most valuable.

As a simple purgative for children, it may also be employed with the best results, in doses of from 1 to 5 grains.

PREPARATIONS AND DOSE.—Jalapa, gr. v.—xv. (.30—1.); Tinct. Jalapæ, ʒ ss.—jss. (2.—6.); Resina Jalapæ, gr. ij.—vj. (.12—.36); Pulv. Jalap. Co., gr. x.—xxx. (.65—2.).

---

[1] If treated with alkalies, convolvuline is converted into convolvulinic acid, a substance analogous, as mentioned above, to jalapinic acid.

ZYGOPHYLLACEÆ.

GUAIACUM (Guaiacum Officinale).

ACTIVE INGREDIENTS.—Guaiacum resin has a slightly aromatic and balsamic odor; on being chewed, it softens, and communicates a slightly bitter and acrid taste, followed by a peculiar burning and prickling sensation in the back of the throat. The sp. gr. is 1.25. Not more than nine per cent. of it is soluble in water; but alcohol dissolves about 91 per cent., acquiring a deep brown color, and from this the resin is again precipitated by water. It is soluble also in ether and in alkaline solutions. The nature of guaiacum resin is that of an acid. It forms soluble salts with the alkalies, and is precipitated from alkalies by acids.

The constituents are about 10 per cent. of guaiaretic acid, $C^{20}H^{26}O^4$, which is crystalline; and about 70 per cent. of guaiaconic acid, $C^{19}H^{20}O^5$, with other subordinate matters.

PHYSIOLOGICAL ACTION.—Guaiacum, given in small doses, has the effect simply of stimulating the vascular system, and often produces diaphoresis. Given in large doses, it produces dryness of the mouth, burning in the throat, a sensation of heat in the stomach, loss of appetite, heartburn, flatulence, nausea, vomiting, and purging.

These symptoms, which vary according to the dose and the length of time that the patient is kept subject to the influence of the drug, are accompanied by palpitation of the heart, headache, apparently of a congestive type, confusion of mind, giddiness, fainting, and general lassitude. Stiffness, of a rheumatic character, is felt at the same time in the nape of the neck and the small of the back, with pains in the bones of the legs, the limbs feeling as if swelled; darting pains, apparently of a rheumatic neuralgic character, extend also from the feet to the knees. The results described are further attended, in many instances, by profuse perspiration, and are sometimes followed by an exanthematous eruption, and casually, by ptyalism. When guaiacum fails to act upon the skin, it often operates as a diuretic.

THERAPEUTIC ACTION.—Our first knowledge of guaiacum was obtained through the Spaniards, who brought it to Europe from St. Domingo, about the year 1508, with the reputation of its being antisyphilitic. The name of "lignum vitæ" was given to it almost immediately, and by 1519 many thousands of patients are said to have been cured.

The mode of exhibiting the medicine was in decoction, this being frequently made with port wine. It was employed in every form of the disorder, and in every stage of its progress, the physicians prescribing at the same time the use of purgatives, baths, and appropriate diet. Resort to mercury was discontinued, and even censured, and for a period of two centuries the new medicine enjoyed the highest repute. Boerhaave went so far as to assert that guaiacum-wood was competent to expel the venereal poison from the system. Yet those who most strongly believed in its efficacy were constrained to admit occasional failures; and, on the other hand, we can now see that many conditions which guaiacum was supposed to cure probably did not belong to the venereal class, so that the drug acquired a celebrity it did not deserve.

The ancient reputation of guaiacum as an antisyphilitic remedy was in great part due to the celebrated name of one of the earliest patients who benefited by its use. This was no less a person than Ulric von Hutten, the satirical and military champion of the early Reformers, and author of the "Epistolæ Obscurorum Virorum," a book which set all Europe laughing at the monks. This famous personage has himself related the story of his cure, which appears to have been effected by means of a compound decoction of various woods, of which guaiacum was assumed to be the most important.

Catarrh and Gout.—In chronic pulmonary catarrh, especially when occurring in gouty subjects, there can be no doubt that guaiacum manifests a beneficial action. Even in gout itself it once had great renown. Cullen speaks of it in terms of commendation.

Tonsillitis.—Dr. Hannah and Mr. Bell[1] advise the use of guaiacum in tonsillitis. They state that it reduces the pain and inflammation with marked rapidity.

Menstrual Disorders.—I may add that guaiacum is useful also in amenorrhœa and in dysmenorrhœa.

PREPARATIONS AND DOSE.—Resina Guiaci, gr. x.—xxx. (.65—2.); Tinct. Guiaci, ʒss.—iss. (2.—6.); Tinct. Guiaci Ammoniata, ʒss.—iss. (2.—6.); Decoct. Sarsaparillæ Comp., ʒss.—j. (15.—30.); Syrupus Sarsaparil. Co., ʒij.—iv. (10.—20.)

MYRTACEÆ.

THE CLOVE (Caryophyllus Aromaticus).

Active Ingredients.—These, according to Trommsdorf, consist chiefly of volatile oil, tannin, a gum-resin, extractive, and lignin. Upon analysis, the volatile oil resolves into a hydrocarbon, $C^{10}H^{16}$, in which eugenic acid, $C^{10}H^{12}O^2$, is dissolved; caryophylline, $C^{10}H^{18}O$, a substance isomeric with camphor; and, thirdly, a body called eugenine, probably isomeric with eugenic acid. The caryophylline presents itself in crystals of satiny brilliance, destitute of taste and smell, fusible, volatile, and soluble in alcohol, but insoluble in water. The oil, though clear and colorless when fresh, gradually acquires, by keeping, a dark and reddish-brown hue. The taste is aromatic and somewhat acrid; the odor is that of the spice, and very strong. It is soluble in alcohol, ether, and the fixed oils, and has a sp. gr. of 1.034 to 1.060, being heavier than water.

Physiological Action.—Used as a spice, and in moderation, cloves stimulate the digestive organs; but taken in excess, or too continuously, they exhaust the susceptibility of the stomach, and induce loss of appetite and constipation.[2] The oil, taken in large and undiluted

---

[1] London Medical Gazette, 1840, p. 202.
[2] It is probable that the oil of cloves partakes of that power of lowering reflex irritability which has been shown to pertain to other volatile oils (*vide* chamomile, etc.)

doses, acts as a powerful irritant; but if in smaller quantities, as a diffusible stimulant.

Therapeutic Action.—Cloves are generally considered to be the most stimulating of the aromatics. They are employed more as a condiment than as a medicine, and often to season food of a somewhat indigestible character. The infusion, being a warm and grateful stomachic, is advantageously employed to relieve the sense of coldness in the stomach which attends certain forms of dyspepsia, especially such as arise from the abuse of ardent spirits, from chronic gout, or from flatulent colic.

In Holland, oil of cloves is combined with cinchona and supertartrate of potash, and administered in ague.

It is seldom, however, that the infusion or any other preparation of cloves is found useful, or at all events expedient, *per se*, or as the basis or principal medicine. When given alone, it is chiefly as a carminative.

Infusion of cloves is exhibited for the relief of nausea, vomiting, flatulent colic, and other allied complaints. Cloves are valuable, also, for the purpose of imparting a pleasant flavor to medicines of a distasteful character, and for correcting the irritant properties of drastics. As a local excitant, they were formerly recommended to be chewed in particular cases of paralysis of the tongue.

The essential oil is a popular remedy for toothache, being applied on cotton-wool to carious cavities in the teeth. By diffusing the essential oil in water, with some mucilage, an agreeable draught may be prepared.

Cloves are employed, also, in the preparation of an aromatic syrup, which is afterwards colored with cochineal. They are used likewise as an adjunct to purgatives.

Preparations and Dose.—Infusum Carophylli, ℥ss.—ij. (15.—60.); Oleum Carophylli, ♏j.—v. (.05—.25).

PIMENTO (Pimenta Vulgaris).

Active Ingredients.—The virtues of pimento mainly reside in the pericarp and the seeds, from which are obtained two essential oils, one volatile and the other fixed. The former of these (*oleum pimentæ*) constitutes about six per cent. of the dried berries, and is considered to possess all the active properties of oil of cloves; it is separated by distillation, and on it mainly depend the strong fragrance and the warm and aromatic flavor. The fixed oil is greenish, and has an acrid, burning taste, and a rancid, but somewhat clover-like odor.

Physiological Action.—Very little is known with regard to this, beyond the fact that pimento is an aromatic stimulant, stomachic, and carminative, and that it holds an intermediate place between pepper and cloves. Externally employed, it is a rubefacient.

Therapeutic Action.—Pimento is used as a condiment by persons suffering from exhaustion of the digestive system, and especially by those living in tropical regions. It relieves nausea, flatulency, and griping pains in the bowels. It increases the effects of vegetable tonics, and prevents the griping of purgatives; it is also used to cover the taste of nauseous medicines.

PREPARATIONS AND DOSE.—Pimenta, gr. v.—xxx. (.30—2.); Oleum Pimentæ, ♏ i.—iij. (.05—.15).

CAJEPUT OIL (Melaleuca Cajeputi).

ACTIVE INGREDIENTS.—Cajeput-oil is very fluid and transparent, light green in color, and possessed of a strong but agreeable odor, and a warm, aromatic, and camphoraceous taste, soon followed by a sense of coolness in the mouth. The sp. gr. is 0·914 to 0·927. It is wholly soluble in alcohol, and boils at 343°. Distilled with water, there first passes over a colorless oil, which is the hydrate of cajeputine ($C^{10}H^{14}H^2O$), and constitutes about two-thirds of the crude oil.

PHYSIOLOGICAL ACTION.—When swallowed, cajeput-oil causes a sense of warmth in the stomach, accelerates the pulse, and increases the perspiration and the urine.

THERAPEUTIC ACTION.—According to Ainslie, "cajeput-oil is a highly diffusible stimulant, antispasmodic and diaphoretic, and may be efficaciously given in dropsy, chronic rheumatism, palsy, hysteria, and flatulent colic." In India it is used with much success in particular choleraic affections; also in nervous vomiting, nervous dysphagia, and for the destruction of ascarides.

Muscular rheumatism and nervous headaches are often quickly removed by the oil being well rubbed into the parts twice a day; when thus applied externally it is a strong stimulating rubefacient.

PREPARATIONS AND DOSE.—Oleum Cajeputi, ♏ iij.—x. (.15—.50); Spts. Cajeputi (B. Ph.), 3 ss.—j. (2.—4.).

SMILACEÆ.

SARSAPARILLA (Smilax Officinalis).

ACTIVE INGREDIENTS.—Sarsaparilla contains, besides a varying amount of starch, a volatile oil, and a white crystalline principle, Smilacine, of which the chemical formula is not yet settled. Smilacine is nearly insoluble in cold water; boiling water dissolves it, and the solution, when shaken, lathers like soap. The reaction is neutral; the taste bitter and acrid.

PHYSIOLOGICAL ACTION.—This is in a state of complete uncertainty. I am not aware of any experiments with the volatile oil; and the results of researches upon smilacine are conflicting, but, on the whole, negative. Boecker, whose authority as an experimenter is very high, concludes, from researches which he made with the assistance of Groos, that smilacine is at least entirely devoid both of diuretic and of diaphoretic power, and that it does not cure syphilitic maladies. On the other hand, the experiments of Pallotta seem to show that in some persons at any rate, large doses (7 or 8 grains) of smilacine produce gastric uneasiness, slowing of the pulse, vomiting and perspiration, with faintness. There have been no methodical experiments with smilacine upon animals.

THERAPEUTIC ACTION.—Of the *modus operandi* of sarsaparilla it must be confessed that nothing is known; even those who are warmest in its praise will admit this fact. As to the supposed sudorific and diuretic effects, although it seems a comparatively simple point, there are very contradictory opinions, many believing that it is only when large quantities of a warm decoction are taken that any sweating or diuresis is produced, and that these effects are in reality due merely to the warmth of the fluid, and to the quantity of water which accompanies the medicine. I am convinced, however, that diaphoresis is one of the most common and genuine results of sarsaparilla.

Sarsaparilla was introduced into medicine as early as 1560, when it was employed in Venice as a cure for syphilis. It long enjoyed high repute, but gradually became more and more neglected, till at last, Cullen allows only eight lines of description to its history and qualities together, and declares that he cannot give it a place in the Materia Medica, never having found it useful in any disease! Its restoration to favor, if not to a great place in medicine, we owe to Dr. William Hunter. It must be admitted, however, that even at the present day, there are many who believe Cullen to have been right. Two or three explanations may be given of the disfavor of sarsaparilla: the difficulty of explaining its action, the possibility of removing certain forms of venereal disease without resorting either to mercury or to sarsaparilla, and the probability of a spurious article being mingled with or substituted for the genuine one, either by accident or design, before it comes into the hands of the practitioner.

The general properties of sarsaparilla are alterative and tonic, and the most common result of its administration is diaphoresis. The continued use of this medicine is often followed by improvement of the appetite and the digestion, and consequent increase of flesh and muscular power. Should there be eruptions, ulcerations, or pains of a rheumatic character in any part of the body, these are often mitigated, and in some cases entirely removed. The best effects are seen in those depraved conditions of the system which are popularly attributed to the presence of some morbid poison, or to a deranged condition of the fluids, whence its familiar repute as a "purifier of the blood." It is of special service in secondary syphilis, either alone or in combination with other remedies, evincing powers of a very serviceable kind when the disease resists the action of mercury, or when it has been aggravated by the use of mercury.

For Simple Debility, and for the cure of intermittents, sarsaparilla is useless, or, at all events, not adapted. It differs in several respects from the bitter vegetable tonics, and possesses very little of the bitterness.

In Constitutional Syphilis, after many changes of fortune, sarsaparilla still holds its ground as a most important remedy, and more especially when employed as an auxiliary to mercury, and in cases which resist the action of that drug, or in which mercury having been freely used, the constitution requires to be rescued from sundry evils which are the result of the mercurial treatment, or of the mercury and the syphilis combined. With respect to this point, it will be interesting to cite first the older British authors, and then to quote at length one of the latest summaries of the action of the drug as given by a highly competent German authority. Sir William Fordyce recommended sarsaparilla more particularly as an auxiliary to mercury, and as well adapted to purify the system after a long course of mercurial treatment. Pearson supports this view, remarking that the contagious matter of syphilis and the mercury together may, in certain habits of body, co-operate to produce a new set of symptoms which, properly speaking, are not venereal. These secondary symptoms (which are sometimes more to be dreaded than the simple and natural effects of the venereal virus) cannot be cured by mercury. Some of the most formidable of them may, however, be removed by the use of sarsaparilla—the virus still holding its ground, though partially giving way—and when by the renewed use of mercury the virus has been completely subdued, sarsaparilla frees the patient from what may be called the *sequelæ* of the mercurial course.

Dr. Good says that he has found sarsaparilla to succeed chiefly in chronic cases, when the constitution has been broken down, and this whether resulting from a long domination of the disease, or from protracted and apparently inefficient mercurial treatment. In connection with a milk diet and country air, Dr. Good continues, sarsaparilla is of essential importance. The majority of medical men coincide with the opinions of Dr. Good and Mr. Pearson. Quite recently, Dr. Clifford Allbutt, of Leeds,[1] has given strong testimony to the good effects of sarsaparilla, as observed both by himself and by his colleagues at the Leeds Infirmary, provided that the remedy be administered in large quantities —not less than a pint of the compound decoction in twenty-four hours. He speaks especially of its restorative effects in old and broken-down cases of constitutional syphilis.

Modern German opinion is summed up by Nothnagel.[2] After mentioning that sarsaparilla is always given in conjunction with other vegetable substances which are supposed to act in an analogous and auxiliary manner, he says:—"That this method of treatment often produces good results is positively certain; the particular cases will be presently indicated. In what way sarsaparilla and similar remedies arrest (zum Schweigen bringen) syphilis, has not been explained. The old view, that sarsaparilla exerts a specific operation against the syphilis poison, appears to be much more incorrect even than the corresponding phrase as applied to mercury; there is no trace of a proof of such specificity. It is now generally assumed that the vegetable remedies (sarsaparilla, guaiacum, etc.) produce a cure by means of such increase of natural evacuations (diuresis, diaphoresis, purgation) as hasten the metamorphosis of tissues, and thereby the natural elimination of the morbid matter which is the origin of the syphilitic manifestations. This view has much in its favor. It is supported, for instance, by the fact that in many cases a patient suffering from syphilis rapidly gets well by the use of simple warm baths, and the drinking of a warm tea which promotes sweating and urination. These latter cases also favor the opinion, so frequently put forward, that sarsaparilla *per se* is superfluous in the cure, and that the menstruum is the only active agent. This opinion cannot be directly disproved, since it is not possible distinctly to *prove* the great activity of sarsaparilla.

"Experience teaches us the following facts respecting the use of the

---

[1] Practitioner, 1870, vol. i.
[2] Arzneimittellehre, p. 489.

vegetable remedies for syphilis:—the wood-drinks (Holz-trünke) neither can nor ought to be exclusively employed against syphilis, any more than mercury should be so used. History shows that physicians have repeatedly abandoned the exclusive use of each of these. We have already spoken of the advantages and applicability of the mercurial treatment, but it is also known that syphilis, under favorable circumstances, may disappear spontaneously. This natural cessation may be assisted by a methodic use of the vegetable cure; this mode of procedure is also indicated in the simple, ordinary secondary phenomena, either in robust, or more especially in scrofulous, tuberculous, or scorbutic individuals; in the former mercury is usually superfluous, in the latter it is usually hurtful. Sarsaparilla is also appropriate in the inveterate syphilis of persons who have already undergone various mercurial courses without success; here the vegetable cure is often strikingly effective; also in obstinate and severe secondaries and tertiaries; in the last usually best in conjunction with iodine. Sarsaparilla is superfluous in primary indurated chancre, for it hardly at all hinders the occurrence of secondary symptoms; it is also almost always useless in bone affections; and finally it is inapplicable, from its slow operation, in cases where rapidity of effect is essential (Iritis, brain symptoms).

"We cannot enter into any detailed discussion of the much disputed question of the advantages and drawbacks of the non-mercurial treatment. One point, however, may be brought forward; it does seem certain that the average time required for the vegetable is longer than that required for the mercurial cure; it is not true that relapses are rarer after the former than after the latter; on the other hand, it seems certain that they appear earlier and oftener after sarsaparilla, but in a milder form; though indeed cases have been seen in which a thorough vegetable cure has been followed by no relapses whatever. The vegetable cure really does appear to have the advantage of not being followed by the fearful tertiary symptoms to any extent so frequently as happens with the early and forced treatment by mercury; but it must be admitted that in some cases, even of the vegetable cure, tertiary phenomena have been observed, and in others there have been repeated recurrences of secondary symptoms."

Miscellaneous Uses of Sarsaparilla.—This drug is employed in affections of the stomach which appear to arise from its own morbid secretions. Also in chronic rheumatism, in conjunction with powerful sudorifics or anodynes, such as opium and hyoscyamus, especially when there is reason to suspect that venereal taint is lurking in the system.

Scrofula, elephantiasis, and some other cutaneous disorders, are likewise usefully treated with sarsaparilla, the value of the medicine depending in these cases chiefly upon the tonic and alterative effects. The diaphoretic influence must be assisted by the use of diluents and warm clothing.

In chronic abscesses, attended by profuse discharge; in obstinate ulcer; in diseases of the bones; in chronic pulmonary affections, where there is great wasting; and in many other complaints which indicate a depraved state of the system, sarsaparilla is a very useful medicine.

Again, in the irritable condition of the system which often ensues upon severe operations, or which arises from long continued suppuration, sarsaparilla is employed with marked success. Such, at all events, was the experience at Guy's Hospital, under Sir Astley Cooper. Sarsaparilla increases the appetite, brings down the pulse, augments its tone, and, conjoined with milk, is both food and medicine.

Mr. Lawrence used to say that physicians had no confidence in sarsaparilla, but that surgeons had a great deal. This is very likely still to be true up to a certain point, physicians being less frequently called upon than surgeons to deal with the class of cases in which sarsaparilla is specially beneficial.

PREPARATIONS AND DOSE.—Extractum Sarsaparillæ Fluidum, ℨ ss.—j. (2.—4.); Ext. Sars. Fluid. Co., ℨ ss.—j. (2.—4.); Syr. Sars. Co., ℨ ij.— ℨ iv. (10.—20.). Decoct. Sars. Co., ℨ ss.—j. (15.—30.).

SIMARUBACEÆ.

QUASSIA (PICRÆNA EXCELSA).

ACTIVE INGREDIENTS.—The wood of the picræna owes its qualities to a peculiar neutral principle called quassine, or quassite, $C^{10}H^{13}O^3$. When separated it appears in the form of small, white, prismatic crystals, destitute of odor, but intensely bitter; fusible readily, soluble in alcohol, but very slightly so in either water or ether. There is no tannin or gallic acid in the bark.

PHYSIOLOGICAL ACTION.—On this subject nothing is accurately known, except that quassia is a poison to flies and to fish. The experiments that have been made on warm-blooded animals are too equivocal and self-contradictory in their results to be of any value.

THERAPEUTIC ACTION.—It must be owned that at the present day there are not many cases in which quassia is likely to be especially useful.

As a Vermicide Enema, to be injected into the rectum of patients who are infested with ascarides, it is very useful indeed; in fact there are many cases in which alternate injections of lime-water and of infusion of quassia will suffice for the cure.

As a General Tonic, and as a Stomachic Tonic, quassia is now generally believed to be inferior to gentian.

As an Antipyretic, and as an Antiperiodic, it is but a feeble agent, and not to be compared with the cinchonas.

In the Dyspepsia of Intemperance it is, perhaps, one of the best remedies, but here again it is doubtful whether quinine is not superior.

In Hysteric Affections the unquestionable repulsiveness of quassia is probably a powerful aid to its medicinal effects.

PREPARATIONS AND DOSE.—Extract. Quassiæ, gr. j.—v. (.06—.30); Infus. Quassiæ, ℨ ss.—ij. (15.—60.); Tinct. Quassiæ, ♏ xv.— ℨ j. (1.—4.).

LEGUMINOSÆ.

CALABAR BEAN (Physostigma Venenosum).

ACTIVE INGREDIENTS.—The efficient element of Calabar bean is the alkaloid called eserine or physostigmine, $C^{15}H^{21}N^3O^2$: it is present only in the cotyledons of the seeds. Physostigmine[1] is a colorless varnish-like stuff which dries with heat into thin scales, but at ordinary temperatures soon softens again. It is said, when exceedingly pure, to form crystalline crusts, or even glittering rhombic scales. It is tasteless, and decidedly alkaline in reaction: easily soluble in alcohol, ether, benzol, and chloroform: not very soluble in water. If heated for a long time, at 212° F., it alters to a reddish color, and its solutions in acid are now red. It can be identified at once by these characters, together with the physiological action on the pupil to be presently described.

PHYSIOLOGICAL ACTION.—The most curious and characteristic action of Calabar bean is one that renders it of extensively useful application in ophthalmic medicine and surgery; this is contraction of the pupil and of the ciliary muscle. It induces a condition of short-sightedness, and occasions sympathetic dilatation of the pupil of the other eye. A difficulty is found, however, in regard to the preparations, since the alcoholic extract, which contains the whole of the poisonous principle, can be only imperfectly dissolved in water (the alcoholic solution being of course, *per se*, inadmissible), while the flow of tears, which instantly follows such an application, greatly reduces the amount placed in contact with the membrane, or at any rate renders it very uncertain. Hence it is found useful to employ the extract diffused in bibulous paper after the manner that has been recommended for the application of atropine; while another mode is to use a solution of the extract in glycerine, the latter in no way interfering with its operation. The proportion is 2½ grs. of extract to 100 minims of pure glycerine. Another very useful plan is to impregnate gelatine with calabar extract and to place a minute disc of this material within the eyelids, where it dissolves.

Calabar bean has, however, much wider actions on the body than have yet been mentioned. The researches of Christison, Fraser, and others in this country, and of a number of German and French observers, have illustrated these actions with much fulness. Perhaps the best summary that has been given of the action of the Calabar bean on the heart and spinal nervous system is that of Roeber:[2] "1. The chief action of the bean consists of a depression and final annihilation of the excitability of the ganglionic elements of the spinal cord; and its operation especially affects the groups of cells in the anterior horns of the gray matter which conduct impulses from the brain to the periphery, and then also attacks the elements of the gray matter in the posterior horns which transmit

---

[1] While this is actually going through the press, I am informed, on the best authority, that physostigmine has been discovered to possess one property which is entirely opposed to the action of the bean itself. It paralyzes, instead of stimulating, the terminals of the vagus. [1874.]

[2] Ueber die Wirkung des Calabar-extractes auf Herz u. Rückenmark. Inaug. Diss. von Hermann Roeber. Berlin, 1868. In Practitioner, 1869, vol. ii.

sensations of pain to the brain. 2. By this functional lesion of the gray matter a complete loss of the motor and reflex activity of the spinal cord is produced, likewise a loss of sensibility to pain; while the sense of touch, and the so-called muscular sense, are retained till the death of the animal. 3. Besides this action on the cord, Calabar bean possesses a special power over the movements of the heart, which in small doses it merely retards, but in large doses completely arrests. 4. The interference with respiration, which is especially produced by small doses, is either the consequence of a sudden interference with the heart's action, or is produced by a destruction of the motor power of the respiratory muscles from paralysis of the spinal cord. 5. The poison increases the secretion of tears and of saliva. 6. The increase of defecation observed in poisoning with Calabar bean is the result of a tetanus of the stomach and intestines, the cause of which is not yet fully determined. 7. The motor and sensory nerves are not affected at the commencement or in the development of the affections of the cord: at a latter stage there follows a paralysis or hastened death of the intra-muscular termini of these nerves. 8. The fibrillary muscular twitchings occurring soon after the administration of the poison, which are especially striking in mammalia, may be explained by a local irritation of them, caused by paralysis of the motor nervi-termini. 9. The pupils are strongly contracted both in the external and in the internal use in large doses of Calabar bean extract: but as to the cause of this it will be necessary to institute more exact inquiries.

But the most interesting properties are those, perhaps, which place physostigma in opposition to belladonna, or, more accurately speaking, to atropia, upon which subject Dr. Thomas R. Fraser has recently printed a most valuable and copious monograph, which is also to be seen in the Transactions of the Royal Society of Edinburgh (vol. xxvi. p. 529, 1872). Dealing first with the general idea of antagonism in medicines, he shows it to be necessary, when the action of a particular poison or other energetic substance has to be counteracted, that the physiological functions of the affected organism should be modified;—a principle early, though very imperfectly, recognized in the original employment of alexipharmics. More exact knowledge has led to the discovery that organic modifications of a definite and constant character can be established; that the modifications produced by certain drugs are exactly the reverse of those produced upon the same part by certain other drugs; such, for example, is the case in the action exerted upon the iris by opium or morphia on the one hand, and by hyoscyamus, belladonna, and stramonium on the other. The contrary and opposite action of physostigma and atropia Dr. Fraser shows to be exactly of this character. His researches, commenced in 1868, have been followed up steadily to the present time, and have consisted of a vast number of carefully tabulated experiments, chiefly upon rabbits; and the conclusive deduction is that the lethal action of physostigma finds a powerfully counteracting one in the influence of atropia. That is to say, the fatal effects of a lethal dose of physostigma are neutralized by the employment of atropia (under certain conditions of administration), the atropia so influencing certain structures as to prevent the occurrence in them of the modifications induced by physostigma, which, if not so dealt with, would result in death. The action of physostigma is cancelled by that of atropia, and so decided is the antagonism that so large a quantity even as $3\frac{1}{4}$ times the minimum lethal dose of the physostigma is rendered inoperative. Dr. Fraser, however, had in some degree been anticipated as early as 1864, when Kleinwachter narrated a case of belladonna-poisoning, which had been successfully treated by physostigma. In Paris, also, a case of tetanus had been much ameliorated by the internal administration of the powdered kernel of physostigma—enough, it was believed, to cause death—followed up by the subcutaneous injection of a small quantity of atropia. Dr. Fraser's own experiments disclose facts of a very singular character. For instance, when the two substances (physostigma and atropia) are administered simultaneously, they induce certain actions of intensity sufficient to cause death; whereas, if the administration of the physostigma be delayed for 25 minutes after that of the atropia, those actions cease to be fatal ones. To delay the administration of the physostigma for only 5 minutes, or, with the same respective doses, for only 10 minutes, is still insufficient to ward off death: recovery from the influence of the atropia takes place only at a minimum of 15 to 20 minutes. The counteracting effect of atropia in regard to the lethal action of physostigma, is successfully exerted, moreover, only within a definite range of doses, determinable by experiment.

As to the accompanying symptoms, it does not appear that successful antagonism necessarily brings with it, or implies any particular class of these. The antagonism may be attended by a greater prominence of the effects either of the atropia or of the physostigma, and no special action belonging to either of these two substances requires to be prominently and obviously produced in order that the antagonism shall be successful.

THERAPEUTIC ACTION.—The earliest, and still the most frequent, use of Calabar bean, is in eye affections, as already described. But it has been used, with more or less success, in various nervous affections.

In Tetanus the results obtained by some observers have been so remarkable as to excite great hopes for the future treatment of that terrible disease; but there are very conflicting statements. Dr. Fraser has borne energetic testimony to its value,[1] and Dr. Eben Watson[2] has collected a considerable number of cases of its use, and shows that the majority recovered. Moreover, he gave enormous doses of the alcoholic extract: one patient, who recovered, took (in forty-three days) 1,020 grains, and as much as 72 grains in one particular twenty-four hours. He believes that the dangers of poisoning are to be avoided if care be taken to adapt each dose to the necessities of the moment, and to support and stimulate the patient freely. Mr. McNamara,[3] of Calcutta, has also had excellent results in a large number of cases. On the other hand, many observers have tried the remedy in isolated cases of tetanus, with no good effect. In strychnia-tetanus there is a case of recovery under Calabar bean recorded by Dr. Keyworth.[4]

In Chorea there is a considerable amount of evidence in favor of the bean, but it is scarcely needful to say that it is much more difficult to be sure that the remedy was the real cause of recovery in chorea than in tetanus. Harley,[5] Ogle,[6] and McLaurie,[7] all testify favorably. Gubler,[7] on

---

[1] Practitioner, August, 1868.
[2] Practitioner, April, 1870; Lancet, April 4 and 11, 1868; Glasgow Med. Journal, Nov., 1868.
[3] Practitioner, 1872.
[4] Glasgow Med. Journal, Nov., 1868.
[5] Med. Times and Gaz., Jan. 16, 1864.
[6] Med. Times and Gaz., Sept. 2, 1868.
[7] Commentaires de Thérapeutique.

the contrary, reports its failure in one case, and other persons have had similar negative results.

In **Delirium Tremens** Calabar bean has been tried a few times; but there were no good physiological grounds for expecting success, and it does not seem to have been obtained.

In **Epilepsy**, also, Calabar bean has been tried in a few isolated cases; but there is no positive evidence of any value, and it is not easy to see how the drug could be expected to do much, except, possibly, in a temporary manner, by reducing excited spinal reflex irritability. The latter condition, however, does not exist in the majority of cases of epilepsy.

In regard to the special applications of calabar-extract and of pure physostigmine to the treatment of eye affections, there are two or three matters that should be particularly mentioned. There is a large amount of evidence to show that Calabar bean directly stimulates the third nerve; this is proved by the evidence not only of the ophthalmic surgeons, but of Gubler. Mr. Wharton Jones[1] states some facts which tend to show that physostigma contracts the circular fibres of the iris, and the veins, as opposed to atropia, which acts on the pupil through the sympathetic, and which contracts the arteries. From these actions it results that paralysis of the circular fibres and of accommodation, such as are liable to follow cold, or to follow diphtheria or other debilitating constitutional diseases, can be successfully treated with the bean; also excessive atropinization of the pupil can be neutralized (but as the action of the bean is rather fugitive, it needs to be repeated): again the pupil may be reduced in size for the purpose of shutting out the irritant effects of light upon an inflamed choroid or retina. It is generally employed, for ophthalmic purposes, in a topical manner.

PREPARATIONS AND DOSE.—Pure physostigmine is seldom used, being difficult to prepare; but Veé[2] recommends a solution of 1 in 2,000 or 1 in 1,000 of the sulphate of physostigmine for local use, one or two drops to be applied; for internal use, about $\frac{1}{64}$ to $\frac{1}{18}$ grain for a dose; for subcutaneous injection, about one-half of the latter quantities. But ophthalmists now most frequently employ the divided squares or discs of impregnated paper or gelatine, each one of which, when placed in the eye, is sufficient to produce the required effects. The dose of the bean itself is said to be from 1 to 4 grains, and of the pharmacopœial extract, $\frac{1}{16}$ to $\frac{1}{4}$ grain or more; but the severity of the affection which we may have to deal with must be taken into consideration, and bad cases of traumatic tetanus may require (as shown by Dr. Eben Watson) an extremely frequent repetition of the dose, so that large quantities are taken in each twenty-four hours. The subcutaneous injection of calabar-extract should always be done in the manner recommended by Dr. Haining, viz. by mixing the dose with 10 to 15 minims of water, and neutralizing its acidity with carbonate of potash. But Dr. Eben Watson states that he has not found good results from the subcutaneous injection, and much prefers to give the drug either by the mouth or by the rectum.

One final remark must be made, as to the proper manner of using physostigma in cases of poisoning by belladonna or atropia. The dose and repetitions must be regulated of course in great part by the apparent effects upon the pupil; but it must be remembered that there are two actions

---

[1] Practitioner, Sept., 1869.
[2] Recherches Chim. et Physiol. sur la Fève de Calabar. Paris, 1865.

of physostigma, the more formidable of which is its tendency, in large doses, to paralyze the heart. Very close watch must therefore be kept on the circulation, and the remedy should be given in fractional doses, repeated as often as is found necessary and safe. [Phillips.]

SENNA (Cassia Obovata).

Active Ingredients.—These, notwithstanding some conflicting opinions, appear to be exclusively represented by the Cathartic acid of Dragendorff and Kubly. It is an acid glucoside, amorphous, of a brown color until dried, when it becomes black; its alcoholic solution, when boiled with acids, produces sugar, and Cathartiginic acid, a yellowish brown substance, which is insoluble in water and in ether. The formula of Cathartic acid is $C^{180}H^{95}N^2O^{82}S$.

Physiological Action.—Infusion of senna, when injected into the veins, produces both vomiting and purging. Taken into the stomach in moderate or medicinal doses, it causes, first, a sensation of warmth in that part, and in the abdomen; the pulse becomes slightly accelerated, and a safe purgative action soon ensues, the stools being liquid, and of a yellow color. Even the odor of senna leaves or of the infusion is, in certain susceptible persons, sufficient to cause an evacuation of the bowels[1]. In large doses it produces nausea, vomiting, griping, flatulence, depression, and, after a little excitement of the pulse, severe drastic purgation, with tenesmus. Senna often also produces hæmorrhoids, and in women is apt to increase the monthly discharge. If taken by nurses, the infant is generally purged, a fact which shows that the senna is absorbed, and then thrown out of the system by the excretories.

Although a drastic, the action of senna is remarkable for its mildness. Even when administered in large doses, the effects, though powerful and disagreeable, are never poisonous. Compared with rhubarb, it is more irritant and energetic, but devoid of the tonic properties which belong thereto. Compared with aloes, the action is more powerful and speedy, especially upon the small intestines; the action upon the large one is less decided. When acrid and griping effects are produced by moderate doses, the action is referable probably not so much to the senna as to the leaves of certain plants with which the drug may have been adulterated during collection.

[According to R. and V. senna is a hepatic stimulant of feeble power. It renders the bile more watery.—*Brit. Med. Journ.*, Nov. 6, 1875.]

Therapeutic Action.—Senna being a safe and certain purgative, is employed with advantage whenever there is occasion for prompt evacuation of the bowels, without possibility of failure, and especially when the small intestines require emptying. In all forms of inactivity of the alimentary canal it constitutes a well-known domestic remedy. Ordinarily the infusion is combined with solution of Epsom salts, or tartrate of potash, or with manna, or tamarinds. If a more vigorous action than usual is desired, it is said to be well to combine the senna with guaiacum. In order to cover the nauseous taste, Dr. Cullen recommends the admixture of coriander seeds.

---

[1] Stillé, vol. ii., p. 447.

In cases of constipation, of worms, and of determination of blood to the head, senna is likewise resorted to with great benefit to the patient. Children and elderly persons are especially suitable subjects to derive advantage from it.

The difficulty with senna is that, should inflammation of the mucous membrane of the bowels be present, or if there is a tendency to hæmorrhoids, menorrhagia, or abortion (in the case of pregnant women), it becomes hurtful, by reason of the irritation it is apt to excite. Senna should be employed with caution, also, when there is febrile disorder, the action of this drug upon the pulse being excitant. In such cases as these it is better, perhaps, not to resort to senna at all, but to supersede it with one of the saline purgatives. Again, if the large volume of senna-tea required for an efficient dose be found objectionable, the electuary called "Confectio sennæ" may be recommended as a form more convenient, and to most persons less unpalatable. This latter preparation is not only very successful in its medicinal use, but by no means disagreeable, and may very wisely be prescribed for children and pregnant women, and in cases of habitual costiveness.

Senna has the additional recommendation of not leaving the bowels confined after its purgative effects have passed away. This quality renders it particularly valuable when the bowels are constantly, though not seriously, confined, and when medicine, for the sake of relief, has to be constantly resorted to.

PREPARATIONS AND DOSE.—Senna, ℥ss.—℥ij. (2.—8.); Confectio Sennæ, ʒj.—ij. (4.—8.); Ext. Sennæ Fl., ʒj.—ij. (4.—8.); Infus. Sennæ, ℥j.—ij. (30.—60.).

BROOM (SAROTHAMNUS SCOPARIUS).

ACTIVE INGREDIENTS.—The ends of the young shoots and the leaves upon these parts have a bitter and nauseous taste, and, while fresh, yield a remarkable odor on being bruised. They contain two peculiar principles, one of which is neutral, the other a volatile liquid alkaloid. The neutral principle, called Scoparine, $C^{21}H^{22}O^{10}$, forms yellow crystals, soluble in water and alcohol, and is destitute of bitterness. Dr. Stenhouse believes it to represent the diuretic property of broom-tops. The alkaloid, called Sparteine, $C^{15}H^{26}N^2$, is pale when newly prepared, but on exposure becomes brownish, and has a bitter taste. It forms crystalline salts with bichloride of platinum, terchloride of gold, and corrosive sublimate, and is extremely poisonous, being little inferior in this respect to either conia or nicotia. Extractive matters and salts are likewise found on broom-tops, but they are of no importance.

PHYSIOLOGICAL ACTION.—Broom-tops have long been celebrated for their cathartic and diuretic powers, which are rendered available by infusion in water or in spirit. In small doses, the infusion is mildly laxative; in large ones it becomes emetic and purgative.

As already said, the diuretic effects of broom appear sufficiently accounted for by the ascertained powers of scoparine. The alkaloid sparteine is powerfully narcotic, but exists in such small proportions that it is not easy to judge whether it really takes any important part in the action of broom. Schroff killed a rabbit in six minutes with a single drop of sparteine which he placed in its mouth; there were violent tetanic symptoms. But Mitchell and Stenhouse gave four grains to a rabbit which lived for three hours afterward, and then died in stupor without convulsions. This difference of experimental results, and the fact that sparteine is little soluble in water, make it doubtful whether it goes for anything in the action of decoction of broom-tops.

THERAPEUTIC ACTION.—Broom has been employed in dropsies, and with considerable success. Cullen says he found it more trustworthy than any other diuretic.

But the value of the medicine is dependent of course upon the particular character of the dropsical effusion. In acute inflammatory cases, and when the kidneys are diseased, the use of broom might prove objectionable; and in cases of thoracic dropsy it is expressly declared to be so, especially when the dropsy is accompanied by pulmonary congestion, in which there is inflammation of the lungs, however trifling. In cardiac dropsies, on the other hand, broom is most useful. Whenever employed, diluents should also be freely used, with a view to assisting the action of the medicine. Cullen recommended that half an ounce of the fresh tops should be boiled in a pint of water until reduced to one-half that quantity, two tablespoonfuls of this decoction to be given every hour, until the whole had been taken, or, at all events, until the bowels were moved.

PREPARATIONS AND DOSE.—Infus. Scoparii (B. Ph.), ℥j.—iij. (30.—90.); Succus Scoparii (B. Ph.), ʒj.—iij. (4.—12.).

INDIAN KINO (Pterocarpus Marsupium).

ACTIVE INGREDIENTS.—Kino is inodorous, but has a very astringent taste, which ultimately becomes sweetish. Artificial heat does not affect it, but in the mouth it softens readily, turning the saliva blood-red, and when chewed, clinging to the teeth. In cold water it is partially soluble, in boiling water more so; and in alcohol it is largely or almost entirely soluble. The constituents are a peculiar kind of tannin, called mimo-tannic acid (or Catechu-tannic acid), $C^{13}H^{13}O^5$, and another astringent principle, called Catechine, probably isomeric with Catechu-tannic acid (described as a constituent also of Pale Catechu), with some red gum, and other unimportant ingredients.

PHYSIOLOGICAL ACTION.—Kino operates in the same manner as catechu, but the action is frequently less powerful, kino being less soluble than catechu, when taken into the system.

THERAPEUTIC ACTION.—Kino was introduced into the Edinburgh pharmacopœia in 1774, and into the London one in 1787. It is chiefly employed in obstinate chronic diarrhœa, and then often in combination with chalk or opium. It has, likewise, been found useful in leucorrhœa, and as a tonic in intermittents. As a topical astringent, it possesses a degree of value, and also for employment as a gargle.

PREPARATIONS AND DOSE.—Kino, gr. v.—xx. (.30—1.25); Tinct. Kino, ʒss.—ij. (2.—8.); Pulv. Kino, Co. (B. Ph.), gr. v.—xx. (.30—1.25).

BALSAM OF PERU (Myrospermum Peruvianum).

Active Ingredients.—In consistence this balsam is viscid, and like treacle, with a sp. gr. of 1.15. It is reddish-brown, or almost black; possesses an agreeable odor, somewhat resembling that of vanilla, and has a warm, acrid, and slightly bitter taste. It is inflammable, and burns with a considerable amount of smoke; is entirely soluble in alcohol, and in about 5 parts of rectified spirit; and undergoes no diminution when mixed with water.

Balsam of Peru contains cinnameine (cinnamate of benzyl, $C^{16}H^{14}O^2$) to the extent of about 70 per cent. This is a neutral, acrid, colorless, strongly-refracting oil, inflammable, of sharp taste, and slightly aromatic odor. It is heavier than water, and insoluble in that medium, but soluble in alcohol and in ether.

By the action of caustic potash, cinnameine is converted into cinnamic acid, which constitutes about 6.4 per cent. of the balsam; and a light and oily fluid called Peruvine. [Cinnamic acid may also be formed by exposing oil-of-cinnamon to the atmosphere.] Cinnameine is colorless and soluble in alcohol and ether; slightly, if at all so, in cold water.

About 24 per cent. of the fresh balsam consists of Resin-of-Peru, which is formed by the cinnameine uniting with the elements of water, and forming a hydrate of cinnameine. The resin increases in quantity with the age of the balsam, at the expense of the oil.

Physiological Action.—Balsam of Peru is stimulant, tonic, and expectorant, the action closely resembling that of storax and benzoin. It is not possessed, however, of the remarkable influence over the urinary organs which characterizes the produce of its near botanical ally, the Copaifera; while in tonic properties it is equally inferior to myrrh. Acting most decidedly upon the mucous membranes, it promotes digestion, causes warmth and excitement throughout the system, stimulates the pulse, and increases the secretion of the kidneys and the skin. In large doses it produces pain and oppression in the stomach, with nausea, vomiting, colic, and diarrhœa.

Therapeutic Action.—Balsam of Peru is employed in chronic bronchitis, asthma, gonorrhœa, gleet, leucorrhœa, and other disorders of similar character. Converted into a plaster, it is applied externally for the cure of headache and toothache. In its simple condition it is excellent for closing recent wounds, and for healing ulcers and other lesions of the body. It often proves a valuable external application also for sore nipples, in cases of which, the following preparation, used five or six times daily, rarely fails to be of service:—Balsam of Peru, 120 gr.; oil of almonds, 90 gr.; gum-Arabic, 120 gr.; rosewater.

Preparations and Dose.—Two to ten minims and upwards may be given in an emulsion of almonds, mucilage, or the yolk of eggs, with a little sugar.

BALSAM OF TOLU (Myrospermum [or Myroxylon] Toluiferum).

Active Ingredients.—The composition of Balsam of Tolu is very similar to that of Balsam of Peru, and, like the last-named, is chiefly represented in cinnameine, cinnamic acid, and resin. It is soluble in alcohol, ether, and the volatile and fatty oils. The soft balsam contains a large proportion of oil, but less acid than the dry and brittle kind.

PHYSIOLOGICAL AND THERAPEUTIC ACTIONS.— These are similar to those of Balsam of Peru.

PREPARATIONS AND DOSE.—Syrup. Tolutanus, ℥ ss.—j. (20.—40.); Tinct. Tolutana, ℨ j.—ij. (4.—8.); Tinct. Benz. Co. (external).

LOGWOOD (Hæmatoxylon Campechianum).

ACTIVE INGREDIENTS.—Logwood contains hæmatoxylin or hæmatine, $C^{16}H^{14}O^6$, which substance is sometimes found crystallized in the crevices; tannin also, resin, and the ordinary constituents of wood in general. The taste of this substance is slightly bitter, acrid, and astringent; it is soluble in alcohol and ether, and slightly so in water. Acted upon by alkalies, it gives a red or purplish color; by acids, a yellowish or red one. The coloring and the astringent principles of the wood itself are dissolved out both by water and by alcohol; the solutions are deepened in color by alkalies, and are rendered slightly turbid by acids.

PHYSIOLOGICAL ACTION.—Logwood, in decoction, is a mild astringent. It does not constipate, nor does it disorder the bowels, and its color can be detected both in the urine and the stools. The use of this medicine has been known to produce phlebitis.

THERAPEUTIC ACTION.—Logwood is usefully administered in dysenteries and diarrhœas of long standing, and is specially adapted to diarrhœa in children. It is also employed as an astringent in hæmorrhages from the lungs, the uterus, and the bowels, and has been found efficient in leucorrhœa. Dr. Percival used logwood to check profuse perspiration in phthisis.

PREPARATIONS AND DOSE.—Extract. Hæmatoxyli, gr. v.—xx. (.30—1.25). Decoct. Hæmatox., ℥ j.—ij. (30.—60.).

COPAIVA (Copaifera Multijuga, and other species).

ACTIVE INGREDIENTS.—The constituents which represent nearly the whole of copaiva are two, a resin, in the proportion of about 52 per cent., and a volatile oil, to the extent of 40 per cent. These proportions vary, however, with age and exposure. The resin (copaivic acid, $C^{20}H^{20}O^2$), is crystalline, and resembles common resin or pinic acid. The volatile oil is a colorless liquid, having the smell and taste of copaiva, and is isomeric with oil of turpentine, i.e., it consists of $C^{10}H^{16}$. In addition to these two principal constituents, there is about 2 per cent. of a soft brown resinoid matter, the nature of which is unknown. It is more abundant in old than in recent copaiva; and is soluble in anhydrous alcohol, ether, and the oils, both fixed and volatile.

Copaiva itself becomes darker in color, and more dense, with age and exposure to the atmosphere. Like other oleo-resins it is insoluble in water, but soluble in alcohol, ether, and oils, also in an equal volume of benzol. With alkalies it forms a kind of soap, insoluble in water.

PHYSIOLOGICAL ACTION.—The most complete inquiries as to the operation of copaiva, when taken into the system, are undoubtedly those of Ricord. He shows that the skin, the bronchial tubes, the digestive organs, and the whole of the mucous surfaces, are affected by it; the nervous centres also, occasionally; and the urinary apparatus most markedly and uniformly. The action upon the digestive organs is manifested in its causing heat in the pit of the stomach; in its lessening the desire for food, and giving rise to nausea and wretching, or eructations that possess the peculiar taste and odor of the medicine. When copaiva cannot be tolerated by the stomach, purging of the bowels often follows; the alvine secretions are sometimes bloody, and attended by mucous discharge, also by a sense of burning in the region of the sphincter ani, and by violent tenesmus.

The influence upon the urinary organs is shown by diuretic effects, and by the changed quality of the secretion, which acquires a deeper color, a certain degree of bitterness, and a slightly balsamic odor, or, at times, the smell of violets. This peculiar odor is perceptible, likewise, in the stools. After standing a little while, the urine of persons who have taken copaiva often presents, also, a filmy covering, or delicate pellicle, which is apt to be iridescent; the suspension in it of a certain resinous matter produces, at the same time, a very manifest turbidity. Should the dose have been unduly large, micturition becomes too frequent, and is preceded and followed by itching, smarting, and burning, in the urethral passage, and during its progress is accompanied by heat and tenusmus, and even by hæmaturia and ischuria. The urine deposits, likewise, a sediment resembling albumen, and which properly consists of vesical and urethral mucus.

[There is little doubt that the resin of copaiva is the efficient ingredient of the balsam, especially for diuretic purposes, a fact so little appreciated by the profession, that a distiller of the oil some time since informed the editor that finding no market for the resin, he used it as fuel in his factory.]

While these local results appear, the whole system becomes excited. The pulse is quickened, and becomes fuller, and thirst and headache supervene.

The influence thus exerted upon the urine and the urinary organs proves that copaiva enters the general circulation. This is further indicated by its tainting the breath, and by its action, eruptions being produced which resemble urticaria and roseola, or, in some cases, they are like rubeola. These symptoms, however, are said to be concomitant only with imperfect medicinal action of the drug, or to appear when the system generally is repugnant to it. Even in moderate doses, the action of copaiva upon the mucous membranes, especially those of the urethra, is very striking; and it may be added that among the early symptoms of its operation are increased flow of saliva, and a flatulent rumbling in the bowels, followed by colicky pains and a desire to vomit.

The effects of copaiva upon the respiratory tract are shown in its producing irritation in the larynx and the bronchi; dryness also, in the larynx, huskiness in the chest, and dry and painful cough, in connection with which there is expectoration of a semi-purulent, greenish, and nauseously smelling mucus.

And lastly, in regard to the influence of copaiva upon the nervous system, it is said that the continued use of it has been followed by epileptic convulsions.

The physiological action of the oil of copaiva is nearly similar to what has been described, but it is not so powerful. Micturition is increased; the stools become frequent and watery, and sometimes bloody; the breathing is hurried; there is palpitation of the heart; and, usually, restlessness. Eventually the medicine is excreted by the customary channels.

THERAPEUTIC ACTION.—The great and special use of copaiva is in the chronic stages of gonorrhœa, also in gleet and in fluor albus.

Women who may be suffering from gonorrhœa are less successfully treated with it than men, since in woman gonorrhœa is not confined to the mucous lining of the urethra, but extends to that of the vagina, whence it is obvious that the disease cannot be benefited to the same extent by the local action of the medicine, which, under all circumstances, is its special and paramount recommendation. Copaiva has, likewise, been employed with success in chronic inflammation of the bladder; also, according to Armstrong (1818) and other writers, in leucorrhœa.

Favorable mention is also made of copaiva in cases of **chronic pulmonary catarrh**. But it is only adapted for long-standing cases and for torpid constitutions.

In **chronic inflammation of the mucous membrane of the bowels**, especially of the colon and of the rectum, copaiva has again proved a valuable agent.

Cullen, in his Materia Medica, further commends it for use in **hæmorrhoids**.

Formerly, again, copaiva was used as an application to **wounds and ulcers**.

In the treatment of gonorrhœa—the complaint for which copaiva is so popular a remedy—two methods of exhibition are resorted to. The first is to give the medicine at the very commencement of the disorder, so as to arrest or suppress it; the other is to delay the use of copaiva until the inflammatory symptoms have subsided. The first method is that employed (at all events, to some extent) in America. In Europe, though it has been followed by some practitioners for nearly seventy years, it is disregarded for the sake of the second. This second method is commenced with antiphlogistic and soothing treatment, and when the inflammation has much abated, the copaiva is given with a view to reducing or stopping the discharge. Hunter, Sir Astley Cooper, and Lawrence, all observed this latter method, and there can be no question that it is the safest; for, independently of the usefulness of the preliminary antiphlogistic treatment, instances have occurrred when copaiva given in the early stages of gonorrhœa has actually aggravated the symptoms.

In administration, Dr. Chapman recommends that the copaiva be poured upon half a wineglassful of water, to which is then to be added a small quantity of some bitter tincture. By this means the copaiva is collected into a small globule which may easily be swallowed, the taste, so nauseous to many patients, being entirely masked.

Ordinarily, copaiva is prescribed in the form of emulsion, in doses of from half a drachm to a drachm thrice a day.

It may also be taken upon sugar, and in this way is more disposed to act upon the urinary organs.

Combined with the *liquor potassæ*, its effects in the last stage of gonorrhœa are much increased.

PREPARATIONS AND DOSE.—Capaiba x.—xx. (.60—1.20); Oleum Copaibæ ℥ ss.— ℥ j. (2.—4.); Resina Copaibæ (not officinal), gr. ij.—v. (.12—.30). Unless thoroughly triturated with milk sugar, the resin is apt to pass the bowels unchanged.

CANELLACEÆ.

CANELLA (Canella Alba).

ACTIVE INGREDIENTS.—By distillation, the bark yields a warm and aromatic oil. This is a mixture of eugenic acid, $C^{10}H^{12}O^2$, with two neutral oxygen-containing oils.

PHYSIOLOGICAL ACTION.—This is not at all accurately known.

THERAPEUTIC ACTION.—In England, Canella bark is principally employed as an aromatic adjunct to tonic and purgative medicines. It has also been employed against scurvy, but is not an important remedial agent. Combined with aloes, it forms a popular remedy well known by the name of "hiera picra." In tincture of senna, it might be advantageously substituted for cardamoms.

PREPARATIONS AND DOSE.—Pulvis Canellæ, gr. xv.—xxx. (1.—2.); Pulvis Aloës et Canellæ, gr. x.—xx. (.65—1.25); Vinum Rhei, ℥ j.—ij. (4.—8.).

VITACEÆ.

THE GRAPE (Vitis Vinifera).

ACTIVE INGREDIENTS.—The skin of grapes is properly rejected in eating them: the remaining contents are juice, semi-solid matter, and pips, all of which should be consumed. The important ingredients of grapes are water, sugar (both grape and fruit), gum, tannin, azotized matters, bitartrate of potash, sulphate of potash, tartrates of lime, magnesia, alum, and iron; chlorides of potassium and of sodium, tartaric, citric, racemic, and malic acids.

The relative proportions vary considerably, more especially those of the sugar and of the acids: and these differences of course have important practical bearings. The variations in the proportion of albuminoid matter also appear to be of some consequence.

PHYSIOLOGICAL ACTION.—After what has been said of the complicated chemistry of the grape, it must be evident that a complete account of its physiological action is impossible. Certain leading facts, however, are quite established.

In the first place, the large amount of water in the juice of the grape renders this fruit highly diuretic when eaten in quantities. The action is much assisted by some of the other ingredients, particularly the vegetable acids (which are burnt in the organism, and pass off in the urine as alkaline carbonates), and the salts of potassium, sodium, etc.

With less constancy, but still with much frequency, grapes act in a laxative or even a decidedly purgative manner; and if this irritant effect be carried to excess, particularly in children and delicate persons, there may be excoriation of the tongue, chronic diarrhœa, and an aphthous condition of the whole alimentary canal.

Again, in the early stages of the treatment, grapes are sometimes found to produce excitement of the heart and circulation, but this effect is usually only temporary.

Grapes are undoubtedly nutritious. The small amount of albuminoid matter does not count for much in this way, but is far from valueless; the sugar, on the other hand, is doubtless highly useful, not merely as a combustion-food, but for storing up, in the shape of unburnt excess, as fat, and possibly in other forms of tissue.

THERAPEUTIC ACTION. — This subject, though important, is strangely neglected in most English works on therapeutics. The "grape-cure" is certainly not an exploded quackery, but an important instrument in the hands of the medical man, as testified by a large number of excellent observers. At the present day there are numerous grape-countries where, during the vintage, sick persons are treated on the plan of eating three to six lbs., or even much more than this, per diem. The principal places are Meran, in the Tyrol; Vevay, on the Lake of Geneva; Dürkheim, in Bavaria; Bingen, Rüdesheim, Kreuznach, St. Goar, etc., on the Rhine banks; Grünberg, in Silesia; Aigle, in Savoy. No doubt, like every other cure, the grape-treatment has often been indiscriminately and mischievously applied; but it is a powerful remedy in appropriate cases.

The remarks made on the physiological action of the grape indicate the probability of there being at least two ways in which the treatment may be efficaceous. In fact, according as the grape contains much sugar and little acid, or *vice versâ*, there will be a preponderance either of the so-called "alterative" purgation, or of the alimentary and tonic effects.[1]

Abdominal Plethora.—The conditions which are especially benefited by the laxative action of the grape-cure are those which are somewhat vaguely grouped under the denomination of "abdominal plethora." There are numerous cases in which the biliary secretion is inactive, and the digestion feeble, in consequence of sluggish portal circulation. In these instances it is proper to administer grapes that are not over-ripe, and, as a rule, the comparatively non-saccharine kinds. The same course should be adopted with women suffering from abdominal congestion which has given rise to amenorrhœa, with its accompanying train of symptoms—headache, vertigo, oppression, and palpitation of the heart. In not a few

---

[1] This is well explained by H. Carchod, in his "Cure de Raisins," Paris: J. B. Baillière et fils, 1860, to which the reader may also be referred for lists of nearly all the important works on the grape-cure.

cases it will be found (especially where there has been very marked engorgement, and even a tendency to solid enlargement of the liver or of some other abdominal viscus), that the grape-cure is not a sufficiently energetic remedy to commence with, but that it comes in very effectively after a preliminary course of powerful mineral waters and other medicinal agents. A case which came under the observation of Dr. Bezencenet *père* is in this respect very striking. A middle-aged woman suffered from marked hypertrophy of the liver, for which disorder she had long been treated, when, being in poor circumstances, it was proposed by a benevolent society to contribute funds for her comfort. Dr. B. begged that the money might be applied toward the necessary expenses of a grape-cure, and accordingly the patient was supplied for six weeks with five pounds of grapes daily. At the expiration of that time the engorgement of the liver and all the accompanying morbid symptoms had completely disappeared.

Another case of Dr. B.'s shows what the grape cure may do, even unaided, for the removal of serious hepatic engorgement. A young lady was tapped for ascites; after the operation the liver was found to reach as low as the umbilicus, and was very hard. Being a homœopathist, she would not take ordinary drugs; she was persuaded to adopt the grape-cure, and in a few weeks the liver was completely righted. She married, and bore children, and had no return of hepatic mischief until twenty-five years later.

Chronic Catarrhs of the Mucous Membranes are frequently benefited by the grape-cure. Dr. Cossy reports a typical example of this action, in relation to pulmonary catarrh:—A man aged 30, previously of good health and appearance, got a cough, at first dry, and then accompanied by much greenish-yellow expectoration; no hæmoptysis occurred, but night-sweats, fever, and dyspnœa to a considerable degree came on, and he was much weakened, so that suspicions of phthisis naturally arose. On examination of the chest (eleven months after the appearance of the first symptoms) the physical signs were those rather of pulmonary catarrh than of phthisis. The grape-cure was persevered in for five weeks, and at the end of that time the man was substantially cured.

The mucous catarrhs of the intestines, and even true dysentery, appear to be greatly benefited by the use of grapes. The authoritative names of Pringle, Sydenham, and many others, may be quoted in support of this statement.

Chronic catarrhs of the urinary mucous membrane are also said to be much amended by the grape treatment.

Dyscrasic Maladies, so-called, are sometimes most happily modified by the grape cure. Among these, skin-disorders of the eczematous, ecthymatous, and impetiginous classes, linked to the herpetic or the rheumatic diathesis, have proved most amenable. As regards the phthisical diathesis, the evidence seems much more doubtful; and it may be questioned whether the supposed cures (other than those of mere pulmonary catarrh) were not effected rather by change of climate than by the specific treatment.

As a Tonic Mode of Treatment, adapted to convalescents from acute diseases and the effects of hæmorrhages, and from various states of debility, there seems ample evidence that the reconstituent powers of the grape-cure are often very efficacious. The digestion is strengthened, the appetite is increased, and the patients make a very considerable amount of flesh. For these purposes the sweeter grapes are preferable.

It would be impossible to enumerate all the diseases which have been confidently said to be curable by means of grapes; the above are those respecting which there is trustworthy evidence of real benefit having resulted. At the same time, a few words must be said as to the accompanying diet. Formerly, especially by Germans, it was recommended to keep the patient almost entirely without other food; but except in a very few cases, in which something like a starvation-cure might be necessarily expected to do good, this plan is not a sound one. It is possible that abdominal engorgements, due solely to over-eating, and to intemperance in drink, might be greatly benefited by a short course of almost exclusive grape diet. But proper nutrition for any extended length of time cannot be maintained in this way, and for delicate subjects it is altogether improper. Moderate meals, with Bordeaux wine in moderation, may legitimately be allowed.

It should be added that there are persons who do not well bear the grape-cure. Women do not bear it so well as men, and children least satisfactorily of all.

LILIACEÆ.

SQUILL (Urginea Scilla).

Active Ingredients.—When prepared for employment in medicine, the bulb of the squill is cut into slices. These, clammy while recent, become, when dried, brittle and slightly translucent, and are easily pulverized; but, if exposed to the air, they recover moisture, and become flexible. The scent is feeble; the taste is disagreeable, mucilaginous, strongly bitter, and somewhat acrid.

Squill is said to yield its active principles alike to water, to acetic acid, and to alcohol.

It must be allowed that the active ingredients of squill are not yet thoroughly understood. A bitter stuff, scillitine, soluble in hot water, and a resinous, very acrid body, soluble in alcohol, which is probably the same as that mentioned by Maudet under the name of Sculline, unquestionably divide between them the powers of squill, but there are great contradictions among different experimenters. It is still quite doubtful whether scillitine is or is not the diuretic principle, and also whether (as Schroff thought) it is also a direct narcotic poison in large doses. It seems, at any rate, evident that the resin is the source of the phenomena of irritant poisoning exhibited in inflammation of the alimentary canal, &c. Whether also diuretic, is not decided.

Physiological Action.—The principal action of squill appears to be exerted upon the lining membrane of the excretory organs, particularly the bronchial, the urinary, and the gastro-intestinal. Upon the urinary apparatus the operation is very marked. As a rule, it produces, in the first place, some degree of strangury, and, as a secondary effect, a profuse secretion of water. Should it fail to act upon the kidneys, it often causes perspiration. In full medicinal doses, squill excites nausea and vomiting, and not unfrequently, purging; and in excessive doses it becomes poisonous. Considered with reference to its diuretic powers, squill is sometimes compared with digitalis; but in its primary action it is more stimulating to the kidneys, while, in regard to the general tonic and sedative effects of foxglove, it is less energetic.

THERAPEUTIC ACTION.—As a Diuretic, squill is a valuable medicine, and is administered either in the recent state or in the dry. The dose of the former is from 5 to 15 grains; and of the latter, from 1 to 3 grains; the smaller quantity being used at first, morning and evening, in the form of a pill, and the dose being gradually increased until the diuretic effects have come to pass. By some practitioners it has been recommended to be given in an amount sufficient to induce nausea; but this is very hurtful to the patient, as well as distressing, and, when practised, frequently compels the discontinuance of a medicine of undoubted utility, since, if once the stomach rebels, it is seldom that squills can be again administered. Combined with mercury, the effects of squill, as a diuretic, are sometimes materially increased. The combination in question is particularly adapted to cases of dropsy which depend upon enlargement or other affections of the liver, or which are directly connected therewith. When the mercurial preparations induce purging, the diuretic action of the squill will be suspended. This result must be obviated either by substituting friction with mercurial ointment, or by the employment of counteracting medicines. [A favorite diuretic pill of Prof. A. Clark, of New York, is composed of one grain each of squill, digitalis, and calomel.]

As an Expectorant, squill is employed most beneficially in cases where there is an increased secretion of pulmonary mucus, and is supposed to operate by promoting absorption, diminishing the quantity of fluid effused, and thus facilitating the expectoration of the remainder.

In Whooping-cough, particularly when attended by troublesome emesis, and although the effects of the drug as an emetic are sometimes distressing, we are still called upon to employ squill, seeking, if possible, to keep its emetic action aloof, and endeavoring to secure the full exercise of its tonic power. In coughs, with tickling in the throat, it is best to employ either syrup or vinegar of squill.

In Chronic Bronchitis, when occurring in debilitated patients, and attended by very profuse loose expectoration of a mucous or muco-purulent character, I have always resorted to squill with great success.

PREPARATIONS AND DOSE.—Pulvis Scillæ, gr. ss.—ij. (.03—.12); Acetum Scillæ, ♏v.—xx. (.30—1.20); Syrupus Scillæ, ʒ ss.—j. (2.—4.).

ALOES (ALOE VULGARIS, and ALOE SOCOTRINA).

ACTIVE INGREDIENTS.—The resin of the vulgaris, when pulverized, is of a dirty yellow color, always presenting a dull appearance; even thin laminæ are very opaque. The odor is excessively disagreeable, and is intensified by breathing upon the lump, which in fracture is imperfectly conchoidal. Proof spirit dissolves nearly the whole. Socotrine aloes give a resinous or vitreous fracture; thin layers are sometimes translucent; the odor, instead of being repulsive, is aromatic; when powdered, the color is bright yellow, and in proof spirit the solution is complete. No difference is perceptible in the bitterness of the two varieties.

The most important constituent of aloes of either kind appears to be aloine, $C^{17}H^{18}O^7$ a neutral and very bitter substance, which crystallizes in needles, is insoluble in cold alcohol, and very sparingly soluble in cold water. In either fluid, however, when warmed, it is readily soluble; and if heated to 212°, it rapidly oxidizes and decomposes. Aloine is likewise soluble in alkaline fluids, forming with them a yellow solution which gradually darkens in color.

Aloes also contain a resinoid substance, which differs from all ordinary resins in being soluble in boiling water, and which is formed probably by atmospheric action upon the aloine. Aloetic acid likewise occurs, striking, with the persalts of iron, an olive-brown. The action of nitric acid upon aloes gives rise to the acids named polychromic, chrysammic, and chrysolepic, all of which are crystalline substances. The colors of their solutions are red and purple.

PHYSIOLOGICAL ACTION.—By whatever means introduced into the system, aloes have a laxative action. Whether swallowed, or injected by the rectum, or injected subcutaneously, or rubbed into the skin, the operation is the same; and it may be added that nursing mothers who take aloes communicate the laxative effect to the infants that imbibe the milk. Cullen believed that the specific action of aloes was exercised solely upon the colon and rectum and that the drug was a simple evacuant of the fæces. Wedekind, on the other hand, held that the primary action is to increase the secretion of bile, and that the purgative or laxative effects are secondary. I lean to the opinion that Wedekind was right; and I further believe that it is the secondary action which excites the power of the muscular coat of the colon and of the rectum; and that it is by this increased muscular action that the fæces are expelled, rather than that the expulsion takes place through the hyper-secretion which is thrown out from the mucous surfaces of the intestines mentioned. Aloes, no doubt, cause augmentation of the mucous secretion of the colon and rectum; but what we have carefully to remember is that the good effects of the medicine are owing, as before stated, not to this simple augmentation of mucus, but to the prior increase in the flow of the bile,[1] and the excitation of the action of the muscular coat.

When aloes are resorted to as an habitual laxative they produce dryness of the throat, an unpleasant sense of warmth throughout the abdomen, with uneasiness amounting almost to pain in the hepatic region, and a tightness and throbbing in the right hypochondrium. The whole of the pelvic viscera are brought into a more or less engorged or congested condition, and in the portal system there is likewise a tendency to congestion. These symptoms are accompanied by heat, irritation, and tenesmus in the rectum, often with subsequent hæmorrhoids; the bladder becomes irritated; the urine hot and scanty. In women the menstrual secretion is, under ordinary circumstances, greatly augmented; and when the secretion is absent, or nearly so, the natural flow is restored. Aloes also have the property, when taken in small and repeated doses, of increasing sexual desire (Stillé). The pulse, as a rule, is slightly quickened.

Administered as a medicine, a dose of one to five grains of aloes acts upon the bowels in from eight to twelve hours. A dose of this description does not produce nausea or any general disturbance, although usually

---

[1] It is right to mention that several recent observers, of good repute, deny this action on the liver; and Nothnagel says there are "no convincing proofs of it." I nevertheless hold to the above opinion, from my own observation of the character of the stools.

causing irritation, and a sense of warmth throughout the abdomen. The evacuation of the bowels is generally attended by griping pains; the stools are copious, but not too abundant; there is usually only one, or at all events there are not more than two. They are fæculent, and of a dark brownish yellow color, and have a characteristic odor. The number of motions depends more upon the condition of the bowels, *i.e.* to what degree they are loaded, than upon the strength of the pill; for it is well known that a single grain of aloes often acts as powerfully as five grains will.

THERAPEUTIC ACTION.—Aloes, in small doses, operate as a warm and stimulating purgative, particularly adapted to the temperament called the melancholic. They also exert a tonic power, and hence have a very beneficial influence in chronic affections of the stomach and bowels, such as loss of appetite, flatulence, constipation, and others usually denominated dyspeptic. The operation is slow, but generally effectual; and as before observed, large doses do not appear to exert any greater influence than small ones. Acting more particularly upon the colon and the rectum, aloes are efficacious also in expelling ascarides; but for the same reason they are said to occasionally produce hæmorrhoids: this idea has, however, been rejected by many good observers, who maintain that the aloes never caused the piles, *which existed already;* and on the contrary, aloes often cure piles by removing constipation. In dysenteric diarrhœa, verging into the chronic stage, and attended by tenesmus, I have used aloes with excellent effect; they probably do good by substituting regular and rhythmical peristaltic action for the spasmodic condition which produces tenesmus and prevents the bowel being effectually evacuated. Aloes are well adapted for use in jaundice, especially when there is hypochondriasis.

Chronic Constipation is often very effectively dealt with by means of aloes; and there is a particular plan introduced by Dr. J. K. Spender, of Bath, which deserves notice. Dr. Spender gives a pill containing one grain watery extract of aloes and two grains sulphate of iron; this is administered, at first three times, then twice, and then once daily. We must not be impatient in administering aloes in this way; it may take some days, or even two or three weeks, to produce a decided effect, but the desired result is usually attained in the end.

As an Emmenagogue, aloes have had a great reputation, but have often been indiscriminately employed in a routine way which is not justified by what we know of the physiological action. Given at chance times, aloes (as Graves pointed out) are no more likely to restore defective menstruation than any other ordinary purgative. Indeed they are likely to do special mischief; inasmuch as they certainly do tend more than other aperients to engorge the pelvic viscera, and it is very undesirable that this should be done except at stated times. Graves points out that in nearly all cases of amenorrhœa there are abortive efforts at the performance of the function, which may be traced in the periodical recurrence of pain in the loins, thighs, and hypogastric regions, flushing, etc. It is at these times, only, that we should administer such substances as cause a direct flow of blood to the uterus; at any other times the congesting influence will only do harm, because the ovarian disposition to menstruation does not exist. Pregnant women should beware of using aloes as an aperient, though it is doubtless true that aloes have sometimes cured piles in pregnancy, by removing constipation.

The medicinal effects of the two kinds of aloes, the Barbadoes and the

Socotrine, differ but little. The Barbadoes are said to be more energetic, for which reason they are used by veterinary surgeons, who find the action to take place upon horses in from twelve to twenty-four hours.

Preparations and Dose.—Aloe, gr. ss.—v. (.03—.30); Pilulæ Aloës, No. 1—3; Pil. Aloës et Assafœtidæ, No. 1—3; Pil. Aloës et Mastiches, one after meals. Pil. Aloës et Myrrhæ, No. 1—3; Pulvis Aloës et Canellæ, gr. i.—v. (.06—.30); Tinct. Aloës et Myrrhæ, 3 ss.—j. (2.—4.); Vinum Aloës, 3 j. (4.); Suppositoria Aloës.

ROSACEÆ.

THE ALMOND (Amygdalus Communis).

Active Ingredients.—The almond yields by expression a considerable quantity of oil, which, when compared with olive oil, is found to be more pure, and less liable to become rancid. An essential oil is likewise procurable by distilling almond-water with barytes, so as to separate the prussic acid. In close vessels it is very volatile; exposed to the air it becomes solid, crystalline, and inodorous.

Physiological Action.—Sweet almonds, as the name would imply, have a bland and agreeable flavor. They are nutritive and emollient, but somewhat difficult of digestion through containing so much oil. They have been known to cause nausea and urticaria. Almond oil (usually expressed from the bitter variety, on account of its cheapness, and the higher value of the residual cake) possesses the dietetic and medicinal properties of the other fixed oils, and is, in its local action, emollient. Taken in large doses it acts as a mild laxative. It is also demulcent.

Cases are recorded of alarming symptoms having been produced in the human subject through eating bitter almonds, and others in which death has occurred. Orfila relates the cases of two children, in whom the symptoms of poisoning by these almonds were very marked and rapid. Within a quarter of an hour after eating them, pallor and collapse of the features set in; the pupils became dilated; the respiration sighing, with somnolence and muscular relaxation.

In another case, a man fell down dead very soon after partaking freely of the almonds. He frothed at the nose and mouth, and the eyes became fixed and glistening.

An eruption resembling urticaria often comes out on the skin from the effects of taking bitter almonds.

The volatile oil of bitter almonds is a deadly poison. Taylor says that in one hundred parts of this oil there are nearly thirteen parts of anhydrous prussic acid, one drop of which has been known to kill a cat.

A case is reported by Mertzdorf of a man who swallowed two drachms of the essential oil of bitter almonds; his features became spasmodically contracted; his eyes fixed and upturned, and starting from his head; his breathing jerking and hurried; and death followed in half an hour. On the other hand, several cases are quoted by toxicologists in which large quantities have been taken without fatal results.

Therapeutic Action.—Dr. Pavy has proposed, as a substitute for bread or starchy food for patients suffering from diabetes, cakes made of sweet almonds.

The "mistura" is useful in cough, and as a lotion to allay itching of the skin. It was once a favorite vehicle for the administration of tartarized antimony, in doses of $\frac{1}{8}$ gr., to subdue inflammatory action of the lungs, and to relieve cough. As a demulcent and emollient in pulmonary affections, it is certainly good, as well as in inflammatory affections of the alimentary canal and of the urinary organs.

Almond-oil may be employed for the same purposes as olive-oil. Combined with an equal volume of syrup of roses, or syrup of violets, it is a suitable laxative for infants.

It is useful also in the preparation of emulsions, and of certain kinds of linctus.

Bitter almonds, combined with decoction of cinchona, have been extolled as a remedy for intermittent fever.

Six or eight blanched almonds are said to relieve heartburn.

Preparations and Dose.—Aqua Amygdalæ Amaræ, ʒ ij.—iv. (.8—15.); Oleum Amygdalæ Amar., ♏ $\frac{1}{4}$—$\frac{1}{2}$ (.01—.03); Mistura Amygdalæ, ʒ j.—iv. (4.—15.); Syrupus Amygdalæ, ʒ j.—iv. (5.—20.); Oleum Amygdalæ Expressum, ʒ j.—iv. (4.—15.)

CHERRY-LAUREL (Cerasus Lauro-cerasus).

Active Ingredients.—The leaves possess a bitter and somewhat acrid flavor, and the odor evolved, when they are bruised, is offensive and characteristic. Drying dissipates the odor, but the taste remains, and then exhibits a certain amount of astringency. Their most important constituent is amygdaline, the formula of which, according to Wöhler and Liebig, is $C^{20}H^{27}NO^{11}$; a substance identical with the glucoside of that name which is found in bitter almonds, although not crystalline, like the latter, but amorphous. Amygdaline is capable of a variety of chemical transformations, but the most interesting is that which it undergoes when mixed with certain albuminous matters and with water, which is also present in the almonds. In bitter almonds it unites, under these circumstances, with emulsine, and the immediate result is the formation of hydrocyanic acid, bitter almond oil, sugar, and (probably) formic acid. A similar kind of transformation takes place when laurel leaves are distilled with water, although it is not known what is the albuminoid body that here performs the functions of the emulsine in bitter almonds. The consequence is that laurel-water contains a small (but unfortunately a varying) proportion of prussic acid, and, so far as is known, this is the ingredient which alone gives it power to affect the system.

Physiological Action.—The leaves and the kernels of the fruit possess poisonous properties. "Laurel-water," prepared from the former, is similar in action to prussic or hydrocyanic acid, being capable of inducing sudden insensibility, and even of producing death within a few minutes. The parts of the body particularly affected by the swallowing of an overdose are the brain and the true spinal system. The cerebral affection is indicated by pain in the head, insensibility, and coma; and disorder induced in the spinal system has corresponding evidence in tetanic convulsions.

Taken in non-poisonous or medicinal doses, laurel-water occasions nausea, accompanied, perhaps, by vertigo, and pain in the stomach; the first effects being increased secretion of saliva, irritation in the throat, nausea (very frequently), disordered and laborious respiration, pain in the head, giddiness, obscured vision, and sleepiness. If the dose be increased, there is decided vertigo, faintness, and perhaps rapidity of pulse, with sickness, and a sense of constriction at the præcordia; while a dose that stops short only of being fatal, occasions insensibility and extreme feebleness of the action of the heart.

Swallowed in poisonous quantity, other symptoms quickly show themselves. The respiration becomes difficult and spasmodic; the pupils are usually dilated, though sometimes contracted; the pulse becomes small or even imperceptible; and presently there are tetanic convulsions, with insensibility, and death quickly follows. The proximate cause of death, in most cases, is obstructed respiration; but in some instances it is referable to stoppage of the action of the heart.[1]

Regarded as a medicine, cherry-laurel water, like prussic acid, is said to be narcotic, sedative, and anti-spasmodic, but it does not possess the power of lessening pain in general; nor has it the property of inducing sleep in a direct manner, in the way, for instance, that is effected by opium. Neither, again, has it the power of controlling the pulse, such as we find possessed by digitalis.

It is remarkable that the pulp of the fruit is not only free from the poisonous properties of the leaves and kernels (as happens also with the fruit of the yew-tree), but is wholesome and palatable.

THERAPEUTIC ACTION.—Laurel-water may be employed in all cases where it is customary to resort to prussic acid; and emphatically when there is need of a sedative narcotic. The results of the administration are rendered uncertain, however, by the varying strength of the medicine, the energy of which is greatest when moderately young leaves are employed, or when the water has been prepared quite recently from dried ones. It is a useful remedy in gastralgia, pyrosis, hiccough, and some other forms of dyspepsia. Palpitation, depending upon dyspepsia, is, in common with the other symptoms, greatly mitigated by the use of it. The powers of this medicine are again very conspicuously shown in cases of vomiting unconnected with inflammation of the stomach. Frequently, as with pregnant women, it has been found to arrest, at the first few doses, vomiting which had existed for several weeks, and which arose merely from morbid irritability.

As with prussic acid, laurel-water will not, however, mitigate pain felt in the intestines. Not being a general anodyne, its employment for the relief of pain is proved by experience to be often useless; though the attacks of pain which occur in angina pectoris, and which, in situation and course, so closely resemble gastrodynia, are often more quickly alleviated by laurel-water, or by prussic acid, than by any other agent at command. It must be added that laurel-water sometimes very quickly allays and even removes *tic-douloureux*.

Hooping-cough, etc.—I have frequently employed laurel-water, with

---

[1] The most interesting case of fatal poisoning with laurel-water recorded is that of Sir Theodosius Broughton, who was murdered in this way in 1781.

advantage, in hooping-cough, and in the affection termed "spasmodic cough," in both instances as a substitute for prussic acid; also, in cases of inflammation of the chest, after the subsidence of the acute symptoms. In repeated catarrh, and in chronic bronchitis, small doses of the same medicine often prove eminently serviceable; it will also relieve some of the more distressing symptoms of phthisis, and is valuable therefore in its competency to afford temporary comfort.

Prurigo, etc.—In prurigo, impetigo, inveterate psoriasis, and other cutaneous affections, attended by severe itching and tingling, laurel-water again affords great relief to the patient. The powdered leaves of the tree, mixed with flour or linseed-meal, have been employed as a poultice for ulcers.

But for the procuring of sleep, for the relief of pain in general, or as a remedy for diabetes, laurel-water, like prussic acid, though often recommended, will be found a poor substitute for other drugs.

PREPARATIONS AND DOSE.—Aqua Lauro-cerasi (B. Ph.), ♏v.—xxx. (.30—2.). The strength of this preparation being very uncertain, caution is required in its use. The Codex directs that the "Eau distillée de Laurier-cerise" shall be adjusted by analysis so as to contain .05 per cent. of hydrocyanic acid.

THE PLUM (Prunus Domestica).

ACTIVE INGREDIENTS.—Prunes contain a large amount of water, with about 20 per cent. of solid matter. The latter consists of malic acid, gum, sugar, pectin, lignin, and other unimportant substances, with a purgative principle, the exact nature of which has not yet been accurately determined.

PHYSIOLOGICAL ACTION.—Prunes, like most other kinds of plum, are gently laxative, but if eaten in excess, they are apt to occasion flatulence, griping, and diarrhœa.

THERAPEUTIC ACTION.—Employed as an article of diet by persons of costive habit, prunes often induce beneficial effects. Hence, too, they constitute an ingredient of the *Confectio Sennæ* of the Pharmacopœia. They are also considered useful during convalescence after febrile and other disorders.

PREPARATIONS AND DOSE.—Confectio Sennæ, ʒi.—ij. (4.—8.); A favorite laxative with many ladies is made by stewing an ounce each of prunes and senna with a pint of water. The requisite dose is adjusted by trial.

KOUSSO (Brayera Anthelmintica).

ACTIVE INGREDIENTS.—It still remains doubtful whether the crystalline, acidly reacting substance, koussine, is the only active element in kousso; or whether a peculiar resin, a volatile oil, and two species of tannin, which it contains, are also poisonous to intestinal worms. So far as I know, the only positive evidence is in favor of koussine, $C^{26}H^{44}O^5$, which occurs in very minute white crystals that have a bitter and irritant taste. It is very insoluble in water, easily soluble in alcohol; concentrated sulphuric acid dissolves it yellow, and, if water be added to this solution, white flocky masses are separated. The alcoholic solution gives a brown precipitate with perchloride of iron, and a grayish yellow one with acetate of lead.

PHYSIOLOGICAL ACTION.—Kousso produces no observable effect on the human body, excepting occasional sickness from the nauseousness of its taste. But it is a perfect and direct poison to intestinal worms, especially tape-worms. And as regards the active ingredients, it may be mentioned that koussine has been successfully employed by various physicians, in a dose of about 20 grains, against tape-worm; this sometimes causes vomiting and diarrhœa, and it has been found needful to give it in three or four divided portions, with some aromatic substance added.

THERAPEUTIC ACTION.—As a vermicide, against both varieties of tape-worm, and to a less extent against round worms, kousso has received the support of very large and varied experience. So long ago as 1850 the *Lancet* gave details of its successful employment by several London physicians: and in Switzerland, where tapeworm (especially Bothriocephalus latus) is very common, from the habit of eating certain meats raw, the value of kousso has been well established. It has also been freely used both in France and Germany, though in the latter country it is not much employed at present. Its nauseous flavor is certainly an objection, but the same defect belongs to the Filix-mas, the value of which no one of any experience could now permit to be disputed on that ground.

PREPARATIONS AND DOSE.—For administration to an adult, about one oz. of the powdered flowers should be mixed with lukewarm water, and allowed to infuse for a quarter of an hour. A little lemon-juice is then swallowed by the patient, and subsequently the infusion, flowers included, is added to some tea without either milk or sugar, a short interval being allowed after the lemon-juice. Should the worm not come away in three or four hours, a castor-oil or saline purge should be administered, kousso being in itself scarcely at all purgative. The dose of koussine has been already mentioned (20 gr.); its taste may be covered by peppermint.

THE WILD CHERRY (Prunus Virginiana).

ACTIVE INGREDIENTS.—The leaves of this tree were ascertained by Procter to yield on distillation both hydrocyanic acid and a light straw-colored volatile oil; the former in quantity sufficient to allow of the product being substituted for cherry-laurel water. Neither of these substances exists in the plant ready formed. They come, as with the cherry-laurel, of the reaction of water with the element amygdaline, during the process of distillation, the changes being induced by another principle also present, the albuminous one called emulsine.

The bark both of the stem and of the roots, especially the latter, also yields hydrocyanic acid. In the fresh state, when alone it should be used (since by keeping its quality deteriorates), it evolves an odor resembling that of peach leaves. The taste is agreeably bitter and aromatic, and resembles that of bitter almonds. An infusion of it, either in cold water or hot, is possessed of all the active properties, but, by boiling, these become in some measure dissipated. There appears also to be some other principle present, of an energetic character, but not yet determined, since an extract of the bark retains its bitterness even when the amygdaline has been wholly withdrawn. Besides the ingredients named, there are resin, starch, tannin, gallic acid, and others.

PHYSIOLOGICAL ACTION.—Wild cherry bark is tonic and sedative. The first result, when it is taken in doses not exceeding thirty grains, is quickening of the pulse, which is also rendered fuller and stronger. In certain constitutions, a desire for sleep is induced. To insure the sedative action, it is needful, however, that rather copious use should be made of the medicine.

Taken in large quantities frequently repeated, it impairs the energy of the digestive organs, produces nausea, and has a depressing effect upon the action of the heart and arteries. There are several cases on record in which the frequency of the beats has been reduced from 75 to 60; and in one reported by Stillé (he himself being the subject of the experiment) the reduction, effected in a remarkably short time, was from 75 to 50. Both classes of result are attributable of course to the hydrocyanic acid.

The oil is very similar in general properties to the oil of bitter almonds. Two drops administered to a cat caused death in five minutes.

THERAPEUTIC ACTION.— The value of this medicine has been proved in many ways. It restores tone after the system has been reduced by inflammatory disease, and particularly if the patient suffers from irritation, either general or local. By its use in doses not too large nervous excitability is calmed. In various forms of dyspepsia it is resorted to with advantage, and also in intermittent fevers where the fever lingers on from day to day; here, however, the operation is decidedly not equal to that of cinchona. In America, wild cherry bark has been found efficacious in the hectic fever of scrofula and consumption. It has been extensively employed, and there seems no reason to doubt what is asserted as to its value in this disorder. Good effects are produced also when there is occasion to allay irritable and nervous cough.

In Hæmorrhoids wild cherry has been found to be of considerable service, probably from its tonic effect on the intestinal canal.

In the Dyspepsia and Dyspnœa of Cardiac Disease, this remedy has often proved very efficacious. In this country it has been recommended principally by Dr. Clifford Allbutt, who states that many medical men have assured him that, in consequence of his recommendation, they have employed it, and with marked success. But Dr. Allbutt himself remarks that experience has taught him that it is only in comparatively early stages of cardiac disease that Prunus Virginiana acts with decided effect. In the later and more confirmed conditions of this complaint it may be usefully replaced by the subcutaneous injection of morphia.

PREPARATIONS AND DOSE.— Extractum Pruni Virginianæ Fluidum, ℨ ss.—j. (2.—4.); Syrup. Pruni Virg. ℨ ij.—iv (10.—20.); Infus. Pruni Virgin. ℥ j.—ij. (30.—60.).

[HAMAMELACEÆ.

WITCH HAZEL (Hamamelis Virginica).

Active Ingredients.—The chemistry of this plant has not been fully studied. The older analyses discovered a very large proportion of tannin. It contains in addition a volatile principle, the nature of which has not been accurately determined. The so-called "hamamelin" of the eclectics is an uncertain preparation of indefinite composition.

Therapeutic Action.—Rafinesque[1] speaks of Hamamelis as being in use among the Indians for ulcers, tumors, sores, etc., but for twenty years after, the name hardly, if at all, occurs in medical literature. The drug was first brought fairly into notice by Dr. Jas. Fountain, of Peekskill, in 1848,[2] and his observations were confirmed a year later by Dr. N. S. Davis.[3] By these writers it was recommended in hæmoptysis and hæmorrhoids. It subsequently appeared in homœopathic literature in an article by Preston,[4] as a remedy in epistaxis, uterine hæmorrhage, varicosis and phlegmasia dolens. Since then, among homœopaths, it has enjoyed a very high repute, but has been but little used by the mass of the profession. The cause of its neglect by the profession is easily accounted for when we consider that at the time of its introduction by Dr. Fountain, it could not be obtained on prescription at the drug-stores, and consequently the city physician must seek it himself, or take some special pains to procure it, which was altogether out of the question. The country physician, finding nothing said in its favor by his metropolitan brother, treated it with equal neglect, although the plant was growing at his very door.[5] If it was employed at all it was probably carefully dried (and a part of its virtues dissipated in the process) according to the spirit of the U. S. Pharm., and on trial was not found to possess all the valuable qualities attributed to it. The only other preparations attainable for many years were the homœopathic tincture and certain proprietary "extracts," the use of which, for obvious reasons, was not countenanced by the profession. Later, the wholesale druggists began to put on their lists fluid-extracts made from the dried plant, which, so far as we have been able to learn, have not attracted (nor merited, we suspect) much confidence. The history of hamamelis is that of many other indigenous remedies. The only modern regular text-book that contains a favorable notice of hamamelis is that of Ringer, who has found it useful in hæmaturia, hæmorrhoids, and varicocele.

The writer has employed it with satisfaction in hæmorrhoids, varicose veins, varicose ulcers, and as an anti-pruritic (locally applied) in cases of eczema.

---

[1] Medical Flora, Phila., 1828.
[2] New York Journal of Medicine, p. 208.
[3] Trans. Am. Med. Ass'n, Vol. i., 1840.
[4] Phil. Jour. of Hom., Jan., 1853.
[5] It is exceedingly abundant in New York and New England, and considerable capital is invested in the manufacture of preparations intended for direct sale to the laity.

It will be seen from the above that the sphere of action of hamamelis is mainly confined to the vascular system, and to the venous rather than the arterial, in fact its influence on the former is as decided as that of aconite on the latter. There is no evidence, however, to show that it in any way influences vessels of the viscera, but, so far as is yet known, limits its effects to vessels distributed to the skin and mucous membranes. It covers a portion only of the ground occupied by Ergot in this respect, but within its own proper field it does not yield to this latter in efficacy. The hæmostatic and vessel-shrinking power of hamamelis has been attributed to the large amount of tannin that it contains. This is probably not the case, for the double reason that tannin, by itself, is not capable of filling the place of hamamelis, and, second, because the so-called "extracts" (more properly distilled waters) contain no tannin, and yet are by no means inert.

PREPARATIONS AND DOSE.—Non-officinal. In the market are found fluid extracts from the dry plant, which we do not recommend; second, a variety of proprietary "extracts" made by distillation from the fresh plant. These latter contain the volatile principles only, and do not represent the entire properties of the drug; third, the homœopathic tincture, made by macerating one part of fresh hamamelis in two parts of alcohol. This contains both the fixed and volatile principles. The only objection to it is its cost, which is much greater than that of the fluid extracts, though these latter are nominally twice as strong. The parts used are the leaves and young twigs gathered late in the fall. The dose of the fresh tincture is from two to ten minims, given in chronic complaints once or twice a day, but in acute cases (hæmorrhage, etc.) repeated with greater frequency. In hæmorrhoids it is well to employ it also in enema or suppository, and in varicose veins and ulcers, as a lotion or an ointment. For local use the tincture should be diluted with five to ten volumes of water.]

SANTALACEÆ.

SANDAL-WOOD (SANTALUM ALBUM).

ACTIVE INGREDIENTS.—I am not acquainted with any accurate analysis of the oil, which was introduced to the notice of the English profession either by Dr. Henderson, of Glasgow,[1] in 1865, or by the late Dr. E. Miller, of the same city, at an earlier date. At any rate Dr. Henderson mentioned that it was in use among the natives of India as a remedy for gonorrhœa; and in the appendix to the Pharmacopœia of India we find it stated that the wood of white sandal yields about 2.5 per cent. of the oil, which is of a pale yellow color,[2] and has a resinous taste and a sweet peculiar smell, best appreciated by rubbing a few drops on the warm hand.

---

[1] Glasgow Med. Jour., April, 1865.
[2] [*Oleum Santali flavum*, not *Oleum Santali flavi*, as often seen. The *oil* is *yellow*, not the plant.]

PHYSIOLOGICAL ACTION.—On this subject no accurate knowledge exists; but the following remarks may be of interest. They were communicated to the Practitioner by Mr. Robert Park, of Great Stanmore:[1] " Its physiological effect is generally to constringe all the mucous membranes of the body. Thus the immediate effect of a dose of it is to cause dryness of the fauces and thirst. During its digestion and absorption this thirst is kept up and intensified in some cases; in all, if a large dose be given (*i. e.*, gr. xv. to xx.), and in many, with a merely legitimate dose, it produces a feeling of 'drawing together of the bowels,' sometimes verging on colic. Its action is somewhat more powerfully felt in the kidneys. At a period varying from two to three hours after it has been taken, a sense of fulness is experienced in both (renal) regions. This sense of fulness, or weight, or tension, lasts for a period varying from ten to twenty minutes."

THERAPEUTIC ACTION.—There is a good deal of concurrent evidence to show that yellow sandal oil is a powerful remedy for gonorrhœa. The principal authorities are Dr. Henderson,[2] Mr. Berkley Hill,[3] Dr. Purdon, of Belfast,[4] Dr. H. W. Beach,[5] Mr. Robert Park (already mentioned), and Dr. G. Pridie.[6] Mr. Park thus sums up the result of a very considerable experience of his own on this subject: " The cases in which it is most efficacious are not those of sickly patients with thin muco-aqueous discharges, but those of full plethoric subjects with thick purulent discharge, and even considerable scalding. Given in a case of this kind as soon as possible after the appearance of the discharge, the effect is simply very gratifying both to patient and doctor. The discharge disappears, and the scalding is no more felt; and if the medicine be continued and collateral treatment attended to, in from two to five days the cure is complete; that is to say, there will have been no discharge seen for twenty-four hours, and there is no pain or abnormal symptom whatever. The patient must not be dismissed now, however, without a caution to abstain rigidly from all stimulants for at least a week, nor must he within a similar period take a purge, else he will assuredly bring back the discharge; then this redeveloped discharge is, for some reason I have never been able to fathom, more difficult to deal with, nor have I been able to find any attempt at explanation in the works of either Cullerier, Ricord, Lancéreaux, or any of our own authorities. . . . I will merely add that complications are to be treated separately, though the exhibition of the oil is by no means contra-indicated by the presence of orchitis, prostatis, or phymosis. It may be given perfectly well while the appropriate treatment for each of these is being carried out. To this rule there is one exception, viz., in orchitis. If it be requisite, as it very often is, to give a purge, let it be given at once, and before the oil is administered; as purges and stimulants (including coffee and, I strongly suspect, tea also) are contra-indicated during its administration. I have never once, however, seen any of the above-mentioned complications occur during a course of the oil,—I mean idiopathically."

---

[1] Practitioner, 1869, vol. ii., p. 260.
[2] Loc. cit.
[3] Brit. Med. Journ., July 6, 1867.
[4] Med. Mirror, Sept. 1866.
[5] Boston Med. and Surg. Journal, Nov., 1868.
[6] Quoted in Appendix to "Pharmacopœia of India."

PREPARATIONS AND DOSE. — [Not officinal.] The oil itself should be administered in mucilage. Authorities differ as to the dose, some recommending 20 minims, some 30 or 40. On the other hand, Mr. Park, whose knowledge of the remedy seems very practical, thinks that larger doses than 5 minims are useless and even injurious; but this quantity should be given as often as every four hours. [Phillips.] The editor has used the yellow oil of sandal since 1866, at a time when it was not found in the general drug trade, and cost from two and a half to three dollars an ounce. Of late years it has greatly suffered in reputation in this country in consequence of the great difficulty of getting a pure article. Most of it is adulterated with oil of cedar, a body of little commercial importance until brought into use as an adulterant. The present New York market price of sandal oil is from about six to sixteen dollars a pound, depending on the amount of cedar oil and turpentine that the purchaser is willing to include in the weight. Under these circumstances it is manifestly unreliable in practice, and is very properly being abandoned. Its principal use at present, we believe, is in the form of proprietary capsules and pastes sold direct to the consumer who hopes to "cure himself" without the aid of a physician. The writer has used it very little for the last few years, having obtained better results in the treatment of gonorrhœa with other drugs (notably *Cannabis sativa* var. *Americana* and corrosive sublimate) than he has ever had with sandal, even when at its best.

UMBELLIFERÆ.

HEMLOCK (CONIUM MACULATUM).

ACTIVE INGREDIENTS.—The special and characteristic medicinal substance contained in hemlock is a peculiar alkaloid, called *coniine*, and known also by the names of conin, conia, and cicutine. It resides in every portion of the plant, but is chiefly abundant in the pericarps of the full-grown but still green fruit, and is separable by distillation with water and caustic potash. It exists in combination with coniic acid. When the leaves are dried, all the conia volatilizes, so that, in this condition, they are useless.

When pure, the alkaloid is, at all ordinary temperatures, an oleaginous, colorless, and volatile liquid, of sp. gr. 0.89, possessed of a strong, penetrating, and stupefying odor, and an acrid flavor, somewhat resembling that of tobacco. The volatile character is made apparent by dropping a small quantity of the alkaloid upon paper, which it will render, in that part, translucent, and as if greasy, the spot disappearing upon subjection of the paper to a slight degree of heat. It burns with a bright flame, giving off, during combustion, a good deal of smoke; the vapor also is inflammable; the boiling point is between 340° and 413° F. At ordinary temperatures, coniine is but slightly soluble in water, requiring 100 times its own volume; but in ether and in alcohol it dissolves readily. With a quarter of its weight of water, it combines, forming therewith hydrate of conia.

The composition of coniine is $C^8H^{15}N$. It is a strong base, and resembles ammonia, not only in being devoid of oxygen, but in many of its reactions. With the vapor of hydrochloric acid it produces copious white fumes, completely neutralizing the acid; and with nitric acid and acetic acid the results are precisely similar. The addition of a small quantity of nitric acid causes coniine to become blood-red. Exposure to the atmosphere resolves it into ammonia and a bitter non-poisonous extractive matter.

A second ingredient in hemlock is the volatile alkaloid called *conydria* or *conhydrine*, $C^8H^{17}NO$, discovered by Wertheim, and distinguished from coniine by being solid and crystalline.

A third ingredient, present only in minute quantity, is a volatile oil, which possesses in a high degree the characteristic odor of the plant, but is scarcely at all poisonous. This plainly shows that it is not the active principle of conium; and, further, that in judging of the pharmaceutical value of different preparations of hemlock, no stress can be laid either upon the presence or the absence of the accustomed odor.

PHYSIOLOGICAL ACTION.—For our knowledge of the physiological action of conium we are indebted to several observers, who have worked out the subject in a very complete manner. We may dismiss from consideration the statements which were made by the older authors as to the action of conium upon the glandular system, as all these effects have been proved to be imaginary. Their incorrectness is especially demonstrable in regard to the sexual organs. It used to be said that the continued use of conium in large doses would cause the mammæ and testes to waste, and destroy sexual power; but the experience of J. Harley, and of many other recent observers, has shown that while conium has the power of repressing unnaturally high sexual excitement, it has no influence upon the natural function, far less any power to injure the glandular structures. The truth is that all statements which were based on the employment of the preparations which were in fashion before the discovery of the alkaloid conia are worthless, for there is every reason to believe that those preparations were quite inert. As to the preparation of hemlock used by the ancient Greeks for the execution of criminals—if our hemlock were really the poison employed (which is perhaps not absolutely certain)—there is every reason to suppose that the fatal drink consisted of the expressed juice, the *succus*, which has been introduced of late years into our Pharmacopœias, and is the only one of our officinal forms which has the slightest value. All preparations in which heat is employed are manifestly worthless, in consequence of the great volatility of coniine, as was long ago remarked by Neligan. In speaking, then, of the action of hemlock, we may disregard all results except those that have been obtained by the employment of the alkaloid itself, or the juice, or, it may be added, the non-officinal tincture of the nearly ripe fruit, which J. Harley has shown to contain more of the active principle than any other part of the plant.[1] Exact research into the action of conium began with Christison,[2] and at once led to important though very incomplete results. Since that time the subject has been investigated in every possible way. Kölliker, Albers, Guttman, Schroff, and many others have investigated the action of coniine upon frogs, upon fish, and upon almost every class of warm-blooded animals; and within the last few years Drs. Fraser and

---

[1] "Old Vegetable Neurotics," p. 92. Also a paper in Practitioner, Oct., 1871.
[2] Transactions Royal Soc., Edin., 1835.

Crum-Brown have made most valuable additions to our knowledge by investigating the modifications which are introduced into the action of coniine by sundry chemical combinations. Upon the human subject the most important observations are those of John Harley.

The combined results of numerous modern researches very clearly indicate that coniine is a nerve-poison, with definite and singularly limited affinities. It is one of the most powerful poisons discovered, yet it operates within a very contracted sphere. Normal (*i. e.*, unmodified) coniine acts as a pure paralyzer; upon the motor system it operates through the nerves alone, leaving the spinal cord intact; small doses act exactly like curara, viz., upon the nerve terminals only, but larger doses affect also the motor trunks. The hind limbs are paralyzed first; but the whole system of voluntary motor nerves is ultimately affected, and death comes by cessation of respiratory movements.

Here, however, it must be remarked that there is considerable variance between Harley and other observers; although perhaps a part of the difference is more apparent than real. Harley states that a full dose of the succus (five to six drachms) acts in the following manner, according to circumstances: if the taker keep up a brisk walking exercise, in half or three-fourths of an hour he feels a sense of weariness in the legs, with some giddiness and heaviness over the eyes, but if he persevere in his walk these symptoms soon pass off. If, on the contrary, he remain at rest after the dose, the eyes become first affected; accommodation is interfered with, and then there are drowsiness and dilatation of the pupil; after these come weakness of the legs; pallor, inability to stand steady, a general diminution of muscular power, especially in the hamstring muscles. From the consideration of these facts, and from the results (hereafter to be mentioned) which conium afforded him in the treatment of chorea, etc., Harley believes that the central motor tract is affected by conium, and that its action (in full and repeated doses) is primarily tranquillizing, and subsequently tonic, to the muscular movements. He particularly states that the vertiginous and quasi-paralytic effects are much more pronounced in those who are quiescent than in subjects who are active; so that a delicate child, if lively and running about, will take with impunity a dose that would muddle a sedentary adult.

As regards the production of dizziness, impaired accommodation, dilated pupil, and other narcotic brain symptoms, although there are some observers who deny them, a great preponderance of evidence is confirmatory;[1] but they appear on the whole to be trifling in comparison with the effects on the motor system.

Upon the sensory nerves coniine appears to exert a direct, but moderate, paralyzing influence, which is local in action. Applied to a nerve laid bare, it causes no pain, but simply interrupts the conduction of sensory impressions. There is no reason to think that it at all affects sensation by any action on the nervous centres. It is said, however, that the first momentary effect of applying coniine to a wound, or to the mucous membrane of the mouth, is a burning feeling; but this sensation is immediately succeeded by a loss of sensibility, and there is no ground for viewing coniine objectively as an irritant.

The question whether conium affects the heart, and, if so, how, remains undecided. The general result of experiments on animals certainly

---

[1] See especially Schroff's account of the experiments made under his control by Dillnberger and Dworzall.

discountenances belief in any direct action on the heart; as to human beings, the evidence is more conflicting. J. Harley declares that there is usually little or no effect; but the pulse may, if anything, become firmer and fuller.

The known general physiological character of the drug entitles us to treat its action on the heart as but of slight consequence.

Conium was formerly credited with diuretic and diaphoretic action; but the evidence is very conflicting. Some weight may perhaps be given to the affirmation by several American physicians of its diaphoretic powers; but the assertion of Fountain that this result is equally produced by coniine and the extract of hemlock is obviously open to great suspicion.

Of the fatally poisonous effects either of hemlock or of conia upon man, we possess very little information; it is desirable, therefore, to take particular notice of any well-observed case. Dr. Hughes Bennett recorded the case of a man who ate hemlock in mistake for something else in a salad; the first symptom observed was weakness of the legs, producing faltering gait; as the weakness increased the man staggered as if drunk, and, at the same time, the arms began to be paralyzed. Perfect loss of all voluntary movement, including swallowing, followed; and, lastly, palsy of respiration put an end to his life. Consciousness and intelligence were unimpaired up to death, but sight was destroyed. It is worthy of notice that in this case there seems to have been no primary brain affection.

The account of the physiological action of conium and coniine would be very incomplete if it did not include the observations of Drs. Fraser and Crum-Brown. They made the very important discovery that, at any rate in Britain, there are two varieties of coniine sold, of which one (Morson's) really contains not only coniine, but a compound of it—methyl-coniine, the properties of which are very different from those of pure coniine. Methyl-coniine was carefully investigated by them, and was found to have the power of influencing not only the motor-nerves, but the spinal cord itself; with large doses the former action is completed before the latter, but with small doses the action on the cord is completed before that on the nerves. It is evident that in proportion to the varying degrees in which the so-called coniine may contain methyl-coniine, the effects vary. Hence the discrepancies of different authors.

Therapeutic Action.—The therapeutics of conium are quite new: in fact, they have been altogether in the stage of discovery and conjecture up to the present time. For it must be remembered that such preparations as the extract—which was chiefly in use till a very few years ago—have been proved by various experimenters to be inert, even in doses greatly larger than it was the custom to employ, and the effects obtained with them must, therefore, have been imaginary. Indeed, the only trustworthy fact among the older observations on conium was the local sedative action.

As a Local Remedy for pain and irritation conium is occasionally very useful: the pain of cancer seems to be, for some reason, particularly amenable to its influence. The best application to a scirrhous breast, for example, which is in the stage of painful tension, immediately before ulceration, or even in the ulcerative stage itself, is a mixture of the succus conii and glycerine in equal parts, or a very weak solution of coniine, or a poultice of the fresh leaves. For cancer of the stomach the succus may be given in doses of two to five drachms or more; and to allay the pain of cancer of the rectum, one ounce of succus may be injected with one ounce of starch-decoction. In the painful "scrofulous" photophobia Dr. Mauthner has employed coniine most successfully; he used a mixture of ½ grain coniine with a drachm of almond oil, a little of which was applied two or three times daily to the conjunctivæ. The irritability and soreness of the air-passages in bronchitis, and in the tickling cough of phthisis, can be admirably calmed by the inhalation of coniine-vapor. Harley's formula is the best: coniine, 1 gr., alcohol ʒiss.; dissolve the coniine in ʒss. of the spirit, and add the remainder with ʒij. of water: ♏ xx. contain $\frac{1}{12}$ gr. of coniine, and form a proper dose to be placed on a sponge in a suitable apparatus.

In Spasmodic and Convulsive Disorders, coniine, though apparently so suited by its physiological action to their treatment, has not yet established any settled rank, notwithstanding the strong statements that have been made in its favor.

In chorea, for example, Dr. Harley has found the succus of remarkable value; he gives it in increasing doses until he discovers the dose necessary to produce the characteristic slighter physiological symptoms above described. One, at least, of Dr. Harley's cases is beyond criticism: for the chorea was of old date, and the boy (æt. six) had also suffered from morbid restlessness from his very birth. It was not till very large doses were given that any decided improvement was observed. He took 104 fl. oz. in twelve weeks; and under this treatment not merely did he lose his chorea, but his congenital restlessness disappeared, and he also gained flesh and strength remarkably.

On the other hand, many practitioners have given patient trial to conium in chorea, and have either had no success at all or else very capricious and uncertain results. Ringer, for instance, speaks of the improvement under conium as apparently only temporary, and adds: "It has yet to be shown that conium will shorten the course of this disease." I have heard many practitioners express a similar view, and many, who at first believed that they had found in conium a specific remedy for chorea, have afterwards lost all faith in it. We cannot, however, lose sight of such cases as are recorded by Harley; and on the whole the reasonable conclusion seems to be that there are cases over which it has a very powerful influence, and other cases for which it does nothing. What may be the pathological difference that causes this varying result, we unfortunately do not know. Dr. Anstie remarks that he has never once seen conium do good in the severe choreas of commencing sexual development; while, on the other hand, it has sometimes appeared strikingly beneficial to young children.

In Tetanus the high anticipations of success from conium which once prevailed have not been sustained. Experimentally, Guttmann found that frogs which were tetanized with strychnia were not at all benefited even by paralyzing doses of coniine; and the cases, if apparently successful, recorded by some authors, must be received with much suspicion. Thus, Mr. Corry's case[1] was treated with the *extract*, in 5-grain doses, and as there is every reason to believe that this is a perfectly inert remedy, the recovery was probably spontaneous. And the cases published by Professor Johnson, of Maryland, are, as Ringer points out, most unsatisfactory, for in both of them other powerful remedies were simultaneously employed. Nor does it seem probable, from the best knowledge we can get of the physiological action of the drug, that

---

[1] Dublin Quarterly Journal of Medicine, Nov., 1860.

conium could antagonize either pathological or strychnia tetanus, notwithstanding Dr. Harley's arguments to the contrary.

In Epilepsy I regret to say that I have obtained no confirmation of Dr. Harley's favorable statements. Dr. Harley limited his commendation of conium in epilepsy to those cases in which self-abuse has been the apparent cause; but even in this special relation I have not found that the drug is effective. It is probable that self-abuse is never more than the exciting cause acting upon a system predisposed to the disease. Consequently, we could hardly expect efficient remedial action from any treatment, except one which should either remove the exciting cause, or should deeply modify the nutrition of the nervous centres; and there is little hope that conium will prove adequate for either of these purposes. Whatever can be done in the way of lowering the susceptibility to peripheral excitement can probably be much more completely effected by bromide of potassium; while the slower, but more radical, benefit which may be hoped for from improvement in central nervous nutrition, is certainly more likely to be obtained by nutritive tonics, such as cod-liver oil.

In Mental and other Cerebral Diseases there seems much more probability that conium will prove efficacious. The practical authority of Dr. Crichton Browne [1] is strongly in favor of its use in acute mania; and Dr. J. Wilkie Burman [2] has published some remarkable results of the combined subcutaneous injection of conium and morphine in the same disease. Dr. Burman experimentally confirmed J. Harley's statement that the union of morphia and conia heightens the effect of each, and that the combined influence is tranquillizing both to the mental and the motor centres. Dr. Burman's estimate of this treatment, if correct, is very important: the question ought to be settled by extensive experiments in our large asylums; and this could very speedily be done. I am bound to state that I have heard of instances where the treatment has been patiently and carefully tried, in apparently suitable cases, without any good effect; but no general decision ought to be arrived at until thousands of well-conducted trials have been made.

In Delirium Tremens, according to Harley, conium is most useful, especially in combination with opium, the action of which it supplements in a very useful way. It may here be remarked that, should it prove true that conium has a direct influence over two varieties of delirium tremens, we might be led to think that it possesses special affinities for the brain-cortex; and further, remembering the interesting discoveries of Professor Ferrier, we might reasonably suspect that if conium influences affections of the motor system at all through the brain, it must be through Ferrier's motor centres in the gray matter, and not through the corpus striatum or the optic thalamus.

PREPARATIONS AND DOSE.—Extractum Conii Fructûs Fluidum, ♏ v.—xxx. (.30—2.); Extractum Conii Alcoholicum; Tinctura Conii, Extractum Conii, Succus Conii. The last four of these are so utterly unreliable that no doses can be assigned. Conia (G. Ph.), gr. $\frac{1}{300}$—$\frac{1}{60}$ (.0002—.001, Binz).

---

[1] Lancet, Feb. 3d, 10th, and 17th, 1872.
[2] Practitioner, Dec., 1872.

ASAFŒTIDA (Narthex Asafœtida).

Active Ingredients.—Besides resin, gum and other unimportant substances, which constitute the chief bulk, asafœtida contains a variable quantity of volatile oil, upon the presence of which its odor and its properties essentially depend. The usual quantity is about 3.5 to 4.5 per cent. The composition of the oil, according to Hasiwetz, is a mixture of 2 (C^2H^{11})S, and $C^4H^{10}S$. When newly distilled, like the essential oils of horseradish and black mustard, it contains no oxygen. It is lighter than water, and at first is colorless, but exposure to the air causes it to acquire a yellow tinge and also to become acid. The odor is very powerful, and, as evaporation proceeds rapidly, it is soon perceived at a long distance. The taste is at first mild, afterwards bitter and acrid. Phosphorus is probably one of its elements, and sulphur certainly, sulphuretted hydrogen being disengaged during the process of boiling. It dissolves in all proportions of alcohol and ether, but requires more than 2,000 times its weight of water for the aqueous solution. When the asafœtida itself is rubbed with water, the gummy matters dissolve, the resin and the volatile oil are suspended, and an emulsion is formed. The resinous matter is soluble in alcohol; and, if the alcoholic solution be mixed with water, a milky fluid is formed, owing to the deposition of the hydrated resin.

Physiological Action.—The first complete investigations into the action of asafœtida appear to have been those made by Jörg. He ascertained that distinct effects followed the administration of a single grain, but that in different individuals there is very various susceptibility to its influence, and that even the characteristic smell of the drug, which has been asserted to be very evident in the secretions after employing it medicinally, is by no means constantly present. The administration of small doses causes alliaceous eructations, which often continue for twenty-four hours, showing what a length of time the medicine is retained in the stomach; the digestion is impaired; there are burning sensations in the fauces; there is pain, fulness, and oppression of the stomach; the abdomen becomes distended with flatus, which, when discharged, is of a very fetid and disagreeable character; there is frequent inclination to evacuate the bowels, and the discharge is thin and watery. The urine is not augmented in quantity, but becomes acrid, and communicates a sense of burning. The pulse at the same time is quickened; the head becomes more or less affected with flying pains, often attended by much giddiness; and various nervous and hysterical phenomena make their appearance. Jörg's statement that the menstrual period is advanced has been confirmed by numerous observations. The sexual desire also becomes excited. Like the pulse, the respiration becomes quickened, and the secretion of the bronchial membrane is promoted.

In larger doses, say ten to thirty grains, asafœtida increases the secretions both of the pulmonary organs and of the abdominal ones, especially those of the liver and bowels, the peristaltic action of which it augments. Appetite and digestion are quickened; morbid secretions of the mucous membrane are improved in quality; and if intestinal worms be present, they are expelled.

To persevere with the medicine too long has a tendency, however, to enfeeble the digestive organs; while larger doses than those indicated cause colic and heat in the abdomen, attended by nausea, vomiting, and diarrhœa.

Upon all parts of the organic system, asafœtida in moderate doses appears, in short, to operate as a wholesome stimulant; and, especially in hysterical subjects, among other good results, it leads to improvement of vision and enlivening of the spirits. That these favorable results do not follow in all cases is quite likely, since a predisposition to benefit by this medicine appears essential; and possibly, in the majority of cases in which it may be employed, it may play, like other nervines, a part very subordinate.

THERAPEUTIC ACTION.—Acting on the nervous system as a stimulant and powerful anti-spasmodic, asafœtida becomes useful in hysterical convulsive affections. Whooping-cough, asthma, and other nervous disorders are likewise usefully treated with it. According to Sydney Ringer, it is valuable in the flatulence of young children, when unconnected with constipation or diarrhœa, a teaspoonful being given every half-hour of a mixture consisting of a drachm of the tincture to half a pint of water. Sydney Ringer states at the same time that when flatulence has constipation or diarrhœa for its accompaniment the asafœtida does little good. In hysterical tympanitis, asafœtida may be administered with much benefit in the form of enema, or it may be taken internally. It is useful again in certain forms of chronic bronchitis and whooping-cough, by reason of its expectorant power; also in globus hystericus, and in hysterical cough. After all, in nervous affections nothing more must be expected of asafœtida than that it shall palliate certain symptoms. Substantial cures can only be looked for from medicines of greater energy. In old chronic catarrhs, especially when accompanied by spasmodic cough and by occasional difficulty of respiration, asafœtida alone, or combined with ammonia, gives decided relief.

Dr. Garrod says that he is "inclined, from the result of much observation, to regard asafœtida as one of the most valuable remedies of the materia medica; far above all other ordinary antispasmodics;" and he thinks "the value of the drug is chiefly due to the sulphur oil contained in it."

PREPARATIONS AND DOSE.—Asafœtidæ, gr. v.—xv. (.30—1.); Mistura Asaf., ʒss.—j. (15.—30.); Tinct. Asaf., ʒss.—j. (2.—4.); Pil. Asaf.; Pil. Aloës et Asaf.—dose of either No. 1—3; Suppositoria Asaf.; Emplast. Asaf.

ANISE OR ANISEED (PIMPINELLA ANISUM).

ACTIVE INGREDIENTS.—The efficient ingredient of anise is the oil obtained from the fruits, which is a mixture of a fixed oil and the volatile anethol, a substance common not only to several Umbelliferæ, but also to the Illicium anisatum and the Artemisia Dracunculus. The formula is $C^{10}H^{12}O$. Anise oil is colorless when fresh, but gradually becomes yellowish; it has the characteristic taste and smell of the seeds. Sp. gr., 0.98 or 0.99. It is very soluble in alcohol and ether.

PHYSIOLOGICAL ACTION.—Anise oil acts as a rapid poison to lice and to itch insects; it is slightly irritant to the human skin; it is a powerful narcotic poison to certain small birds; in half-ounce doses it kills rabbits with symptoms of narcotic-acrid poisoning. It affects cats and dogs but slightly.

THERAPEUTIC ACTION.—Taken internally, anise oil in small doses is a mild stimulant to the stomach and intestines, and to the bronchial mucous membrane; and is useful in atonic dyspepsia, colic, flatulence, and chronic bronchitis. Externally it has been employed with success in the form of an ointment or a soap to destroy lice and itch insects. Various other therapeutic actions have been ascribed to it which it does not really possess. It is unquestionably eliminated both in the urine and the milk, and has therefore been credited with the power of increasing the excretions, but there is no good evidence of this. It is probable that the old epithet of intestinorum solamen, applied to it by Van Helmont, adequately expresses its value. By Galen anise was reckoned among the true anodynes or cordials.

PREPARATIONS AND DOSE.—Anisum, gr. x.—xxx. (.65—2.); Oleum Anisi, ♏ j.—v. (.05—.25); Spts. Anisi, ʒ j.—ij. (4.—8.)

FENNEL (FŒNICULUM DULCE).

ACTIVE INGREDIENTS.—The oil of sweet fennel, constituting the active principle, is of a light straw color, and retains the odor of the fruit. In composition it corresponds with the oil of anise; it contains, moreover, a peculiar camphor-like body with a high boiling-point.

PHYSIOLOGICAL ACTION.—The vapor of fennel oil causes secretion of tears, and occasionally of saliva. Given in large doses to rabbits, it proves fatal, in much the same way as anise oil, but, unlike the latter, it does not escape in the urine.

THERAPEUTIC ACTION.—Fennel is a useful carminative for the relief of flatulence and griping.

PREPARATIONS AND DOSE.—Fœniculum, gr. x.—xxx. (.65—2.); Aqua Fœniculum, ʒ j.—ij. (30.—60.); Ol. Fœnic., ♏ ij.—v. (.10—.25).

CUMIN (CUMINUM CYMINUM).

ACTIVE INGREDIENTS.—The properties of cumin depend on a volatile oil, which is of a pale yellow color, limpid, and lighter than water. It is a mixture of cuminol and cymol. Cuminol, $C^{10}H^{12}O$, isomeric with anise oil, is a colorless oil with a sharp burning taste, and a strong smell like caraway. It is insoluble in water, readily soluble in alcohol and ether. Cymol, $C^{10}H^{14}$, is a colorless oil, strongly refractive of light, with an unpleasant, camphor-like smell. It is insoluble in water, readily soluble in alcohol, ether, and fats.

PHYSIOLOGICAL ACTION.—Cumin oil has acquired an altogether unexpected degree of importance from having been made the subject of the researches of Grisar, already referred to under chamomile oil. Grisar experimented with the oil on frogs, in the same manner as when testing chamomile oil. He came to the conclusion that cumin oil very markedly depresses reflex excitability, though its action is on the whole weaker than that of the other substances experimented with. He also proved the power of cumin oil to antagonize the tetanic excitement of strychnia in a decided manner.

THERAPEUTIC ACTION.—The use of cumin in medicine has heretofore been confined to its employment as a carminative, but the abovementioned experiments of Grisar will doubtless cause it to be tested for the purpose of reducing reflex excitability in its various forms, as mentioned under chamomile.

PREPARATIONS AND DOSE.—Cumin is not officinal; the oil ought alone to be used in doses of from two to eight minims, with sugar, or in an emulsion.

CORIANDER (CORIANDRUM SATIVUM).

ACTIVE INGREDIENTS.—The taste and odor of coriander seeds are referable to a yellowish volatile oil which in flavor renders them moderately warm and pungent. Kawalier believes it to be of the camphine class, and to consist of $C^{10}H^9O$.

PHYSIOLOGICAL ACTION.—Stomachic and carminative.

THERAPEUTIC ACTION.—Coriander seeds are added to infusion of senna, and to other purgatives, with a view to covering the flavor, and checking any tendency there may be to gripe. As an independent medicine they are rarely used.

PREPARATIONS AND DOSE.—Coriandrum, gr. x.—xxx. (.65—2.); Oleum Coriandri (B. Ph.), ♏j.—v. (.05—.25). Coriander is an ingredient of Confectio Sennæ, and of Tinct. Rhei et Sennæ.

CARAWAY (CARUM CARUI).

ACTIVE INGREDIENTS.—Caraways owe their properties to a volatile oil, which is thin, colorless, or straw-yellow, with a well-known penetrating smell, and a hot, bitterish taste. It dissolves in an equal volume of rectified spirit. It consists of two bodies, carvol and carvene. The former has the formula $C^{10}H^{14}O$; it is a thin, clear fluid, with caraway smell; sp. gr., 0.953; and does not boil under 250° C. Carvene is a camphor-like body; formula $C^{10}H^{16}$; it forms a crystalline salt with hydrochloric acid, and boils at 173° C.

PHYSIOLOGICAL ACTION.—Caraway oil appears to be, in large doses, a fatal narcotic poison to rabbits. One case has been recorded where a man took about a drachm; he was attacked with shiverings and heats, congestion of the head, and delirium.

THERAPEUTIC ACTION.—Flatulent colic in children is often satisfactorily treated with preparations of caraway, but the chief use is that of a flavoring agent. Caraway-water is a convenient article for saline purgatives; it is good also for covering the taste of nauseous medicines and preventing the griping action of purgatives in general. For this latter purpose the oil is frequently used in the preparation of cathartic pills.

PREPARATIONS AND DOSE.—Oleum Carui, ♏ j.—v. (.05—.25). Caraway is contained in Tinct. Cardomomi Comp.

DILL (ANETHUM GRAVEOLENS).

ACTIVE INGREDIENTS.—The fruit of this plant owes it properties to the non-volatile and volatile forms of anethol, which has been already described under anise.

PHYSIOLOGICAL ACTION.—Nothing is accurately known in regard to the physiological action of dill, but it is presumably the same as that of anise.

THERAPEUTIC ACTION.—The medicinal uses of dill are identical with those of anise and caraway; the most extensive employment being as a remedy for the flatulence of infants. In India the infusion is administered to women immediately after parturition, as a stomachic and grateful cordial.

PREPARATIONS AND DOSE.—Aqua Anethi (B. Ph.), ℥ j.—ij. (30.—60.); Oleum Anethi (B. Ph.), ♏ ij.—v. (.10—.25).

SUMBUL (EURYANGIUM SUMBUL).

ACTIVE INGREDIENTS.—The analyses of Reinsch show that when the clear pale yellow resin extracted by ether from sumbul root is treated with alcohol (of 75 per cent.), and the solution evaporated, a balsam-like body is left which contains much angelic and much valerianic acid. According to Sommer, the dry distillation of this substance brings over a blue volatile oil containing umbelliferine, $C^9H^4O^7$, a substance which crystallizes in transparent, silky needles destitute of taste or smell, and subliming at a high temperature. Its solution in boiling water is remarkable for possessing a blue fluorescence. Umbelliferine is readily soluble in alcohol, ether, and chloroform.

PHYSIOLOGICAL ACTION.—Upon this subject it is to be regretted that we as yet possess no information beyond the very general idea that sumbul acts on the brain and on the spinal centres with a calming and antispasmodic influence.

THERAPEUTIC ACTION.—Sumbul is a drug of which we probably as yet only very partially know the value. In Russia it is used in low fevers of a typhoid type; also in asthenic cases of dysentery and diarrhœa; and it has even been said to have proved successful in cholera. It is considered by Thielmann, of St. Petersburg, the most trustworthy remedy in delirium tremens. Dr. Granville, who introduced it into England, recommends it in cases of gastric spasm, hysteria, dysmenorrhœa, epilepsy, and nervous disorders. Mr. Murawieff, a Russian physician, recommends it in chronic bronchitis, moist asthma occurring in old anæmic and scorbutic subjects, in atonic dyspepsia, leucorrhœa, hypochondriasis, and hysteria. I can bear witness to its decided efficacy in chronic bronchitis, and in certain stages of phthisis.

As a Remedy for Neuralgia of a certain type, sumbul has probably more value than any other known drug. It is sometimes very surprising to observe the rapidity with which a severe facial, sciatic, or ovarian neuralgia will yield to a few doses of sumbul, though it had resisted very powerful remedies. It is difficult to say exactly what are the characteristics of the neuralgias which prove thus amenable to sumbul, but certainly this kind of pain is most frequently found in women, particularly those of a quick and lively nervous constitution. On the other hand, the dull migraine of hysterical women who have a phlegmatic constitution, and a great tendency to obstinate constipation, does not yield readily, if at all, to sumbul.

In the restlessness of pregnancy, an affection which is exceedingly distressing to some women, though others do not suffer from it, sumbul is often invaluable; a draught of 30 to 40 minims of the tincture, with a little chloric ether, giving quiet nights for a long time together without losing its power.

In the Insomnia of Chronic Alcoholism sumbul often serves a similar purpose very efficiently.

PREPARATION AND DOSE.—Tinctura Sumbul (B. Ph.), ℥ x.— ʒ j. (.60—4.).

LOBELIACEÆ.

LOBELIA (LOBELIA INFLATA).

ACTIVE INGREDIENTS.—The active properties of lobelia are yielded both to alcohol and to ether, and appear to depend principally upon the presence of lobeline,[1] a peculiar alkaline principle which presents itself in the form of a pale yellow liquid, of less specific gravity than water, upon the surface of which it floats. The odor is aromatic; the taste is pungent. It is soluble in water, more readily so in alcohol and in ether; and forms crystalline salts with the mineral acids. Lobelia also contains lobelic acid, which is crystalline, and soluble in water, alcohol, and ether; also a volatile oil, to which the odor of the drug is due; and a resin of acrid taste.

PHYSIOLOGICAL ACTION.—In small doses, lobelia acts as a diaphoretic and expectorant, the latter property being manifested without involving the pain of coughing. In full medicinal doses, say twenty grains of the powder, it acts as a powerful nauseating emetic, causing speedy and violent vomiting, attended by distressing and persistent nausea, and sometimes by purging, copious sweating, and great general relaxation. These symptoms are usually preceded by giddiness, headache, and general rigors. The effects produced are thus very similar

---

[1] Discovered by Bastick and Proctor. See "Journ. Pharmaceut. Trans.," vol. x.

to those of tobacco, only that for their occurrence larger doses are required. All experiments, alike upon men and animals, show that violent emesis and prostration ensue upon the administration of moderately large doses. A single grain, administered to a cat, induced these results in less than two minutes, the creature recovering in about three hours. Administered by the rectum, it produces the same distressing symptoms—sickness, profuse perspiration, and universal relaxation—which ensue upon a corresponding use of tobacco.

In excessive doses, or in full medicinal doses, too frequently repeated, the effects of lobelia are those of a powerful acro-narcotic poison. Extreme prostration is first induced, then great anxiety and distress, followed by convulsions and death. These symptoms are produced also by the alkaloid. The earliest effects are pain in the back of the head, with a feeling of fulness, tightness, and giddiness; these are followed by general tremor, with prickling sensation throughout the body, nausea, and profuse perspiration. Violent emesis, if not already present, soon ensues.

The painfulness and giddiness in the head generally alternate with the nausea; and, on the occurrence of profuse diaphoresis, the head-symptoms generally subside. Great prostration of strength and relaxation of the entire muscular system now set in, accompanied by heavy despondency and fear of death. There is much thirst; the hands and arms are thrown about, and the sufferer rubs or beats his stomach; the secretion of saliva and of mucus is increased; there is dryness, burning, and rawness in the throat, and frequent and difficult deglutition, with irritation of the œsophagus, and oppression of the præcordium. Extreme spasmodic difficulty of breathing attends these distressing conditions, and there is great flatulent distention throughout the abdomen, especially in the neighborhood of the navel, with frequent eructations, and flatulent discharges from the bowels.

At the same time the urine is very profuse, and causes a smarting sensation along its passage. Most other secretions are likewise increased. The pulse is irregular, slow, and feeble; or regular, slow, and full.

Before profuse perspiration sets in, there is generally a feeling of great restlessness, with distention of the abdomen, and irregular and spasmodic respiration.

The cheeks are usually suffused; the pupils are dilated, and the eyes become more brilliant. The senses are rendered more acute; the brain is generally excited; the mind wanders, sometimes lapsing even to wild and furious delirium, although a calm and placid sensation pervades the system generally.

The evacuations of the bowels are seldom increased in frequency. The patient mostly remains quite still, since to move causes return of the sudden and violent vomiting, with additional prostration. After a time he gets short periods of sleep, or sinks into a semi-somnolent condition, and out of one of these sleeps he awakes quite well.

When lobelia does not cause vomiting, its power is expended in profuse diaphoresis or diuresis.

When the dose is extreme, excessive prostration, anxiety, contracted pupils, insensibility, and convulsions set in, followed by death.

["In 1809 Samuel Thompson was indicted in Massachusetts for the wilful murder of Ezra Lovett, Jr., by giving him a poison called lobelia, on the ninth of January, of which he died the next day."[1] The founder of

---

[1] Barton: Vegetable Materia Medica, Phil., 1817.

Thompsonianism escaped conviction on the ground that he destroyed life through ignorance and not by design. If the State laws had declared that the wilful administration of powerful drugs by ignorant persons was a grave criminal offence, and had provided appropriate punishment, it is probable that the almost incalculable damage effected by this agent in the hands of Thompson's followers in New England and in this State would have been avoided. On the contrary, much of our medical legislation has been directly protective of this class of persons, and encouraged an increase in their number.]

THERAPEUTIC ACTION.—In the United States this powerful plant is considered the most active article of the materia medica. When Europeans made their first settlement in that part of the world it was found to be the medicine most regularly resorted to by the aborigines. For a long time, however, it was left in the hands of irregular practitioners, and was only introduced to the notice of the profession by the Rev. Dr. Cutler, of Massachusetts. It was introduced into English practice in the year 1829, by the late Dr. Reece, who published a treatise upon its "anti-asthmatical properties."

The principal value of lobelia, in cautious hands, appears to be that of an anti-spasmodic. In some cases it becomes a useful adjunct to diuretics.

Lobelia has been resorted to as a substitute for tobacco in cases of strangulated hernia, in which it has been employed in the form of enema.

Paroxysmal spasmodic asthma, with extreme oppression, striving to cough, with inability to do so, is relieved by lobelia. It also often mitigates dry cough, with continual tickling in the throat, such as hinders the patient from sleep.

But, in some instances, it has appeared to lose power by repetition; and in any case it should be administered in doses so small as not to cause either nausea or vomiting. If either of these results should ensue upon the administration, the giving of the medicine must be immediately suspended.

Other disorders of the organs of respiration, such as catarrhal asthma, croup, and whooping-cough, when treated with lobelia, though great expectations were once entertained of its utility, appear not to have given encouragement to persevere with it.[1]

As an emetic, lobelia is unfit for general use, the operation being much too distressing for the patient, and by no means free from danger.

Compared with nicotine, it is a less active principle, though it is to tobacco that the general properties of lobelia admit of being most nearly compared. The taste and the sensations excited in the throat are similar; and, in default of any exact chemical tests by which the presence of lobelia may be ascertained, these form, at all events, an approximate clue to it. The analogy between lobelia and tobacco was originally observed by the American practitioners, and was confirmed by Dr. Elliotson.

I agree with many authors in believing that the administration of lobelia is best confined to temperaments remarkable for predominance of the nervous element.

PREPARATIONS AND DOSE.—Acetum Lobeliæ, ℥ x.—xxx. (.60

---

[1] Dr. Ringer, however ("Handbook," p. 484), speaks very highly of its use in uncomplicated whooping-cough, and in all purely spasmodic respiratory affections. He has observed a much greater tolerance of the drug in children than in adults.

—2.); Tinct. Lobeliæ, ♏ x.—xxx. (.60—2.); Tinct. Lobeliæ Ætherea (B. Ph.); ♏ x.—xxx. (.60—2.).

ERICACEÆ.

UVA-URSI (Arctostaphylos Uva-ursi).

Active Ingredients.—The leaves, when dried and pulverized, evolve an odor resembling that of hay or tea; their taste is bitter and astringent. The constituents, which are brought out both by water and by spirit, consist of about 35 per cent. of tannin; 1.5 per cent. of bitter extractive; gallic acid, just a trace; a peculiar substance called arbutine, to which the properties of the plant are probably due; and one or two others of little moment. The infusion of the leaves gives a bluish-black precipitate with perchloride of iron.

Physiological Action.—Taken internally, by the healthy subject, uva-ursi, as a rule, does not appear to act as a diuretic. But when given in disorders of the urinary organs, it would seem both to increase the quantity of the secretion and to affect its quality, lithic deposits becoming lessened by perseverance in the use of it as a medicine. The astringent principle passes off in the urine, giving to the latter a dark color.

In small medicinal doses, it increases the appetite, but rather constipates the bowels. In large doses, it nauseates, and produces vomiting and purging. It also produces inflammatory irritation of the lining membrane of the bladder and of the urethra, accompanied by tenesmus, and often by a bloody discharge and, later on, by a purulent and bloody one.

Therapeutic Action.—The principal value of uva-ursi is shown in chronic affections of the bladder, attended by mucous, bloody, or purulent discharges, with burning in the urethra during urination, and especially when these symptoms are produced by calculus. In nephritic paroxysms, and in irritable states of the bladder dependent on renal disease, uva-ursi is decidedly effective.

In Catarrh of the Bladder, in Leucorrhœa, and in Gonorrhœa, uva-ursi has likewise been administered successfully. For acute cases it is unsuitable.

Contraction of Gravid Uterus.—Dr. Harris, of Fayette (Alabama), and Beauvais and Gauchet, in France, assert that uva-ursi possesses the property of causing contraction in the same way that ergot does; but this requires confirmation.

Preparations and Dose.—Extractum Uvæ Ursi Fluidum, ℨ j.—ij. (4.—8.); Decoct. Uvæ Ursi, ℥ j.—ij. (30.—60.).

PYROLACEÆ.

PIPSISSEWA (CHIMAPHILA UMBELLATA).

ACTIVE INGREDIENTS.—When bruised, the chimaphila evolves a powerful odor; the leaves, while fresh, possess considerable acridity; though in taste warm and somewhat pungent and astringent. In the dried state the plant smells like tea. The chief ingredients appear to be a bitter extractive matter, in which is contained a yellow crystalline principle called chimaphiline, discovered by Fairbank, a resin, also tannin, and, perhaps, a little gallic acid.

PHYSIOLOGICAL ACTION.—The freshly gathered leaves, if bruised and applied to the skin, act as a rubefacient and vesicant. The decoction of the dried plant is tonic, producing agreeable sensations in the stomach, exciting the appetite, and assisting digestion. All the secreting organs, the kidneys in particular, receive a degree of stimulus from it, and it is said to check the secretion of lithic acid.

THERAPEUTIC ACTION.—The chimaphila was found to be in use among the aborigines of North America almost as early as the first settlement there of Europeans; but it was not until 1803 that the notice of the medical profession was directly drawn to it.[1] In 1817 Dr. Wolff, of Göttingen, published a treatise, "De Pyrola Umbellata," but in England it has never received the attention it would seem to deserve.

In several forms of chronic nephritic disease, attended by albuminuria, it unquestionably has power. It is useful, also, as a diuretic in dropsy (especially when the disease is accompanied by loss of appetite and great debility). Chronic catarrhal affections of the urinary organs are likewise amenable to it, as are hæmaturia, ischuria, dysuria, and gonorrhœa.

As a remedy for scrofula, the chimaphila holds a certain amount of reputation. As a tonic, it may be exhibited with advantage. Externally it has been found useful in tumors and ulcers of various kinds.

Chimaphila may be expected to prove a useful remedy in gout and rheumatism, especially if, as above stated, it has the power of checking the secretion of lithic acid.

PREPARATIONS AND DOSE.—Extract. Chimaphilæ Fluidum, ℥ xv.— ℨ j. (1.—4.); Decoct. Chimaph., ℥ j.—iv. (30.—120.).

[[1] This is a mistake: both the Pyrola umbellata and P. maculata are mentioned by Schoepf (Mat. Med. Amer., 1787). Of the latter he says: "*Infusum foliorum, ante annos aliquot, sub nomine Pipsisseva, frequentissime ad Febres intermittentes exhibebatur in Pennsylvania.*" In the "Reports of the Medical Society of the City of New York, on Nostrums or Secret Medicines" (1827) Chimaphila is mentioned as a probable ingredient of Swaim's panacea, a then celebrated nostrum.]

STYRACACEÆ.

STORAX (Styrax Officinale).

Active Ingredients.—" Storax in the tear," the genuine liquid balsam of storax, is brownish-yellow, translucent, and of the consistence of honey. The odor is peculiar and agreeable; the taste is warm and aromatic. When pure, it is soluble in alcohol and in ether. The constituents are a volatile oil, cinnamic acid, styracine, and two peculiar resins, one hard, the other soft. The oil, called Styrole, C^8H^8, is a colorless, translucent, and very volatile liquid, with a burning taste, and yielding the odor of storax. It is soluble in alcohol and in ether; boils at about 295° F.; burns with a sooty flame; and is of sp. gr. 0.924. Cinnamic acid, $C^9H^8O^2$ (a constituent also of the Balsams of Peru and Tolu), is colorless and crystalline; soluble in alcohol, and sparingly so in cold water. It is distinguished from benzoic acid by yielding the odor of oil of bitter almonds on being treated with a solution of chromic acid. The third important element of storax, styracine, $C^{18}H^{16}O^2$, is a crystalline solid, sometimes obtained in a liquid state. It is soluble in ether, less so in alcohol, and insoluble in water.

Physiological Action.—In its operation, storax closely resembles the Balsams of Peru and Tolu.

Therapeutic Action.—Being stimulant and expectorant, storax induces all the usual effects of balsamic substances. The stimulant properties are manifested more particularly when it is employed in cases of irritation of the mucous surfaces. It has been used with advantage in affections of the respiratory organs, such as chronic bronchitis, in which it usefully promotes expectoration.

Chronic catarrhal affections of the genito-urinary membrane are also relieved by the use of storax. It has likewise been found serviceable in amenorrhœa.

Externally, storax has been used in the form of ointment, as a detergent for foul ulcers, appearing to mitigate the offensive character of the matter secreted.

Preparations and Dose.—Styrax præparatus (B. Ph.), gr. v.—xxx. (.30—2.). Storax is an ingredient of the Tinctura Benzoini Comp.

BENZOIN (Styrax Benzoin).

Active Ingredients.—The best Siam benzoin occurs in white or reddish tears, agglutinated together by a darker colored and amorphous portion, the fracture presenting an amygdaloid appearance. The odor is agreeable, to some persons it is fragrant; the taste is balsamic and slightly sweet. When heated, it melts, and gives off benzoic acid in the form of white fumes, with a little empyreumatic oil, and then gradually consumes away. It is soluble in alcohol, and in solution of potash. Ether dissolves the resinous element in part. Sp. gr., 1.092.

The principal constituents are about 80 per cent. of resin, and 10 to 20 per cent. of benzoic acid, $HO\ C^{14}H^5O^3$, or perhaps $HC^7H^5O^2$. This acid was known as long ago as 1608, and is procurable by sublimation. It presents itself in the shape of soft white feathery and flexible crystals, with a pearly lustre, and usually with a slight impregnation of the above-mentioned empyreumatic oil, which confers on it the odor of the original mass. It is soluble in alcohol, also in solutions of lime and of caustic alkalies, from which it is precipitated by hydrochloric acid; but in water it is very sparingly soluble, requiring about 25 parts of boiling, and about 200 parts of cold, for complete solution. It melts at 248°, and boils at 462°, entirely subliming when pure. Of the resin, one portion is soluble in ether and in carbonate of potash, and the other portion is not soluble.

PHYSIOLOGICAL ACTION.—If chewed for any length of time, benzoin causes irritation of the fauces: the fumes produced by heat and combustion are also irritating; and the powder, if inhaled, excites sneezing. Of the action of this substance upon the body, when swallowed, we possess, however, no exact knowledge. Taken in large doses, it is said to quicken the circulation, and to increase the flow of urine and the cutaneous exhalations. Schreiber swallowed half an ounce of benzoic acid, divided into 40 doses, in the course of two days, with the result of prolonged irritation of the throat, and a sense of warmth, first in the abdomen, and afterwards throughout the body; the pulse at the same time became more frequent, though afterwards it subsided gradually. After the first day, the perspiration and the expectoration were augmented, but the quantity of urine was unchanged. The head was somewhat confused, and digestion was temporarily impaired.

Professor Wœhler, in 1831, expressed his opinion that benzoic acid, during digestion, was probably converted into hippuric acid; and Dr. Ure states that the administration of benzoic acid causes uric acid to disappear from the urine, the latter being replaced by hippuric acid. On this ground he recommends it as a remedy for the gouty and calculous concretions of uric acid. Keller, on the other hand, maintains that the uric acid undergoes no change, and that the benzoic is directly converted into hippuric acid. Garrod likewise takes this view, and states also that it renders the urine more acid as well as somewhat irritating. That benzoin is a local irritant there is no doubt. The vapor, when too concentrated, induces violent coughing and sneezing.

THERAPEUTIC ACTION.—Being stimulant and expectorant, benzoin is usefully employed in bronchitis and other chest affections. By reason, however, of its irritant effects upon the mucous membrane, great care must be taken not to administer too large a quantity, such as would over-stimulate that membrane. In the chronic bronchitis of old persons it proves especially useful. Benzoin has likewise been employed with advantage in chlorosis and other uterine disorders.

The compound tincture is an admirable local application for sore nipples, and for chapped lips and hands. When intended to be thus employed, one part of this tincture should be added to four parts of glycerine. Mixed with water, it is used as a cosmetic to remove freckles and other eruptions. The special cosmetic called "Virgin's milk" is a spirituous solution in about twenty parts of rose-water. "Friar's Balsam,"

"Wade's Drops," and "Jesuits' Drops" are nothing more than forms of the compound tincture.

Benzoic acid is used in chronic cystitis, and as a stimulant of the mucous membrane of the bladder. It often removes the fœtor of the urine in cases of irritable bladder induced by enlarged prostate. It is likewise a very useful remedy in cases of phosphatic urine dependent upon irritation of the bladder, being perfectly capable of rendering the urine acid in a short time. Hence, too, it renders alkaline urine acid.

Benzoate of ammonia acts in a manner very similar to benzoic acid, and is thus a very convenient salt for exhibition in cases such as are amenable to the pure acid.

Benzoin, it may be added, is the basis of "Turlington Balsam," a useful preparation for the healing of wounds and ulcers, and especially valuable to travellers and others who cannot easily obtain surgical aid. It is employed likewise in the preparation of "paregoric elixir," and in the manufacture of court-plaster and of pastiles.

PREPARATIONS AND DOSE.—Tinct. Benzoini, 3 ss.—j. (2.—4.); Tinct. Benzoini Comp., 3 ss.—j. (2.—4.); Acidum Benzoicum, gr. v.—xv. (.30—1.); Ammonii Benzoas, gr. v.—xv. (.30—1.); Ungt. Benzoini.

MYRISTICACEÆ.

THE NUTMEG (MYRISTICA FRAGRANS).

ACTIVE INGREDIENTS.—Nutmegs contain about 30 per cent. of a concrete oil, which, when extracted by means of heat and expression, receives the name of butter of nutmegs. In commerce it occurs in solid, oblong, brick-shaped cakes, of an orange color, and possessed of the agreeable and aromatic odor of the nut. Ordinarily they are wrapped in the leaves of some endogenous plant. This primary or concrete oil is soluble in four times its weight of boiling alcohol, and in twice its weight of ether.

Upon analysis it yields a fixed oil, or fat, which is solid and crystallizable, melts at 118° F., and contains myristic acid ($C^{14}H^{n}OHO$) in combination with glycerine. The myristic acid crystallizes in silky needles. In addition to the fixed oil, there is a volatile one (the same as that obtained by distillation), which is colorless or pale yellow, and has the peculiar smell and taste of the nutmeg itself. Sp. gr., 0.95.

PHYSIOLOGICAL ACTION.—Nutmeg, like other spices, if taken in moderate quantity, is cordial and carminative, and has an agreeably stimulating effect. In excess, on the other hand, it proves narcotic, and causes giddiness, oppression of the chest, intense thirst, headache, delirium, and stupor. In particular constitutions, symptoms analogous to those produced by narcotic poisons are developed. Even apoplexy, with a fatal termination, is said to have resulted from the excessive use of nutmeg. Cullen is emphatic in his recommendation that persons who are believed to have a tendency towards apoplexy should abstain from the use of nutmeg as a condiment.

The volatile oil of nutmeg has been shown by experiment to possess rather powerful properties. Applied to the human skin it acts as a decided rubefacient: given to rabbits in doses of two to six drachms, it proved fatal, with symptoms of narcotic-irritant poisoning extending over a few hours or days; the urine had a peculiar smell, which was not that of the oil itself. When the dose was not quite large enough to kill, there was prolonged constipation after recovery from the acute symptoms.

THERAPEUTIC ACTION.—On account of its cordial, carminative, anodyne, and astringent properties, nutmeg has been employed in diarrhœa and in dysentery, and to relieve nausea and vomiting. For these disorders it is best administered in wine or in brandy-and-water, and in the form of simple powder or grated nutmeg. In mild cases of diarrhœa nutmeg proves an efficient substitute for opium.

As an External Stimulant the oil is occasionally applied to the skin, especially in paralysis and chronic rheumatism.

Mace is in India a favorite medicine in low stages of fever, in consumptive complaints, and in humoral asthma; also (in combination with other aromatics) in long-continued and wasting bowel-complaints.

PREPARATIONS AND DOSE.—Myristica, gr. v.—xv. (.30—1.); Spts. Myristicæ, ʒ ss.—j. (2.—4.); Oleum Myristicæ, ♏ij.—v. (.10—.25).

ASCLEPIADACEÆ.

HEMIDESMUS (HEMIDESMUS INDICUS).

ACTIVE INGREDIENTS.—The properties of the root are believed to depend upon the presence of a volatile crystalline substance, called hemidesmine or hemidesmic acid. They are yielded to boiling water, but little is yet accurately known respecting their nature.

PHYSIOLOGICAL ACTION.—Upon this point likewise nothing is yet definitely known, except that the effects of the plant are generally diuretic.

THERAPEUTIC ACTION.—The diuretic virtues of hemidesmus render it useful in certain diseases of the kidneys: also for some syphilitic skin diseases, and for indigestion, when it becomes serviceable after the manner of sarsaparilla, with which reputation it was brought to England by Dr. Ashburner, about the year 1830. In India, where the plant is common, the roots are largely employed as a substitute for sarsaparilla, acting also as a tonic and diaphoretic.[1] In England, however, hemidesmus is principally employed as a flavoring agent. Lindley stated, in 1838, that much of it was then employed in London as a very fine kind of sarsaparilla, and that it was said to be quite as efficient a medicine as the

---

[1] Waring ("Manual of Practical Therapeutics") quotes O'Shaughnessy to this effect, and decidedly concurs with the view of the latter that hemidesmus resembles sarsaparilla, but is superior to it for the above-named purposes.

best sarsaparilla of America. The roots of other species of hemidesmus are probably sometimes mixed with it, or substituted for it.

PREPARATIONS AND DOSE.—Syrupus Hemidesmi (B. Ph.), 3 j.—ij. (5.—10.).

VALERIANACEÆ.

VALERIAN (Valeriana Officinalis).

ACTIVE INGREDIENTS.—The ethereal oil of valerian contains the active ingredients of the plant. There is still much uncertainty as to its precise composition; perhaps the analysis of Pierlot comes nearest to the truth. According to this author, the oil consists of valeren, 25 parts; valerianic acid, 5 parts; valerian-camphor, 18 parts; resin, 47 parts; water, 5 parts. Valeren is a colorless oil, smelling like turpentine, which boils at 160° C., and under the influence of nitric acid is transformed into ordinary camphor. Valerianic acid, $C^{15}H^{10}O^2$, is a colorless, oily fluid, with a smell recalling at once the odor of valerian, and that of decayed cheese; it has a strongly acid and burning taste; the specific gravity is somewhere about 0.945; the boiling point about 170° C. It greases blotting-paper, but the mark gradually vanishes. It is soluble, in all its preparations, in alcohol and in ether; but requires about 30 parts of water to dissolve it.

Valerian-camphor, $C^{12}H^{20}O$, forms white crystals. It is not possible yet to assign their respective shares to these ingredients in the action of valerian oil. Valerianic acid is usually prepared by the oxidation of fusil oil, or amylic alcohol.

PHYSIOLOGICAL ACTION.—Valerian oil, procured by distillation of the root with water, may be taken as the representative of the collective virtues of the plant, and hence the recent researches of Grisar upon its action are full of interest and importance. Grisar states that he experimented with the ordinary volatile oil of valerian employed in pharmacy, and mentions that every specimen used was highly charged with the peculiar odor of valerianic acid. The general result of his experiments with valerian oil was similar to that of the trials with chamomile oil, etc.; reflex excitability was always manifestly reduced, and the specific reflex excitement of strychnia-poisoning was antagonized. Valerian oil appears to be one of the most powerful of the group of oils investigated by Grisar. There are many other remarkable effects of valerian, both upon man and some of the lower animals, which are difficult at first sight to reconcile with those above mentioned. Cats, according to Foy, discover it by the smell at a long distance. They are seen rolling themselves upon the plant, and are heard mewing and purring in the most extraordinary manner. After a while they are seized with spasms and convulsions, and at last expire in a kind of voluptuous frenzy.

In mankind, when in health, the powder of the root inhaled by the nostrils excites sneezing. Taken internally, as a medicine, in moderate and continued doses, it improves the digestion and increases the appetite, without interfering with the action of the bowels. Perseverance in its use for too long a period induces a decided tendency to low melancholy and hysterical depression.

Stillé says:[1] "Two drachms at a single dose generally occasion a sense of heat and weight in the abdomen, eructations, and frequently vomiting, colic, and diarrhœa; and, along with these symptoms, some excitement of the circulation, augmented warmth of the whole body, and either perspiration or an increased flow of urine. The first effects of valerian upon the nervous system," he continues, "are in doses of one to two drachms, to render the mind tranquil, to incline to good humor, and dispose to exertion. But these results are usually accompanied by a lively formication in the hands and feet, and a sensation about the head and spine which has been compared to the *aura epileptica.*"

Berbier, in his "Materia Medica," gives an account of a patient in the Hôtel-Dieu of Amiens, who, after taking six drachms of valerian daily for some time, woke up delirious, fancying that the side of the ward opposite to where he lay was in flames. Other patients who had taken the medicine imagined that flashes of fire darted from their eyes.

When taken in these large doses, there can be no doubt that valerian causes headache, hallucinations, with much mental excitement and giddiness. In a case related by Dr. Abell, the pulse became frequent, tremulous, and irregular, and the pupils were extremely and fixedly dilated; the patient fancied himself beset with dangers; he staggered, and had a constant desire to urinate. In addition to the above symptoms, following the use of large doses, there is restlessness, occasionally attended by spasmodic movements of the limbs. The patient is generally troubled with repeated eructations, tasting of valerian, and often with a disposition to vomit. According to Kummer, there is a sense of constriction in the pharynx, accompanied by headache, which last he describes as a pain in the right frontal region, extending to the vertex, and, more decidedly, to the eyes, causing a sensation of pressure when those organs are used. These head symptoms appear and disappear periodically, continuing for a few hours at a time. The headache sometimes appears to shift to the left side, but retains its original character. Guntz adds to the above that there is a feeling of fulness and heaviness in the epigastrium, accompanied by rumbling and cutting pains in the bowels, followed by stools of a diarrhœic character. The desire to urinate is one of the results of taking valerian in an overdose; micturition not only becomes more frequent and profuse, but the fluid is generally turbid, and deposits a "bran-like sediment,"[2] a "brick-red sediment,"[3] or "a slimy sediment which seems to dissolve after shaking the vessel."[4]

THERAPEUTIC ACTION.—From the time of Galen down to the present day, valerian has had repute in **epilepsy**. Fabius Colonna thought that he cured himself of this disorder by employing the powdered root, when many other medicines reputed trustworthy had utterly failed. Scopoli relates a case (in the "Flora Carniolica") of supposed epilepsy resulting from fright, which had yielded to the same drug. And Marchant, in the "Histoire de l'Acad. Roy. des Sciences," A.D. 1706, mentions several instances of the same kind. But it is probable that some at least of the authors who have commended valerian in this complaint have mistaken the epileptiform convulsions which occur in hysteria for true epilepsy. Convulsions, it must be remembered, are also induced by

---

[1] Vol. ii, p. 34. [2] Guntz. [3] Winkler. [4] Engler.

worms, and such convulsions again often resemble epilepsy. Valerian destroys worms, and might consequently remove convulsions which were referable thereto. Moreover, from all that is known of the general action of valerian, which is not curative, but simply palliative, it would hardly be expected that true epilepsy should be removable by it. Epilepsy, so-called, developing itself from hysteric spasms, is without doubt often relieved, even permanently so, by the use of valerian.

The most likely opportunities for the advantageous employment of valerian are those in which a nervous excitant is apt to do good, and where stimulants are admissible, in which cases it becomes valuable as an antispasmodic. It is especially serviceable in headaches (hemicrania) of a nervous or hysterical character, and in those which follow profuse or painful menstruation; also to persons of a nervous temperament, especially females, who are subject to attacks of hysterical dyspepsia, and to flushing of the face upon the least excitement, and when the dyspepsia is accompanied by temporal or frontal headache, irritation, and distended abdomen. The flatulent distention which comes on without notice and subsides quite as suddenly, after causing more or less discomfort, is peculiarly well dealt with by the use of valerian.

Valerian, in a word, is useful in most cases of nervous affection occurring in excitable temperaments. In hypochondriasis it calms the nervousness, abates the excitement of the circulation, removes wakefulness, promotes sleep, and induces sensations of quietude and comfort; sadness is removed, and the hypochondriac state of mind in general abates. In globus, in all asthmatical and hysterical coughs, and nervous palpitation of the heart, accompanied by dyspnœa, valerian has likewise proved serviceable; also in facial neuralgia of an hysteric type. In fevers, too, of a nervous type, such as have for their symptoms (among others) morbid sensibility, uncertain movements, rapid respiration with sighing, often accompanied by gaping, sleeplessness, restlessness, fidgets, anxiety, and a profuse flow of limpid urine, this medicine has repeatedly done much good. In Germany it is more valued in this direction than among ourselves. Chorea, likewise, is sometimes very successfully treated with valerian, especially when attributable to worms, to which, as before stated, valerian is fatal. Stimulating the nervous system, valerian is again very usefully employed in combination with guaiacum in strumous enlargements of glandular structures. Some physicians believe that it is prudent to combine it with cinchona or with ammonia when administered in hemicrania and hysteria. Ammonia indeed often appears to be a desirable adjunct to valerian.

The anti-hysteric properties of valerian appear to be in a certain measure confirmed by the employment, in the East Indies, of spikenard (Valeriana Jatamansi), which the females of the country find eligible in cases of hysteria.

In addition to these more ordinarily accepted uses of valerian, will probably have to be added in future (to judge from Grisar's researches) the power of mitigating the spasms of strychnia-poisoning. Indeed, it is probable that the ultimate results of those observations will be to lead therapeutists to style valerian a **sedative to reflex excitability**, in preference to the vague appellations of "nervine stimulant," "antispasmodic," etc.

PREPARATIONS AND DOSE.—Oleum Valerianæ, ℳ ij.—v (.12—.30); Extr. Valer., gr. x.—xxx. (.65—2.); Extr. Valer. Fl., ʒ ss.—j. (2.—

4.); Tinct. Valer., 3 j.—iij. (4.—12.); Tinct. Valer. Ammoniata, 3 j.—ij. (4.—8.).

VALERIANIC ACID AND ITS COMPOUNDS.

Something must necessarily be said (though it is not easy to say anything very satisfactory) concerning the action of valerianic acid and the valerianates.

Of Valerianic Acid, $C^9H^{10}O^2$, Reissner[1] is the only author who has made any careful physiological study. As regards its chemical effects he states that it is capable of coagulating blood-serum and milk. Applied to the skin, it acts only very slowly and mildly as an irritant: applied to the tongue, or taken into the alimentary canal, it will kill (in large enough doses) pretty quickly; the symptoms and post-mortem appearances are those of irritant poisoning, but it is remarkable that there is no diarrhœa. The heart beats fast but weakly; the respirations are at first quickened, and subsequently become slow and labored; the lower extremities are weakened, and then actually paralytic; and there are sometimes convulsions before death. The acid undergoes some change in the blood, for it is not eliminated as such in the urine. The acid itself is not used in medicine, but several of the valerianates have been much in vogue from time to time.

Valerianate of Zinc has been much lauded in various spasmodic diseases, especially in epilepsy, chorea, and the convulsions of children, as also in the convulsive forms of hysteria, in neuralgia, etc. It must be allowed by any unprejudiced person who reads over the evidence adduced by Neligan, Tilt, and others, that it is very doubtful whether we have here anything more than examples of the great, though, unfortunately, very irregular and uncertain power of zinc over the nervous system. I consider the efficacy of valerianate of zinc, as such, entirely unproved; and, in reference to its supposed effects in neuralgia, may say that it has proved quite untrustworthy, so as to compel me to the belief that it can only be effective through the agency of the imagination.

In Chorea I have seen no reason to think that valerianate of zinc ever really cures, for the severe cases do not even temporarily yield to it.

In Epilepsy it would be folly to waste time in trying this remedy when we have the bromide of potassium in our hands. Dose, one-half grain to four grains or more.

Valerianate of Soda is even less entitled to credit as a positive remedy than the zinc salt. Dose, one-half grain to two grains or more.

Valerianate of Quinine has attracted a good deal of attention of late years; but there is really no valid evidence in its favor; indeed Gubler has gone so far as to say that the action of valerianic acid is positively antagonistic to that of quinine, and that the only reason why the compound is not perfectly inactive is that it becomes decomposed in the body, and the stimulant effects of the acid passing off rapidly, the quinine is enabled to assert its unchecked influence on the organism. Dose, one to five grains.

Valerianate of Atropine has been very highly recommended; but there is nothing in the accounts by Michéa and others of its effects in

---

[1] "De Acido Valerianico ejusque affectu," etc., Berlin, 1855.

epilepsy which may not be perfectly well explained by the influence of atropine alone.

Valerianate of Iron and Valerianate of Ammonia are also used in medicine. Dose, one to two grains or more.

In short, it may be said with confidence that in the action of the valerianates we trace very little, if anything, of the powers which valerian itself, and the ethereal oil of valerian, undoubtedly possess; and it is quite uncertain whether the acid has any real influence on the nervous system.

CINCHONACEÆ.

CINCHONA.

ACTIVE INGREDIENTS.—The ingredients which confer physiological and therapeutic power upon the cinchona barks are very numerous. They may be divided into four groups, varying in importance, but perfectly well distinguished from one another, and, excepting the last, these are again resolvable into smaller groups. The primary ones are as follows: (1) Alkaloids; (2) Simple Acids; (3) Tannins; (4) the resinoid Kinovine.

1. ALKALOIDS.—These are resolvable into (A) Quinine; (B) Cinchonine; (C) Quinidine; (D) Cinchonidine; (E) Aricine. There is likewise the so-called Quinoidine, an impure residue of manufacture from which is prepared the amorphous quinine of Liebig.

(A) Quinine, $C^{20}H^{24}N^2O^2$, exists in all the medicinal cinchonas, but is most plentiful in "yellow bark," occurring in natural combination with kinic acid and kino-tannic acid. The pure alkaloid is a strong base. It completely neutralizes acids, and produces with them both neutral salts and acid salts which are crystalline. It is very insoluble in water, freely soluble in alcohol, and less so in ether.[1] The solutions are distinguished by a remarkable blue fluorescence. Excess of chlorine-water, and the subsequent addition of ammonia, produce with the salts an emerald green. Quinine *per se*, is never employed in medicine. By far the most commonly used salt is the neutral sulphate ($2C^{20}H^{24}N^2O^2, HO, SO^3 + 7HO$), which substance crystallizes in tufts of fine silky needles, and occasionally scales. These are so light that the aggregate of the mass occupies considerable space. In ordinary states of the atmosphere and of storage, the sulphate contains at least two equivalents of extraneous water, which can be entirely driven off by a temperature of $248°$ F., but speedily becomes reabsorbed. The purely bitter taste of quinine is highly characteristic.

Of late years, however, and especially in Germany, the neutral hydrochlorate of quinine has been preferred by many physicians, particularly by Binz, not only on account of its superior medicinal qualities, but because it is less subject to the fungus which spoils ordinary quinine solutions.

Numerous other salts of quinine will be mentioned under the head of

---

[1] The hydrate, with three equivalents of water, is much more soluble in ether than the anhydrous alkaloid.

Preparations. Here it will suffice to state that the sulphate and the hydrochlorate are probably capable of accomplishing all the good that can certainly be effected by quinine.

(B) Cinchonine, $C^{20}H^{24}N^2O$, is most abundant in the paler varieties of bark. It forms clear, colorless, four-sided prisms, which are soluble in thirty parts of water, and are very insoluble in alcohol and in ether. With acids it forms soluble salts, which do not fluoresce in solution, and are turned lightish brown-yellow by the chlorine and ammonia tests.

(C) Quinidine, $C^{20}H^{24}N^2O^2$, contained in many varieties of bark, is an alkaloid isomeric with quinine, with two equivalents of water, but is less intensely bitter, and less soluble in water and in ether. It gives a similar fluorescence, and the same color with the chlorine and ammonia tests. Sulphate of quinidine is much more soluble in water than sulphate of quinine.

(D) Cinchonidine, $C^{20}H^{24}N^2O$, is isomeric with cinchonine. It occurs in large, shining, striated, rhombic prisms, which are anhydrous, and scarcely at all soluble in ether. The solutions are fluorescent, but do not answer to the chlorine and ammonia tests. In taste, compared with quinine, it is less bitter.

(E) Aricine, $C^{23}H^{26}N^2O^4$. Concerning this it is unnecessary to say anything in detail, since there is no probability of its ever being brought into use in practical medicine.

2. SIMPLE ACIDS.—The simple acids contained in the cinchona barks are the Kinic and the Kinovic.

(A) Kinic acid (or Quinic acid), $C^7H^{12}O^6$, forms large, transparent, colorless tablets, the taste of which is strongly and purely acid. They dissolve very easily in cold water; much less readily in boiling water. They are more soluble in weak than in strong alcohol, and in ether are nearly insoluble.

(B) Kinovic acid, $C^{24}H^{38}O^4$, was for a long time supposed to be obtainable only by artificial (chemical) means from the resinoid Kinovine. It is now known to be a natural ingredient of the raw kinovine, or at all events of the kinovine which is furnished by the cinchonas grown in Java. This acid is probably of far greater importance than the kinic; recent researches, as will be shown presently, have invested it with much interest.

3. TANNIN.—The tannic acids of the cinchonas are two: (A) Kinotannic acid, and (B) Kinovi-tannic acid.

(A) Kino-tannic acid, as prepared by Schwarz's process, is a bright yellow mass, easily pulverized, but very hygroscopic, and possessed of a sour and astringent, but not bitter taste. Friction renders it electrical. It dissolves readily in water, alcohol, and ether. (Formula, $C^{23}H^{30}O^{35}$?)

(B) Kinovi-tannic acid is of a clear, transparent, yellow color, and in taste is somewhat bitter. It is soluble in water and in alcohol, but not soluble in ether.

Whether either of these two tannic acids exists in the officinal cinchonas is, however, somewhat doubtful.

4. KINOVINE (or Quinovine), $C^{30}H^{48}O^8$, is an amorphous resinoid body, which can be rubbed into a smooth white powder, possessed of manifestly electric properties. When warmed, it evolves a feebly balsamic odor. The taste, though very slight, is sharp, and unpleasantly bitter. The reaction is neutral; it is excessively hygroscopic; hardly at all soluble in

water, very soluble in spirit, somewhat less so in ether. Dry distillation with lime develops metacetone and resinoid bodies. Heating with strong nitric acid causes the evolution of red fumes. Concentrated sulphuric acid gradually dissolves it with a dark red color. Hydrochloric acid vapor conducted into an alcoholic solution causes, as above stated, the development of kinovic acid along with kinova sugar.

Whether kinovine is itself an available remedy in disease has not yet been determined. It is employed, however, in the preparation of Kerner's pure and impure kinovates of lime.

PHYSIOLOGICAL ACTION.—The action of the cinchona barks upon the living organism is very complex: indeed, it is not yet possible to specify, in full detail, either the whole of the effects which the bark itself can produce, or the precise shares which are respectively contributed by the several ingredients. I proceed, however, to give some account, first, of the action of cinchona bark itself, and then of the effects produced by the active ingredients above enumerated.

Action of Cinchona Bark.—The first point requiring consideration is that of the chemical differences between the various kinds of officinal bark which are sufficient to cause diversities in their respective effects. The most important part of this question relates to the differences in the percentage (α) of alkaloids; (β) of tannin; (γ) of quinovine; and (δ) of kinovic acid.

(α) As regards the alkaloids, it is found that between yellow bark and red bark there is the following difference: the yellow contains about two per cent. of quinine, and scarcely any cinchonine; while red bark contains upon the whole about an equal quantity of each. Pale bark, in contrast with the other officinal sorts, contains only half the total customary percentage of alkaloids, and what it does contain is chiefly cinchonine. Here it may be noted that the species of cinchona chiefly cultivated in India are the *succirubra* and the *lancifolia ;* and that the latter is reported to contain a larger proportion than the former both of quinidine and of cinchonidine.

(β) As regards the tannin. Of this, red bark contains about 3.2 per cent., and yellow bark about 2.5 per cent. In pale bark the proportion is considerably less.

(γ, δ) The proportions of kinovic acid and of kinovine contained in the different varieties of bark have not yet, so far as I am aware, been exactly determined. The matter is nevertheless one of great consequence. We know at all events from the researches of De Vrij that kinovine is abundant in yellow bark. For reasons which will become evident when the therapeutic actions of the particular ingredients of bark shall have been spoken of, it will be highly interesting to learn whether the officinal pale bark, and other kinds which are weak in alkaloids, are fairly rich either in kinovic acid, or in kinovine,[1] since it may sometimes be desirable to seek to obtain the effects of these two last without material admixture with them of the action of the alkaloids, but still in combination with a small amount of tannin-astringency.

The action of the cinchona barks, as such, has never received the kind of investigation which has lately been pursued in regard to the alkaloids; and in the present state of our knowledge there are no means of deciding

---

[1] It is reasonable to believe that kinovine is decomposed in the gastric juice, with formation of kinovic acid and kinova sugar.

whether the barks of different species operate in ways which would theoretically correspond with the different proportions of the ingredients they contain. The following facts, in regard to cinchona barks in general, are, however, pretty well established: (1) When given in very large doses, they produce the phenomena which are known as "cinchonism," and which will presently be described under the head of quinine. But for the production of these phenomena extreme doses are required, and long before they commence certain disturbances come into play. (2) Upon the mucous membrane of the alimentary canal, cinchona bark produces effects which are probably quite independent of those induced by the alkaloids, and which are attributable to the astringent ingredients. In the provings of Jörg (undertaken with the object of testing the statements of the homœopathists), when powdered bark was taken in two-drachm doses, the symptoms were flatulence and eructation, sometimes accompanied by a little nausea, but more frequently by improved appetite. At the same time there was generally some constipation, but beyond this nothing, except that one of the experimenters was troubled with repeated nocturnal erections. This last-named effect will be more particularly noticed when speaking of the action of the alkaloids.

In passing, I must allude to the statement of Hahnemann and of some of his followers, that cinchona bark can produce intermittent fever. This is quite incorrect. No doubt, when taken in large doses by very susceptible people, cinchona may so upset the nervous system as casually to induce a symptom or two of the kind seen in true intermittents. But it is now admitted, even by homœopathic writers, that no one ever saw cinchona induce a periodic recurrence of attacks such as could in any way fall under the designation of fever. I shall point out, in the section upon quinine, the probable source of this error in observation.

[Concerning this matter, Trousseau and Pidoux speak as follows: "Daily observation, says Bretonneau, proves that cinchona, given in large doses, produces in many persons a well-marked febrile movement. The character of this fever and the period at which it manifests itself vary in different cases. Most frequently tinnitus aurium, deafness, and a sort of intoxication with a slight chill first occur. Succeeding this there is dry heat with headache, which gradually subside with slight perspiration. Under additional or larger doses the fever is increased.

"These physiological effects of cinchona, noticed in the first edition of our *Traité de Thérapeutique*, have been misunderstood and denied by the majority of physicians in this country; since then many observers abroad and at home have testified to the same, and though the authors claimed the honor of a discovery which belonged to Bretonneau, their testimony is none the less valuable; and to-day any physician may, with a little attention, verify the facts on which we here insist."[1]

The occasional occurrence of symptoms, after the use of large doses of cinchona (less frequently after pure quinine), which resemble those which accompany a malarial *paroxysm* is undoubted. The marked peculiarity, however, which characterizes malarial disease is a succession of paroxysms without apparent renewal of the exciting cause. This does not occur with cinchona; Hahnemann himself stating that renewal of the artificial fever only occurred after fresh doses of the drug.[2] It will be seen, there-

---

[1] [Traité de Thérapeutique, 9th ed., 1877, vol. ii., p. 576.]
[2] [As both homœopaths and those who are not have fallen into error as to what Hahnemann did say in this connection, we here give in full the passage, as found in a

fore, that cinchona is not homœopathic to the periodical feature of the disease, though it may sometimes produce symptoms that closely counterfeit an individual paroxysm. Pflüger gives a remarkable instance of this.—(*Berl. klin. Wochenschrift*, No. 37, 1877)].

Physiological Action of Quinine.—This subject has recently assumed vast proportions, and every month adds so much to our knowledge that it is nearly impossible that this article should fail to omit some more or less important facts which will have been published by the time it appears. It is necessary to consider quinine first in its physiological relations to protoplasm, and afterwards as to its action on the various organs of the body. That quinine in large doses is a protoplasm-poison has been lately proved by a number of observers; the fact having been first discovered by Binz, of Bonn, and subsequently verified by many of his pupils, and by various other physiologists. The latest demonstration of it was given by Dr. Buchanan Baxter in an elaborate research published in November last.[1]

As long ago as 1849 Buchheim and Engel had observed that quinine had power to check the progress of alcoholic fermentation; and the interesting researches of Pasteur, which came later, directed general attention to the importance of the low organisms which are present in fermenting liquor. In 1868 Binz published his first researches, which made it evident that quinine exercises a restraining action on the growth and movements of protozoa, and in large enough dose destroys their life. In this and subsequent papers by Binz, and by Scharrenbroich and Martin, it was further shown that this influence applied also to the development and to the migrations of the white corpuscles of the blood; and although there have been disputes as to the degree of saturation of the blood with quinine which is required in order to arrest the movements of the corpuscles, there can be no question now that this action exists, and that it can be carried to an extent which is even fatal to the life of the animal. In the interesting paper by Buchanan Baxter, the process of investigation is described very clearly. The hydrochlorate of quinine was dissolved in 0.75 per cent. solution of common salt; a large drop of this mixture having been placed on a glass, the cut surface of a newt's tail was touched with a clean cover-glass one inch in diameter; a few seconds having been permitted to elapse to allow of coagulation beginning, the cover-glass was carefully inverted on the drop. The excess of the fluid was then drained from the edges of the preparation with bibulous paper, and a ring of oil painted around it to prevent evaporation. Baxter makes the following very practical remarks on the general phenomena to be observed:

---

note to his translation of Cullen's Materia Medica, Leipzig, 1790, Bd. ii., p. 109 : "Ich nahm des Versuchs halber etliche Tage zweimal täglich jedesmal vier Quentchen gute China ein; die Füsse, die Fingerspitzen, u s. w., wurden mir erst kalt; ich ward matt und schläfrig, dann fing mir das Herz an zu klopfen, mein Pulz ward hart und geschwind; eine unleidliche Aengstlichkeit, ein Zittern (aber ohne Schauder), eine Abschlagenheit durch alle Glieder; dann Klopfen im Kopfe. Röthe der Wangen, Durst, kurz alle mir sonst beim Wechselfieber gewöhnlichen Symptomen erscheinen nach einander, doch ohne eigentlichen Fieberschauder. Mit Kurzem : auch die mir bei Wechselfiebern gewöhnlichen besonders characterischen Symptomen, die Stumpfheit der Sinne, die Art von Steifigkeit in allen Gelenken, besonders aber die taube widrige Empfindung, welche in dem Periostium über allen Knochen des ganzen Körpers ihren Sitz zu haben scheint—alle erschienen. Dieser Paroxysm dauerte zwei bis drei Stunden jedesmal, und erneuerte sich wenn ich diese Gabe wiederholte, sonst nicht. Ich hörte auf, und ich war gesund."]

[1] Practitioner, Nov., 1873.

"It is important to define strictly what is meant by the term 'migratory movement' as used in the following pages. If a drop of newt's blood be prepared with salt solution in the manner just described, and examined some hours after, the following will be among the appearances presented by the colorless corpuscles.

1. Migratory movement. The protoplasmic mass is spread out into a thin film, adherent to the glass, and of a most irregular shape. The outline of the film may be well marked in parts, but is never complete; the blurred portions being beset with fine filamentous processes, which are continually changing in size and number, some being retracted, while others are put out. The aggregate result of the changes of form which the corpuscle undergoes is a change in its *position;* it migrates from one part of the field to another.

"2. Change of form without change of position.

"α. The corpuscle is spherical, non-adherent, its surface velvety or shaggy, giving it an indistinct or blurred outline.

"β. Its outline is sharp; it is more or less spherical, non-adherent, and exhibits a peculiar waxy lustre. Its nucleus or nuclei are usually invisible. If carefully watched, it may be seen to change its shape very gradually; one part of its surface rising slowly into knob-like protuberances, while another part sinks into a corresponding extent; the protuberances being limited by definite contours, and exhibiting the same waxy lustre as the main body of the corpuscle. No filamentous processes are put forth. Such corpuscles, though their movements are undoubtedly vital, do not change their place. But they are capable of resuming the migratory condition as soon as the cause of their temporary quiescence ceases to operate.

"γ. The corpuscle may be surrounded by projections of another kind. The main body shrinks gradually into an irregular lump, while minute, translucent spherules of various sizes present themselves round its edge. These spherules do not appear to consist of the same substance as the main bulk of the corpuscle; they may even become detached from it, and float away. The appearances are those which might be presented by a coagulum slowly shrinking and expelling a clear fluid from its interstices; the fluid not being miscible with that in which the corpuscle is suspended. Whether a leucocyte in this state may, as in the two foregoing conditions, resume its migratory powers, I am unable to state.

"3. Permanent repose or death. The corpuscle is spherical or spheroidal, with a distinct outline. It does not exhibit any sort of movement; its protoplasm is very granular, and its nuclei are well marked. This condition is irremediable, and is the immediate precursor of disintegration.

"The first of these conditions is that of the highest vital activity; the second may pass either into the first or into the third. It is possible so to graduate the dose of a poisonous agent as to obtain the second or the third condition at will. In the following experiments, the exact state of the colorless corpuscles is usually indicated; the discordant results of previous observers in the case of quinia being probably due to a neglect of this exactitude. The condition of the red discs is often noted, inasmuch as they furnish the most delicate test of the neutrality of the liquid in which they float—the faintest trace of acid causing the delimitation, a larger proportion the granulation, of their nuclei."

Quinine very rapidly enters the circulation of men and of animals, whether put into the stomach, the subcutaneous tissue, or the cavities.

It is also very rapidly eliminated again, as was proved by Thau;[1] nearly the whole dose taken being discharged as quinine in the course of about twelve hours. Yet it has been proved by the researches of Bence Jones and Dupré that quinine itself (or a body in every respect resembling it) is, in minute proportions, a natural constituent of the body. Under these circumstances it is interesting to inquire whether more quinine is retained in the body when the drug has been given to a fevered, than when it has been given to a healthy subject; but the only difference shown by Thau's researches is, that in the former case the second period of six hours, and in the latter the first period of six hours, is the time during which the greatest elimination occurs.

The power of quinine to reduce bodily temperature, which is mainly a therapeutic and not a physiological effect, must be mentioned in this place, because it stands, or stood until lately, on debatable ground: it was doubtful, that is to say, whether the influence is one exerted through the nervous system by the medium of a supposed heat-regulating centre in the brain, or whether it is a part of a more generalized action on the tissues and fluids of the body. At the present time it is evident that scientific opinion is coming to agree with that expressed by Binz, namely, that the lowering of bodily temperature is produced by means of a general interference of quinine with the oxidation processes of the body in almost every part of it.

The special action of quinine upon the nervous system is most clearly seen in the symptoms which are associated with "cinchonism." When a patient is saturated with excessive doses of quinine, he gets loud ringing noises in the ears, splitting headache, vertigo, amaurosis, sometimes even delirium. In animals, a fatal dose of quinine has often produced convulsions and paralysis of the hinder extremities. It would appear that the lowered sensibility of parts, and the diminished muscular action, are due not to direct paralysis of nerves or to any interference with muscular irritability, but to a diminution of reflex action. For man it is not easy to say what would be a fatal dose of quinine, since enormous doses have been lately given with only temporary bad effects. Still it is certain that, if the stomach could be got to retain a sufficient quantity, we should have a fatal result, preceded, in all probability, by convulsions.

There are other local poisonous actions of quinine which are less constant, and which we can but little explain. One of the most singular of these is its very powerful and disagreeable action on the skin of many patients. Most medical men have met with one or two individuals in whom any dose of quinine, but especially a large one, produced irritation of the skin, followed by free desquamation; cases have not unfrequently been seen in which the whole skin of a hand or even of a limb has come off like a glove or a stocking.

Another occasional effect, more common than the last named, is the influence of large doses of quinine in accelerating the heart's action and the respiration. Many persons seem quite insusceptible to this influence even when very large doses are given to them; but in others the palpitation and hurried breathing are so pronounced as to cause much distress.

Physiological Action of Cinchonine.—This alkaloid, formerly supposed to be next in activity to quinine among the cinchona alkaloids, is now known to be the feeblest of them all. The experiments of Bernatzik, in 1867, showed that upon dogs the fatal dose of cinchonine was

---

[1] Practitioner, 1869, vol. ii.

one-fourth larger than the quantity of quinine which would kill other animals of the same weight.

Still more recent observations [1] confirm this estimate, by showing that the influence of cinchonine as a protoplasm-poison, though resembling that of quinine, is weaker not only than the quinine action, but than the action of either of the other alkaloids. It will be seen presently that this physiological position of cinchonine corresponds to its place in the therapeutic scale.

Physiological Action of Quinidine.—There is every reason to believe that this corresponds exactly in every particular with that of quinine. Baxter ranks quinidine just equal to quinine as a protoplasm-poison; and in all other respects it appears to behave precisely like it in the animal organism.

Physiological Action of Cinchonidine.—The great powers and activity of this alkaloid have only of late been appreciated. As a protoplasm-poison, and probably in every other physiological relation, it comes next to quinine and quinidine, and decidedly above cinchonine.[2]

One additional remark may be made before closing this account of the physiological action of the bark-alkaloids. It has been mentioned that bark itself in large doses has occasionally induced sexual excitement; and there is some reason to believe, though the matter is not cleared up, that each of the four alkaloids is capable of producing this effect under certain conditions.

THERAPEUTIC ACTION.—The curative effects of quinine and the other alkaloids of cinchona are manifold, and may be distributed into four chief divisions; (α) Antimiasmatic, (β) Antiseptic, (γ) Antiphlogistic, and (δ) the special actions on morbid conditions of the nervous system. Probably indeed there should be a fifth class including the (ε) so-called oxytocic action of quinine on the uterus, and some other more doubtful examples of its immediate action on muscular viscera.

α. The antimiasmatic action of cinchona alkaloids, by means of which they put an end to the morbid processes induced by the so-called paludal poison, is still the most interesting as it was the most anciently known therapeutic effect of bark. It is a fact as familiar to the public as to the medical profession that intermittent and remittent fevers yield promptly to the influence of bark or quinine; but it must be added that the *modus operandi* cannot be stated more than conjecturally, though it is now years since this remarkable specific effect was made known to European physicians.

The most probable theory is that which has been advanced chiefly by Binz and his followers, namely, that the paludal poison consists of, or is conveyed by, low organisms which enter the body, and increase and multiply in the blood; and that the anti-miasmatic action of quinine consists in its restraining power over the development and multiplication of these organisms.

The fact that quinine has such a power over the lowest organisms is now so amply demonstrated that there is no occasion to reopen that part

---

[1] *Vide* Buchanan Baxter: Loc. cit.

[2] It is a singular circumstance that, just as cinchonidine is obtaining its due recognition as one of the most powerful alkaloids, evidence has been procured which shows that the Countess of Chinchon (whose cure from intermittent fever originated the name "cinchona") was probably treated with a species of bark particularly rich in cinchonidine.

of the question. But the dependence of malarial poisoning upon the entrance of such bodies into the human organism is still among the vexed questions of pathology; and while increasing numbers of observers are inclined to affirm it, I see no grounds even now for a more positive statement than the following, by Nothnagel, three years ago: "How far these researches with their concordant results—that quinine checks putrefaction and fermentation by destroying infusoria and fungi—will be able to be utilized for pathology, is still uncertain. It is evident that at a time when people are everywhere discovering infusoria and fungi in pathological processes, it is tempting to apply Binz's researches to explain the value of quinine in malarial, puerperal, septicæmic, and (typhous) febrile processes, but this is impossible so long as the fungoid theory of these affections rests upon so unstable a basis as at present."

Without venturing to reverse this judgment, it may be admitted, however, that there is an increasing probability that the parasitic theory of the above-mentioned disorders will receive a more satisfactory confirmation in the future. But it must be remembered that very able pathologists are entirely opposed to the germ-theory of zymotic diseases; and those who wish to see the opposite view very ably presented in its most absolute form may be referred to Dr. Bastian's arguments.[1]

It is at least uncertain whether the germ-theory will not have to be modified in the sense of supposing that neither infusoria nor fungi, but the rapidly multiplying matter of human origin for which Beale pleads, or the analogous substances which are imagined in the graft-theory of Dr. Ross, are the true elements of infection. But in the latter case there would still be room to suppose that Binz's explanation would apply, for it would still be probable that the alkaloid acts curatively by its power to limit protoplasm formation.

We have now to consider the practical employment of quinine in the malarial fevers. The simplest case is that of the typical **Intermittent fevers in the acute stages**; but there are conflicting opinions as to the best method of administration even in these diseases. Probably almost any method which insures the daily absorption into the blood of some five or six grains of quinine would, in the end, cure any not extremely severe case of intermittent fever; but strong statements have been made by different authors in favor of different modes of distributing the dose. By some it has been laid down that we ought to give a single large dose of twenty or even thirty grains immediately before an attack is expected; others say that a smaller dose should be given, but either immediately after a paroxysm, or even while the final (or sweating) stage of the fit still lasts. Others, again, advocate the continuous use of large doses, both during the paroxysm and also during the periods of intermission; while another, and this perhaps the largest group of authorities, maintains that quinine should be given only during the intermissions, and then in repeated small doses (two to four or five grains or less).

I have found that in treating the ordinary agues of Great Britain, there is a decided advantage in employing large doses given only in the intermission. In the case of an ague of moderate severity, and which has not yet lasted very long, a tertian for instance of not more than two or three weeks' standing, the best plan is to order one twenty-grain dose to be taken on the day on which the fit is expected, and about an hour before the time at which it habitually occurs. In quotidian ague, especially

---

[1] See especially Appendix E to the "Beginnings of Life," vol. ii. Macmillan, 1872.

when it has lasted for some time, or has become quotidian from being previously tertian or quartan, at least twenty grains should be given every day, also about an hour before the expected attack is due. In quartan ague it is not always sufficient to give one large dose on the day when the fit is due; for some unknown reason the quartan type is frequently more obstinate than would have been supposed from the rarity of the fits; and I have usually found it best in these cases to give one ten- or fifteen-grain dose every day, some time before the hour when the fit occurs.

In the severe forms of intermittent fever which are met with in tropical countries, it is probably always well to employ large doses; and it is not unfrequently necessary to give the medicine during the paroxysm, as well as in the periods of intermission. The pernicious remittents still more often require the use of large and continuous doses of quinine; thirty to fifty or sixty grains[1] may be given in very severe cases several times a day, and without any special regard to the presence or absence of exacerbation. At the present day we have no reason to dread giving quinine during a paroxysmal period; on the contrary, we reckon quinine among our most active reducers of febrile temperature, and there can be no possible reason for withholding it when other considerations point to the advisability of speedily impregnating the system with it. Whether the fungoid theory of malaria be true or not, it is certain that something like the saturation which is necessary to produce an effect upon low organisms is by far the most effective way of dealing with any malarial fever which is heavily oppressing the system. It will be best at once to go on to the production of decided though not severe cinchonism. As soon as the ringing noise in the ears is fairly set up, we shall in almost every case observe a marked abatement in the fever; and a gentle perspiration is often the immediate forerunner of entire or partial convalescence.

It is not only in cutting short the typical symptoms of malarial fever that quinine is valuable; the majority of the complications of these diseases, of whatever kind they may be, yield better to the treatment by quinine than to any measures addressed immediately to the organ that appears to be suffering. Among these complications of ague, one of the most remarkable is the tumefaction of the spleen which forms the so-called "ague-cake;" and in a less degree the liver is liable to a similar change. I have rarely seen the slightest good effected by measures locally or specially addressed to these organs when so affected; nor indeed will quinine always produce an impression; but, at least in the majority of such cases, the action of quinine in reducing enlarged spleen is not less decided than its influence in cutting short the course of the general feverish symptoms. It is so likewise with the intense gastric irritability, causing incessant vomiting, which is almost universal in bad remittents; nothing quiets the stomach half so effectually as the production of full cinchonism. And the distressing pain and heat of the head, which drives such patients almost frantic, is also far better treated by the influence of quinine than by any local or other remedies for the mere symptoms. In regard to all these matters the profession owes much to Dr. Maclean and other recent Indian writers for the decisive courage with which they have put aside all temporizing measures, and have both practised and taught the radical treatment by quinine in sufficient doses to affect the system fully.

---

[1] It is right to say that so experienced an authority as Dr. Maclean deprecates the use of larger doses than fifteen or at most twenty grains. But the limit is too strict to suit all cases.

Although we may have full trust and confidence in the benefits conferred by a bold and unshrinking use of quinine in the acute stages of fever, there is no reason, however, why we should be reckless in the doses we employ, and more especially in the continuance of large doses beyond the necessary period. This remark gives me the opportunity to notice what I believe is the sole fact which keeps up the belief in the truth of Hahnemann's assertion that bark can produce intermittent fever. A great many, perhaps the majority of those persons who have ever been seriously affected with malarial fever, and have experienced the marvellous benefits of quinine, get into the habit of taking it upon the smallest indication of illness, and in quantities which are often extravagant. This cannot be done with impunity; it throws the nervous system at least into a state of great commotion. And we know very well that great nervous disturbance, from whatever cause arising, is a powerful influence to bring back the phenomena of ague in a person who has once suffered from them; for there are great numbers of persons who live quite comfortably on condition that they pass tranquil and carefully guarded lives, but who immediately suffer from a relapse of aguish symptoms when they happen to be exposed to harassing emotions, or anything else which much distresses the nervous system. And thus it may be said with truth that though quinine is quite unable to produce intermittent fever in a healthy person, its untimely and improper use by a person formerly aguish may very easily reproduce the paroxysm with greater or less severity. I have witnessed many examples of this occurrence, more especially in Americans and Indians who come from districts where malaria appears to be a more serious matter than in England.[1]

In those affections which indirectly result from an original malarial poisoning, it is of much consequence to determine the manner in which quinine may be most usefully employed.

In Neuralgias of malarial origin there is generally a much more regular periodicity of the attacks than in the neuralgias which are independent of this cause; and advantage should be taken of this fact to apply the principles already laid down in speaking of the intermittent fevers. Instead of continuing a number of small daily doses, it is far better to reserve the drug for employment at a time when a paroxysm may be confidently expected shortly to occur. About an hour before the time of the anticipated attack, a heavy dose (five to fifteen or even twenty grains, according to the degree of the patient's previous familiarity with the drug) should be administered; it is not a bad plan to give the quinine in a glass of sherry. Seldom does it fail to produce decided effects; the paroxysm either ceases altogether, or is much weaker than usual. The same plan is pursued before the next expected recurrence; and the neuralgia rarely fails to disappear entirely after a very few such repetitions of the dose.

The above plan is most suitable to severe cases with short intervals; in milder examples of the disease it may doubtless be sufficient to impregnate the system with a daily amount of three to six grains in divided doses.

Epilepsy of malarial origin is very effectively combated with quinine;

---

[1] During the whole period of Messrs. Howards & Sons' carrying on the manufacture of quinine they have never known a case of ague or intermittent fever amongst their workmen which could by any possibility be referred to the nature of their employment, nor have they ever heard of such a case. Messrs. Howards & Sons' is the oldest and largest firm in that branch of trade.

few things are more remarkable than the uselessness of this drug in common epilepsy, compared with its magical operation in the malarial form of the disease. Here we cannot put in force the principle of single large doses in anticipation of an attack; the proper method is to give regular moderate doses (about six grains a day), for several weeks in succession. So important are the benefits secured by the quinine treatment in cases where malaria has had any share in the production of the disease, that even a small probability that such has been the case ought to induce us to try this remedy; and the mere fact that the patient has resided for some length of time in a malarious country, even though he may not have had distinct malarious symptoms, is ample ground for essaying a cure by quinine.

β. **In the so-called Septicæmic fevers,** a large and not very accurately determined class, but of which the principal representations are well recognized, quinine has a part to play only second to that which it fulfils in the malarious fevers; and here again much depends on the mode of administration. With regard to the individual members of this group of diseases, it is true that there are other remedies that frequently prove more directly and specially effective; thus in erysipelas, at any rate in the early stages, the perchloride of iron is the true, and usually successful remedy; and in puerperal fever, as we have already mentioned, the action of aconite is frequently so serviceable as to leave nothing to be desired which drugs have any chance of accomplishing. But in the whole series of this variety of blood-poisonings, quinine holds a permanent place as a remedy; for whether in erysipelas, in surgical pyæmia, or any other of the infections which are specially connected with the absorption of putrid matters or of their emanations, when once a certain gravity of organic disturbance is reached—when the fever is excessively high, and the nervous system profoundly agitated and depressed—there is scarcely anything medicinal which offers the same chance of reducing the pyrexia, of relieving the inflammatory complications, and of sustaining the vital powers during the struggle, as quinine given in large and repeated doses. It is important to remember that both the reduction of actual temperature and the excessive combustion processes which lead to the generation of abnormal heat, are aided in high degree by the simultaneous administration of alcohol; and Socin has recently shown that alcohol helps the organism to tolerate doses of quinine, which would otherwise produce inconvenient or even dangerous toxic effects.

In the treatment of the ordinary Infectious fevers quinine holds a more doubtful place than in that of the so-called septicæmic fevers. In all of them the administration of very large doses of the drug (five to twenty grains or more, frequently repeated) will often produce a notable reduction of the symptoms, and especially of the febrile temperature. But we cannot be sure that any considerable influence is exerted on the general course of the disease; often a striking, immediate, anti-pyretic effect, even many times repeated with unvarying certainty, seems to count for little or nothing in the final event. The example of typhoid fever serves well to test this matter. In our own country the use of quinine in this disease was introduced but a short time ago by the late Dr. Fuller, and both here and abroad it has been fully tested in recent years. That the treatment is at least harmless seems to have been fully established; and in regard to dosage and mode of administration, it may be stated that as much as three or four drachms have been given in divided doses during the twenty-four hours, without injury, and often with benefit. Upon this point it is interesting to quote the recent authority of Dr. Clifford Allbutt, who has continually used quinine largely in febrile diseases, ever since its recommendation by Dr. Fuller some fifteen years ago. In typhoid fever, however, neither Dr. Allbutt nor any other high authority seems to claim the proof either that the disease is shortened in course or that its more serious consequences are materially hindered. Neither those who place their main reliance on the removal of heat *per se*, nor those who think it most important to strike at the combustion processes, which are the origin of abnormal heat, appear now to believe that quinine alone will effect any very material purpose. The same may certainly be said to be even more true of typhus fever. There are, however, some of the acute fevers which apparently benefit far more by the use of quinine; this is especially true of scarlet fever, as was pointed out some years ago by Dr. Peter Hood. In this disease it does appear to be the fact that quinine used in moderate doses from the first has a remarkable effect in averting the graver complications. It is probably not a merely accidental coincidence that quinine thus proves most beneficial to that member of the group of contagious fevers which has some of the closest relations, and even a partial convertibility, with such septic disorders as puerperal fever.

Besides the direct action which quinine may exercise upon uncomplicated fevers, we must remember that its use may be specially called for in those septic infections which so frequently occur as complications; as familiar examples I may mention the mischief caused by the absorption of putrid matters from the throat in scarlet fever, or from the bowel in typhoid fever. It can scarcely be doubted by any experienced person that in these circumstances quinine, with more or less aid from alcohol, affords one of our best sources of hope. Here, however, it is again necessary to employ very large and repeated doses.

In Whooping-cough we have a disease which can as yet only doubtfully be reckoned as of septic origin; but after the recent observations of Binz it must be considered at least probable that this is the case, and that quinine in large doses is one of the most appropriate remedies.

In Hay-fever there appears to be still better ground for supposing that quinine may prove useful in virtue of its antiseptic powers. The observation was first made by Helmholtz, and subsequently confirmed by Binz, that the injection of a solution of quinine into the nares checked the irritative catarrhal discharge and also the spasmodic symptoms.

The Hectic fever even of some chronic diseases is probably in a certain number of instances at least partly due to septic infection. In many examples of phthisis, with lung softening, one can hardly doubt that the rhythmic recurrence of fever is at least as much due to the absorption of putrescent matter from the cavities as to the exhaustion of the nervous system. Still less can we doubt that the hectic attending very large chronic abscesses connected with carious bone partakes of the character of a true pyæmic poisoning in minute but incessantly repeated doses. In both these last-quoted examples the influence of quinine can ordinarily only be effective when it is given in considerable quantities; from twelve to twenty grains daily is by no means an excessive amount.

I freely admit, however, that in many of the examples of septic fever in chronic disease there may be reasons which abundantly outweigh the arguments for the administration of quinine. Irritability of the stomach and intestines will very often prevent one from using it, especially in large doses; but the hydrochlorate will occasionally be tolerated when the sulphate cannot be borne.

γ. **The employment of quinine in Inflammations** is a very large subject, and one which, as yet, has only been incompletely studied. From time to time some author has strongly asserted its merits in regard to a particular inflammatory disease, such, for instance, as peritonitis, in which is was so warmly recommended by Trousseau. Other authors again have occasionally expressed in general terms the opinion that quinine was antiphlogistic; but nothing like a scientific basis for this belief existed previous to the researches already mentioned, in which the multiplication and the wandering of white blood-corpuscles was plainly observed to be checked by the action of the remedy. Henceforth it must be considered as established that quinine is naturally indicated in most inflammations, especially of the acuter sort; and that the question whether it shall be given or withheld in any particular case can only arise in consequence of accidental peculiarities in the phenomena of the disease. That objections of a most practical kind do exist, however, to its use in many cases of inflammation, no experienced practitioner can doubt; and it seems advisable to give a general glance at the character of the various obstacles to its employment which are likely to present themselves.

The primary objection which is most likely to occur is doubtless the incapacity of the stomach to bear with doses that would be sufficient to materially influence progress. In these cases, supposing the need of quinine to be urgent, we have still some chance of being able to use it. As already mentioned, the hydrochlorate will often be borne by the stomach when the sulphate will not; indeed it is probable that the latter ought to be entirely superseded by the former. If not successful, we may try injection per rectum, though for obvious reasons this is inconvenient. Finally, there is the subcutaneous method, which, however, has difficulties of its own as to which I must speak presently. Another objection is the ease with which some patients are cinchonized, making it nearly impossible to give them doses large enough to affect the inflammation. In a certain proportion of such cases it will be possible to mitigate or even prevent this effect by the simultaneous use of alcohol in large doses; this is equally true of severe double pneumonia, as Socin found it true of traumatic erysipelas.

It is also to be remarked that not merely ordinary cinchonism, but even fatal results have occasionally ensued upon the use of doses by no means enormous in comparison with the quantities which have frequently been given; this seems to depend upon an idiosyncrasy which is probably somewhat rare, and it is difficult to see how it can be perfectly guarded against. It is probably less likely to occur if the patient be simultaneously treated with alcoholic stimulants; but there might of course be objections to these.

As regards its fancied tendencies to aggravate delirium, it does not appear that in inflammatory fever quinine produces this effect, except very rarely. If the patient escapes the phenomena of ordinary cinchonism, it will be found nearly always that his nervous centres are in all respects far more tolerant of the drug than they would be in health.

It is perhaps scarcely worth while to dwell upon the still rarer inconveniences which quinine produces in a very small number of persons with peculiar tendencies, e. g., the very singular effects upon the nutrition of the skin which have been mentioned above.

To say, however, that because the difficulties now described do not seriously hinder the administration of quinine in a large majority of inflammatory cases, therefore it should nearly always be administered, would be to go much too far. For there are several other antiphlogistic remedies which, in particular instances, will respectively be found superior to quinine.

In **Pneumonia**, at any rate in the stage of dry, pungently hot skin, it will be much better to commence with aconite, even if we afterwards employ quinine. The action of aconite in rapidly throwing open the cutaneous capillaries and inducing a large transpiration and evaporation from the skin has already been described; and at any rate as a first step in treatment the use of this drug is more appropriate than that of quinine. When the action of the skin has discharged a quantity of superfluous heat from the body, and thus relieved the distress of the nervous centres (including the respiratory), it may perhaps be found that we have already arrived at the end of the active inflammation, and that resolution is about to take place. There are, however, very many cases, especially those of double pneumonia (and more particularly the double pneumonia which is only one phase of a general blood-poisoning, of renal disease, etc.), where no such simple course can be expected; the inflammatory fever will soon regain and maintain its activity, and here the employment of quinine with alcohol will very probably become our best chance. But this is not always the case; and in a good many instances where I have seen a severe pneumonia complicate an acute or subacute nephritis, the use of digitalis afforded relief of a far superior kind to any effect of quinine that I have witnessed, even in more favorable cases.

In cases of pneumonia which are markedly asthenic from the first, it has been shown by Sir Dominic Corrigan that the use of quinine throughout is most valuable; about five grains every three hours are to be given. He states, however, that in young persons with evidence of general capillary congestion, the administration of quinine should be preceded by local depletion—an assertion which is open to much doubt, for these patients are usually in a state of great depression, and while the abstraction of blood from the cutaneous vessels of the chest may very probably cause an undesirable loss of strength, it is difficult to see how it can seriously affect the vascular engorgement of the lungs.

Much more might be said respecting the varying circumstances under which quinine is more or less useful in acute inflammations, but it will be better to pass on to the next division of the present subject.

δ. **The use of quinine in Nervous diseases** is a subject with regard to which every one seems to assume that a great deal is known; yet, when we come to test this knowledge, much of it will be found to be very vague, and a good many of the popular views are certainly incorrect.

In **Neuralgia** quinine is to this day considered the sheet-anchor by most practitioners, although the erroneousness of the opinion has been pointed out by many high authorities. If we exclude the cases which are partially or wholly due to a malarial influence (and which form a very small proportion of the neuralgias actually encountered in practice), it will be found that quinine has not an extensive sphere of curative action. It has some hitherto unexplained preferential influence on the neuralgias of the ophthalmic division of the fifth cranial; but upon neuralgias of other nerves it frequently—perhaps most frequently—fails to produce any decided impression. In fact, not only is this the case, but we often meet with patients in whom much harm has been done by attacking an ordinary neuralgia with repeated and increasing doses of quinine; the nervous system gets seriously upset; there is more or less marked cinchonism; and the pain, so far from being relieved, is aggravated. It is a stereotyped remark in medical works that the more exactly a nervous (or indeed any other) disease conforms to a regular type of periodic exacerbations, the more surely will quinine prove useful. As a general proposition this is untrue. It is only when the neuralgia is due to actual malaria that the rule holds; and in the case of recurrent inflammations, or of hectic, there must be either malaria or else septic poisoning at work, or else we shall find this maxim fail us.

Among the non-malarial neuralgias there are none so often typically periodic as migraine; very often the attacks occur with the strictest regularity every month or every fortnight; yet there is no disease in which quinine is less of a specific. Its failures are many times more numerous than its apparent successes.

In **Epilepsy** I have already said that quinine is highly effective when malaria has been a cause; and by various authors it has been asserted that it is applicable to the simple non-malarious disease. No such pretensions, however, would at present be supported by our principal authorities on nervous diseases. Epilepsy can, unhappily, be studied on an immense scale at the London hospitals and dispensaries; and no one who had had such experience would now entertain the least hope of doing good with quinine, except in cases of a special nature, and which occur but rarely. Even of these it is likely that a considerable proportion are in truth obscurely malarial in origin, though the fact escapes the practitioner's observation.

In **Tetanus**, also, quinine has quite sunk in the estimation of practical physicians and surgeons from the high place which it once seemed destined to fill. There is, no doubt, one variety of the disease in which malaria plays a part, and here quinine will certainly be useful. But of the ordinary English or European cases, traumatic or otherwise, those that recover are now supposed with much probability to owe their escape from death to purely natural causes, independent of remedies. I should be loth to express myself as hostile to all attempts to benefit patients in tetanus with quinine, but I cannot but think that, at any rate in severe cases, it would be better to lose no time in applying one of those agents which are known to have a direct action in reducing reflex irritability, with which property there is no ground for crediting quinine.

In **Chorea** quinine has never appeared to me to exert more than a subordinate and almost accidental curative power. We now and then seem to hit, quite by chance as it seems, some ill-understood want of the organism, and in this way get an apparent quinine-cure. But to any one who has been elated by a success of this kind, there is more than likely to come disappointment, from the total failure of the medicine in a long series of subsequent cases.

In **Chronic Alcoholism**, on the contrary, and in some varieties of chronic **Insanity**, quinine is much more frequently useful; and it may indeed be said that in all cases where the problem is to slowly raise the nervous tone, quinine has a steady and permanent value. It is useless to administer it in large quantities; from two to six grains daily is enough.

The above is a fair summary of the more important uses of quinine which are called for by the occurrences of every-day practice. There remains a more miscellaneous group of disorders, in some of which it is likely that quinine will prove more beneficial than is at present generally supposed. In others it is already known to be of great value.

In all **Gangrenous conditions**, whether the result of ordinary inflammation badly complicated, or of a specific or septic poisoning, quinine has often proved itself to be of value. It is commonly said that cinchona in infusion acts better than quinine in those conditions, but such is not my experience, even when using the sulphate, while it is probable that when the hydrochlorate comes into use in this country we shall still further get rid of any disturbing influence upon the stomach.

In **Laryngismus Stridulus** we have a most valuable remedy in quinine, and it is probably an erroneous idea that the action of the drug in this disease is simply a nervous one. Laryngismus is part of a general abnormal state in which the whole organism is affected in a peculiar manner, and in which, with great frequency, we observe some of the symptoms of rickets. In what way it comes to pass that quinine so benefits this general state we are not able to tell; but it has appeared to me that we can trace, in many cases, a rapid alteration of that fault of assimilation by which the phosphates, instead of being applied to the nutrition of the tissues, are thrown out uselessly in the urine. I do not hold with some authors that it should be given only in the intervals of the paroxysms and in very large doses. I believe it better to give it in small repeated doses, and without interruption, using enemata if the child cannot swallow.

In **Asthma**, which is also usually considered, relatively to any beneficial influence of quinine, a merely nervous disease, it is probable that the action of the remedy is not so simple or so limited. We have the high authority of Dr. Hyde Salter for its use as the best of all tonics; but he usually gave it with iron and mineral acids, and this agrees with my own belief that the good effect of the drug is exerted not only on the nervous system, but also, and probably more decidedly, on the general nutrition of the body and on the digestive process. It is one of the greatest troubles of asthmatics that they cannot take more than a very small quantity of food at once without the danger of exciting a paroxysm; under the influence of a quinine or a quinine and iron tonic this disability often much diminishes, to the great benefit of the patient.

In **Erythema Nodosum** quinine has long been noted for its beneficial action; indeed it is usually the only remedy needed in addition to a few days' rest; and its use certainly renders the course of the affection much shorter than when left to nature. We should probably be justified in reckoning this as one of the examples of the action of quinine upon the nervous system, although we are unable to prove the dependence of erythema nodosum on purely nervous causes.

In **Urticaria** quinine is only useful in special cases. The tendency to nettlerash is very personal; in the commoner kind of cases the individual is always liable to an attack after eating freely of some particular food, such as shell-fish, salmon, or strawberries. To such patients, when they have thus brought on an attack, we need only give an emetic, or an emetic and a purge. But there is a kind of chronic urticaria in which, though an attack may be more easily brought on if indigestion be present, the true source of the disease lies evidently in some peculiarity of the nervous system; and medicines directed to the digestive organs cannot be expected to do much good.

In some of these cases arsenic is the best remedy; but in most of them the long-continued use of quinine and mineral acids overcomes the tendency more surely.

In **Insanity** it would be wrong to speak of quinine as exerting any special effect; but there are many cases in which its tonic effect, especially when combined with that of iron (as Dr. Maudsley observes) is very useful.

As a General Tonic it is universally agreed that the condition in which quinine produces most decided benefits is that in which the flesh is flabby and the skin too perspiring.

As a Tonic to Digestion in the numerous cases of atonic dyspepsia, quinine is probably inferior, on the whole, to cinchona in conjunction with mineral acids. But there are affections of the stomach in which the dyspepsia is accompanied by dilatation of the stomach and symptoms of fermentation of the retained food, which after a time is passively vomited, often in immense quantity: in this disease there is much reason for thinking that the fermentation is not merely an accidental consequence of the primary affection, but that it directly aggravates the mischief to such an extent as to become almost the more important part of the morbid condition. Upon this supposition various antiseptic and antifermentative remedies have been proposed, but it is probable that none of these has a more powerful action of the kind required than quinine, for reasons which have already been stated.

ε. **As an Oxytocic**, or stimulant of the uterine movements, quinine has yet to give its absolute proofs, but a large amount of evidence has been collected, during the last few years, in favor of the belief that it does possess this action. The latest facts on the subject are those related by Dr. Rancillia.[1] Remarking that it has often been observed by French and Italian physicians that quinine administered to pregnant women suffering from intermittents caused abortion, but that this had been attributed by other authorities to the direct influence of the malaria, and not to the quinine, Rancillia states that in his practice as a veterinary surgeon at Caen he had often found the labor pains of bitches brought on very actively, even when ergot had failed, by the administration of one and one-half grain doses of quinine at short intervals. This is a matter which certainly deserves to be practically tested by all accoucheurs.

The Therapeutic Action of the other Cinchona Alkaloids needs but very few words after what has been said of the result of recent inquiries. The extensive comparative trials which were made by the Indian Commission, already mentioned, brought out the general fact that all the four alkaloids, quinine, cinchonine, quinidine, and cinchonidine, act essentially in the same manner against intermittents; and as regards activity, the only point of consequence is that cinchonine is a good deal weaker than either of the other three. When we take these facts in conjunction with the valuable observations of Dr. Baxter, which show that the same relations exist between the purely physiological activity of the respective alkaloids, it seems legitimate to regard the question as practically settled; and all further questions concern only the matter of expense, or such special points as the facility with which one or other alkaloid can be prepared for hypodermic use.

The Therapeutic Action of Kinic Acid is still in a very unsettled position. Of course its primary interest arises from the fact that, in cinchona bark, quinine exists in combination with it, and that this kinate of quinine is a particularly soluble salt. At present we cannot say that there is any other virtue in kinic acid beyond this.

The Therapeutic Action of Kinovic Acid is probably much more important. The researches of Kerner (1863) first brought this acid prom-

---

[1] Practitioner, January, 1874; Union Médicale, 1873, p. 800.

inently into notice, although it had been thoroughly investigated by Hlasiwetz, from the chemical side, in 1859. Kerner directed attention to the fact that kinovate of lime is the active principle in Deloudre's Extract, a preparation successfully employed in various European countries and in India in diarrhœa and dysentery. It has been further stated that kinovate of lime is useful even in true intermittents; but this requires confirmation. Since the year 1869, when kinovate of lime was introduced into London by Messrs. Hodgkinson, a few physicians have occasionally employed it; but there has been no general adoption of it, and it seems to have been forgotten. I cannot speak from personal experience, but it appears that this remedy may be given in doses of from ten to thirty grains. One circumstance is worthy of special mention: it is suspected by some of the continental authorities that kinovic acid is especially extracted in the making of the cold infusions of bark, and it is known that many good authorities have preferred this preparation to any other, especially in febrile conditions where the alimentary canal was irritable.

The Therapeutic Uses of the Tannic Acids of Bark have not been studied with any care, although they well deserve it. It is evident that in them resides the powerful and very peculiar kind of astringency which the cinchonas possess, and which enables us at times to administer bark where we could give almost no other tonic. The separate trial of cincho-tannic and kinovi-tannic acids, as remedies for diarrhœas and other fluxes, is much to be desired.

PREPARATIONS AND DOSE.—Cinchona Flava, gr. v.—xx. (.30—1.20.); Extract. Cinchon., gr. ij.—xv. (.12—1.); Extr. Cinchon. Fluid., ℳv.—xx. (.30—1.20.); Tinct. Cinchon., ℨss.—ij. (2.—8.); Tinct. Cinchon. Co., ℨss.—ij. (2.—8.); Decoct. Cinchon. Flavæ., ℨj.—ij. (30.—60.); Decoct. Cinchon. Rubræ, ℨj.—ij. (30.—60.); Infus. Cinchon. Flav., ℨj.—ij. (30.—60.); Infus. Cinchon. Rub., ℨj.—ij. (30.—60.); Quiniæ Sulphas; Quiniæ Valerianas; Pilulæ Quiniæ Sulphatis.—No definite dose or number, as the circumstances under which they are given vary so greatly.

Of Quinine, the form most commonly employed is the neutral sulphate, $2C^{20}H^{24}N^2O^2,7HO$, which, from its insolubility in water, has to be dissolved by the acid—a mineral acid when the medicine is to be given in liquid shape. Whenever we can give quinine in the solid form we may quite trust the stomach fluids to dissolve it, and for most persons this answers very well. Any one who cannot conveniently swallow pills can either take one of the convenient preparations of chocolate which are now made, with a definite dose in each piece, or can effect the same purpose in a rough and ready fashion by enclosing the dose of solid quinine between the two halves of an ordinary chocolate "drop" and eating it up. Chocolate is exactly the right thing for covering the taste of quinine.

By patients who are severely ill with acute disease we often cannot get sweet things taken, and if we have any difficulty in getting the requisite doses taken by the mouth it is better to administer them in enema. But the sulphate is not a good form in which to administer quinine in some acute diseases, as it is not sufficiently well borne by the stomach.

The hydrochlorate of quinine is evidently the best preparation, especially for the purposes last named. The formula of the neutral salt is $C^{20}H^{24}N^2O^2$, $HCL+2H^2O$. It is soluble in as little as 24 parts of water, and is a good deal less irritating to the stomach than the sulphate. It is fully as active, probably rather more so, than the latter; hence its dose as an ordinary tonic may be reckoned as one to two grains thrice daily; as an antineuralgic, from three to twenty grains (according to the circumstances described above); and in the acute septic diseases, according to Binz, Socin, and others, who have come to employ it to the exclusion of all other quinine salts, it may be given up to the quantity of sixty to one hundred and twenty grains in the twenty-four hours. The neutral salt, it must be remembered, should be crystalline; the acid hydrochlorate forms a gummy amorphous mass, and is not a desirable drug.

If we wish to produce the specific effects of quinine there is no reason to suppose that any of the other very numerous salts of that alkaloid have any advantage over those already mentioned. With regard to the hypodermic injection of quinine, which is becoming important, I am sorry that it is not easy to give any thoroughly satisfactory directions. Ordinary sulphate of quinine, in order to be made sufficiently soluble to be got into the necessarily small quantity of water that can be used for this purpose, requires some other powerful solvent; and, whatever solvent may be chosen, the resulting liquid will to most persons prove so irritant that the formation of an abscess, or at least a hard, painful swelling with a black cicatrix in the centre, is pretty certain. Neither sulphuric, nor acetic, nor (in spite of Bricheteau's assertion) tartaric acid can be used for this purpose, without at least an even chance of producing painful local inflammation.

Probably a plain aqueous solution of the neutral hydrochlorate is the best form for giving quinine subcutaneously. It is true that in order to inject as much as one grain we shall have to use about twenty-five minims of distilled water, but the operation will be entirely painless, provided the needle terminates in a flat-bladed steel point (Buzzard's); and one grain given in this way is at least the equal of three, if not of four, grains given by the mouth. It will therefore be understood that even motives of economy might not unfrequently induce us to adopt the subcutaneous method, and there is also the advantage of not irritating the stomach.

Of the other salts of quinine I shall simply mention the names, since I distrust the statements as to their special virtues.

Citrate,	Ferrocyanide,
Acetate,	Valerianate,
Kinate,	Stearate,
Hydrocyanate,	Nitrate,
Tannate,	Phosphate,
Urate,	Bisulphate,
Tartrate,	Sulpho-tartrate,
Lactate,	Carbolate.
Acid Lactate,	

Besides these there are other salts of quinine which are rather to be considered as preparations of some other substances. The arseniate, the antimoniate, the hydriodate, etc., must be thus looked upon. In short, the ingenuity of chemists has been somewhat unprofitably exercised (so far as mere pharmaceutical purposes are concerned) in multiplying preparations of this alkaloid without any good reason or result.

It should have been mentioned just now that quinine can be very readily given by inhalation of solutions, and that this procedure is very useful in many conditions of the larynx, bronchi, and air-cells in which a local antiseptic effect is required. A convenient method is to dissolve eight grains of hydrochlorate or sixteen grains of sulphate of quinine in twenty ounces of distilled water: this solution gives a very good spray for inhalation.

With regard to one particular preparation of quinine, the so-called citrate of quinine and iron, a few words must be specially said. It is to be regretted, but is certainly true, that this article is so irregularly made that few physicians are justified in prescribing it. There are many wholesale chemists who sell two qualities of this drug ("A" and "B"), the cheaper of which is simply an imposture, containing little if any quinine.

In fact, if it be wished to administer iron and quinine together, we shall do better to mix for ourselves. Sulphate of iron and sulphate of quinine can be given together, in mixture or in pills; and citrate of iron can be given with quinine in effervescence with citric acid and potash or ammonia.

Of Quinidine it is known, especially from the late Indian researches, that the sulphate acts much in the same way and with the same energy as sulphate of quinine. It is also very fairly soluble in water (1 in 108), so that for administration by the stomach it is a convenient salt. The doses are the same as those of quinine.

It is necessary, however, to be sure that we really get the article wished for, as (at any rate till quite lately) the so-called sulphate of quinidine of the shops was in many cases sulphate of cinchonidine. The quinidine salt should yield the tests above mentioned.

Cinchonine is met with here almost exclusively in the form of the sulphate. The neutral sulphate, $2C^{20}H^{24}N^{2}O, SH^{2}O^{4} + 2H^{2}O$, is soluble in 66 parts of cold water, and does well for internal use; the dose must be at least twice as large as that of quinine or quinidine salts. The acid sulphate, $C^{20}H^{24}N^{2}O^{2}, SH^{2}O^{4} + 4H^{2}O$, is far more soluble still, and is a very convenient substance for hypodermic injection in neuralgias, as well as in intermittent fevers, etc. From one to five or six grains may be thus very easily injected with water. Still, like all the cinchona alkaloids, it has some tendency to irritate the subcutaneous tissues.

Cinchonidine Sulphate, as already mentioned, has very often been sold in the shops for quinidine sulphate. Its action is in every way similar to that of the latter, but it is slightly (perhaps one-sixth or one-eighth) weaker than quinine or quinidine salts.

Quinovic Acid, under the form of quinovate of lime, already referred to, can now be readily procured in London. The dose in powder is two to eight grains every hour, for diarrhœa and dysentery. For solution, Messrs. Hodgkinson & Co. recommend the following formula:

℞ Calcis quinovatis, gr. 96.
Pulvis tragacanth., gr. 10.
Acid. phosph. dil., q.s.
Aqua, ℥ vj. Misce.

Of the various other cinchona substances, more especially the bark-tannins, I have no sufficiently accurate knowledge to say anything about the form of preparation or the dose.

PALE CATECHU (Nauclea Gambir).

Active Ingredients.—Pale catechu contains between thirty-six and forty per cent. of tannic acid, and the important substance "catechine" or "catechuic acid," $C^{13}H^{12}O^{6}$ (Rochleder).

The first named, catechu-tannic acid, is soluble in cold water. Exposed to a moist atmosphere, it turns dark red, and becomes insoluble.

Catechine is a modification of tannin. It changes the persalts of iron to grayish green; crystallizes in snow-white silky needles; is soluble in boiling water, alcohol, and ether, but very sparingly soluble in cold water, which apparently simply softens it, and causes it to swell up. Heated, it becomes fusible. It very slightly reddens litmus. From tannic acid it differs in not affecting a solution of gelatine. The sweet taste of catechu is said to depend upon the presence of the catechine.

PHYSIOLOGICAL ACTION.—Pale catechu appears to act upon the human organism as a simple and mild astringent.

THERAPEUTIC ACTION.—Catechu is a very excellent astringent, both for taking internally and for application to the surface. As an internal astringent, it is employed in affections of the mouth and throat, such as relaxed uvula and the troublesome cough induced thereby, and relaxed sore throat. For these purposes it may be sucked or chewed, or employed in the form of lozenges. Catechu may also be chewed as a remedy for atonic dyspepsia, especially when much trouble is given by pyrosis. It may be given, too, in cases of diarrhœa and chronic dysentery, but should be avoided if there be acute inflammatory symptoms, hæmorrhages, mucous discharges, and incontinence of urine. Catechu, in a word, may be employed in most cases where astringents are indicated. It is often prescribed in conjunction with chalk-mixture, opium, and other narcotics. The best time for administration is half an hour or so before meals.

The powder of pale catechu is sometimes sprinkled upon indolent ulcers with good effect.

The infusion injected up the nostrils often stops epistaxis.

PREPARATIONS AND DOSE.—None officinal. The following, from the B. Ph.: Catechu Pallidum, gr. x.—xx. (.60—1.25); Pulvis Catechu Comp., gr. xx.—xl. (1.25—2.50); Tinct. Catechu, ʒj.—ij. (4.—8.); Infus. Catechu, ʒj.—ij. (30.—60.); Trochisci Catechu, No. 1—3.

IPECACUANHA (Cephaelis Ipecacuanha).

ACTIVE INGREDIENTS.—The woody portion of the root of the ipecacuanha plant is nearly inert; that which gives value to it in medicine being the bark. Both portions contain the essential alkaloid, emetine, but in an impure condition.

Pure emetine constitutes about one per cent. of the best ipecacuanha root; when separated it is a white and amorphous powder, inodorous and of bitter taste, and is found to consist of $C^{15}H^{15}O^{5}N$. It is soluble in alcohol; scarcely soluble in ether and oils, very sparingly soluble in cold, but much more soluble in hot water. It dissolves in acids, is precipitated by tannin, and is fusible at 122° F. Perfectly pure emetine is procurable only with difficulty.

The impure form is distinguished by its grayish-yellow color, and occurs usually in transparent scales. In taste it is bitter and acrid; it is nearly destitute of odor, very soluble both in water and in alcohol, but insoluble in ether.

In addition to the peculiar alkaloid emetine, ipecacuanha contains an astringent acid, $C^{14}H^9O^7$, somewhat akin to catechine, and called cephaelic or ipecacuanhic acid. There is also a trace of volatile oil, with some fatty matter, and a small proportion of starch and gum. The cephaelic acid was formerly mistaken for gallic acid. With the persalts of iron it strikes a green color.

PHYSIOLOGICAL ACTION.—In all the more decided effects which ipecacuanha can produce in the animal organism, its powers are fairly represented by the alkaloid emetine, though the latter is not convenient as a remedy. The experiments of Merat and De Lens, of Magendie and Pelletier, and of Schroff with the pure alkaloid, and of Magendie and Pelletier and others with the impure, bring out substantially the same results with those obtained by the action of powerful doses of ipecacuanha itself, the only question being one of degree. 1. It is a local irritant both to the skin and the mucous membrane. 2. In large doses it is a special irritant after absorption (for instance, from the subcutaneous tissue quite as well as if swallowed) to the gastric and intestinal mucous membrane. 3. It has a powerful and, in large doses of the alkaloid, a fatal, depressing action upon the nervous centres.

How far the vomiting, which is so characteristic an effect of full doses of ipecacuanha, and the striking increase of secretion from the mucous membrane of the bronchi and the stomach which are similarly produced, are consequences of the direct action of the drug upon the vagus-terminals, or are brought about by action upon the nervous centres, is not at all known. There is a well-known phenomenon produced by the inhalation of ipecacuanha dust; in ordinary persons this produces merely cough and sneezing, and might be attributed merely to the local irritant action; but in certain subjects—and they are probably more numerous than is generally supposed—the sneezing, coughing, and running at the eyes and nose are supplemented by an amount of dyspnœa and feeling of anxiety which is comparable to that caused by an attack of spasmodic asthma or of the more spasmodic form of hay-fever. In these points of view there is great interest in the experiments of Pécholier with emetine, on rabbits; a dose .072 grain caused immediate excitement, followed by retching and (in two minutes) a remarkable feebleness, depression of circulation and breathing, together with a very curious lowering of temperature in the whole anterior half of the body, while the rectal temperature was raised from local congestion. If this dose were repeated daily for some time, it was found that the singular temperature-phenomena became more marked, while the excitement and the retching were less so, and death at last took place.

Dissection showed congestion, intense in the stomach and milder in the upper part of the intestines; the lungs were anæmic; the liver contained no glucose. The same results were produced when the remedy was given by the skin.

It must, after all, be confessed that, as Nothnagel remarks, we are in an astonishing state of ignorance of the physiological action of a drug which medical men have so largely employed for centuries past.

THERAPEUTIC ACTION.—Fortunately it happens that our information, though in a great measure empirical in its source, enables us to say with confidence that ipecacuanha is highly useful in various ways.

As an **Emetic**, ipecacuanha is unequalled in its value for a large number of cases. It is not violent, nor swift in action, but very certain; it is mild, and the vomiting it causes usually empties the stomach, and is neither repeated nor followed by the collapsed state which antimony and some other emetics often leave behind them.

Its freedom from any disgusting taste, like that of sulphate of zinc or sulphate of copper, renders it still further advantageous for administration to children. Twenty grains of the powder, or half an ounce of the wine for an adult, and from one-fourth to one-half these quantities for a child, will rarely fail to act, but if necessary may be repeated in fifteen or twenty minutes.

As an **anti-Catarrhal remedy**, ipecacuanha has extensive and varied uses; for there is perhaps no part of the mucous tracts the catarrhal affections of which may not be beneficially treated with this drug.

Common Catarrh of the throat and bronchi in the dry stage may be beneficially treated with nauseating or nearly nauseating doses of the drug; thus from one to two grains may be given every two or three hours. The smaller dose will induce diaphoresis, and probably also give some relief to the dryness of the mucous membrane; the two-grain doses will cause a decided increase in the flow of mucus, and in most patients a feeling of nausea, or at least of squeamishness; but at the same time much relief will be obtained for the distressing dry soreness.

There is a different use for ipecacuanha in the later stages of bronchial catarrh, where the secretion is profuse and troublesome. It is then to be given in much smaller doses, and thus given it unquestionably possesses power to limit the mucous secretion and to improve its quality, diminishing the nasty ropiness which makes it so offensive and also so likely to keep up irritation. The doses I now refer to are one-quarter to one-half grain of the powder, or five to ten minims of the wine. Smaller doses must be given to children, but not in the exact proportion of their age, for children usually bear ipecacuanha very well.

In **Catarrh which mainly affects the stomach**, ipecacuanha can also prove very useful, but it needs to be employed with more precaution. Anything like the emetic, or even the nauseating doses, would be almost sure to aggravate the inflammatory tendency and produce gastritis of an unmanageable kind. But in the smaller doses (five to ten minims of the wine) ipecacuanha is highly beneficial; in many cases of chronic catarrhal dyspepsia, and indeed in atonic dyspepsia which is independent of any catarrhal cause, this remedy is most valuable. **In children, even acute catarrh** is much benefited by ipecacuanha.

In **Vomiting of several kinds**, ipecacuanha is demonstrated to be an excellent medicine; but it here requires to be taken in very small doses. One minim of the wine (equal to $\frac{1}{38}$ grain) is taken in water and repeated at short intervals till a decided effect is produced. It is now well established that this effect of ipecacuanha is especially, if not solely, produced in cases where the vomiting is **sympathetic** in character. It is seen very remarkably in the **sickness of pregnancy**, for which there is probably no antidote of equal value; and in the morning **sickness of chronic alcoholism** it is scarcely less useful. On the contrary, it is useless in sickness arising from organic affections of the stomach, such as cancer, chronic ulceration, etc.

As an **Arrester of Hæmorrhage**, ipecacuanha possesses considerable energy; a fact which is the more curious because in poisonous doses it has frequently produced hæmoptysis and other forms of bleeding.

The best doses to employ are those which just stop short at the production of slight nausea; one grain repeated frequently enough to produce a very gentle effect. To produce absolute emesis is to run a grave risk of heightening the very mischief which we are treating. **In the hæmoptysis of early phthisis**, ipecacuanha may dispute with ergot of rye the reputation of being the most valuable styptic we possess. **In the passive hæmorrhages from merely engorged bronchial mucous membrane**, it is equally effective. **In hæmatemesis** it is of value when the hæmorrhage is vicarious of menstruation. **In menorrhagias** it is scarcely worth while to mention it as a remedy fit to be put in competition with ergot of rye and several others.

In Dysentery, ipecacuanha has of late years proved itself more than worthy of the old reputation, implied in the name "radix antidysenterica," under which Leibnitz introduced it to Europe. In acute tropical dysentery there is now a vast amount of Indian experience to show that the bold use of large doses is most successful. At the earliest opportunity twenty-five or thirty grains of powdered ipecacuanha are given in a very small quantity of fluid; some physicians order a full dose of laudanum or a few doses of chloroform half an hour previously. The patient is kept quite still in bed, and may suck a little ice or take a teaspoonful of water, if thirsty. Mustard or turpentine is applied to the belly.

It is surprising how seldom either vomiting or any considerable amount of nausea occurs; and in eight or ten hours the remedy can be repeated in diminished dose. Under this treatment the purging of blood and slime, the abdominal pain, and the tenesmus soon disappear, the motions becoming feculent; and there is often a profuse sweat, followed by refreshing sleep. The smaller doses of ipecacuanha are continued for some days, taking care to keep intervals free for the administration of food. Even when the motions have become healthy, ten or twelve grains should be given at bedtime for two or three nights.

This treatment is applicable to the great majority of patients, but even in dysentery we sometimes meet with evidence of the greater sensitiveness of some persons than of others; and in such exceptional cases two or three grain doses, with opium and gray powder, will often prove very successful.

The general utility of the large doses above described is conclusively shown by Dr. Maclean, who ably sums up the modern evidence, including his own large experience, in his article on Dysentery in Reynolds' "System of Medicine," vol. i. The credit of reintroducing the plan is due to Mr. Docker.[1]

It is very interesting to notice that while ipecacuanha is, *par excellence*, the treatment for acute dysentery of the ordinary type, which so commonly alternates, epidemically, with outbreaks of intermittent and remittent fever, the other great cinchonaceous drug, cinchona (with its alkaloids), is equally successful in that variety of dysentery which is actually complicated with the symptoms of genuine malarial poisoning. In the latter case large doses of quinine perfectly replace the ipecacuanha treatment.

In Diarrhœa, more especially when dependent on nervous irritation, and hence particularly in the **diarrhœa of young children**, ipecacuanha is a powerful remedy, especially when there is also vomiting or retching.

---

[1] Lancet, vol. ii., 1858.

Vinum ipecac. in three-minim doses every two or three hours, will quickly put a check to the disorder, even when it has begun to take on pseudo-dysenteric characters with blood and slime.

In various spasmodic Diseases of the Respiratory Organs ipecacuanha is of great value. In the paroxysms of **nervous Asthma** it was especially praised by the late Dr. Salter, who gave it, however, in one large emetic dose at the outset of the paroxysm. I myself prefer to give repeated five or ten minim doses of the wine every ten to thirty minutes during two or three hours, or until substantial relief is obtained, and this will often happen, even in intense cases, very quickly. In **Whooping-cough**, doses of one minim for children under five years, or two minims for older patients, may be given every one, two, or three hours with the greatest relief.

In simple Inflammatory Croup and in simple Pneumonia, ipecacuanha has only a comparatively limited value; it does not so quickly produce nauseating effects as tartar emetic; it is also inferior to the latter as a diaphoretic, while both are inferior to aconite.

In concluding the general estimate of the therapeutic powers of ipecacuanha, it is necessary to give a caution respecting the extreme inconvenience which may be occasioned by giving this drug, even in very small doses, to patients possessed of the unfortunate idiosyncrasy of suffering irritative catarrh or asthma whenever they take it. If any sick person declares from past experience that he is liable to this, the practitioner (whatever his desire to administer ipecacuanha on any particular occasion) should pause before venturing to give even the minutest dose; for so great is the misery which it causes to such persons, and so keen often is their annoyance, that he may suffer great discredit for his persistence in prescribing it.

[The following is from R. and V., *Brit. Med. Journ.*, June 9, 1877:

"1. Sixty grains of powdered ipecacuan, mixed with a small quantity of bile, and placed in the duodenum, powerfully stimulated the liver. Even three grains had an effect on a dog weighing 6.8 kilogrammes, very nearly as great as the effect of sixty grains on a dog weighing 27.2 kilogrammes, the amount of bile secreted per dog being nearly the same in both cases. 2. The bile secreted under its influence was of normal composition as regards the biliary matter proper. 3. No purgative effect was produced, but there was an increased secretion of mucus in the small intestine. The composition of the bile did not afford any evidence of an increased secretion of mucus in the small intestine. The composition did not afford any evidence of an increasd secretion of mucus having taken place from the glands of the bile-ducts.

"The increased biliary flow that followed ipecacuan could not in these experiments be ascribed to any relaxation of spasm of the bile-ducts; for that no such thing existed was clearly shown by the free flow of the bile before the substance was given. Nor could it be owing to contraction of the gall-bladder, for the cystic duct was clamped. Nor can it be ascribed to contraction of the bile-ducts, for the increased flow was far too prolonged to be attributable to any such cause. It is therefore certain that this substance, like the others, has the power of stimulating the *secreting* apparatus of the liver. This being now proved as regards the dog, it can scarcely be doubted that the *modus operandi* is the same in man. The results of these experiments will, therefore, lead to new speculations regarding the pathology of dysentery; for *every step towards greater accuracy of knowledge regarding the modus operandi of any therapeutic agent is certainly calculated to advance our knowledge of the true nature of the pathological condition that is relieved or cured by it."*]

Preparations and Dose.—Ipecacuanha, gr. j.—xxx (.06—2.); Pulv. Ipecac. Co., gr. x.—xv. (.65—1.); Extract. Ipecac. Fluid., ♏j.—xxx. (.06—2.); Syrup. Ipecac., ♏xv.— ʒ ij. (1.—8.); Vinum Ipecac., ♏xv. — ʒ ij. (1.—8.); Trochisci Ipecac., p. r. n.

Besides the officinal members of the Cinchonaceæ, there is one plant, Coffee, which is of great importance, and cannot be passed over in a treatise on therapeutics. On account of the identity, or close relationship, of their active ingredients, I must also notice the tea-plants, Paraguay Tea, and Guarana, though these belong to different orders.

COFFEE (Coffea Arabica).

Active Ingredients of the seeds vary accordingly as it is raw or roasted. In the raw seed, Caffeine, $C^8H^{10}N^4O^2$, is probably the only important ingredient; it crystallizes in pure white silky needles, which can be sublimed by a heat of about 350°; its vapor has no smell; it has a neutral reaction, and a weak, bitter taste; it is moderately soluble in water, in alcohol, and in ether. It combines with the stronger acids to form acidly reacting salts. It dissolves, colorless, in concentrated sulphuric and nitric acids; but, if the solution with the latter be evaporated, there is a reddish-yellow residue, which gives, with ammonia, a brilliant purple color.

In the raw seed, caffeine exists as a tannate. In the process of roasting, as was first shown by M. Personne,[1] a portion of this tannate of caffeine becomes transformed into the alkaloid methylamine, a substance which, in the form of the acetate, has been proved by M. Behier[2] to possess stimulant qualities.

The volatile and aromatic ingredients of coffee are developed in the roasting; it is not accurately known either what they are, nor how far they influence the physiological action of coffee. It is possible that the whole of the undoubted increase in physiological power which the roasting confers upon the seed is due to the change of part of the caffeine into methylamine. M. Aubert has recently remarked that coffee-extractive free from caffeine is far from being inert.

One more active ingredient must be mentioned, namely, the coffee-tannin, or caffeic acid, which possesses somewhat astringent properties.

Physiological Action.—The infusion of coffee is celebrated for its stimulating and refreshing effects; but there is a certain number of persons who know it only as a more or less powerful poison, and it is probable that even less susceptible individuals would suffer from similar symptoms if they took it in very large doses. Coffee undoubtedly exalts the general reflex excitability: one of the consequences of this is its effect in banishing sleep; another is the production of an irritable state of the heart, in which the slightest excitement is sufficient to bring on violent palpitations.

---

[1] Practitioner, October, 1868.   [2] Ibid.

[The physiological effects of unburnt coffee differ decidedly from those produced by the roasted bean. For many years we have used a wine or a tincture made from green coffee for its diuretic and anti-lithic effects, which are quite marked, and find it specially useful as an eliminating agent in gouty conditions of the system, to be used during the intervals between the acute manifestations of the disease. Gubler has quite recently ascertained that caffeine is an efficient diuretic.]

The action of caffeine has been very sedulously investigated by many observers; its action on the nervous system is specially illustrated by the recent researches of Aubert, already referred to. According to the latter author, caffeine, when injected in sufficient quantities into rabbits, rats, cats, and dogs, always exalts reflex excitability and produces tetanus. The latter originates in the spinal cord, the nerve-trunks remaining intact. It also causes a rigidity of the extremities, which Aubert thinks is dependent upon a direct action on the muscular tissues. The poisonous action can be overcome by the use of artificial respiration. In frogs, caffeine has but little action on the heart. In rabbits it remarkably quickens the pulse, but there are periodic intermissions in which the heart appears to be distended. In dogs the frequency of the pulsations increases, while the blood-pressure is invariably lessened; the former effect Aubert thinks due to action on the excitory apparatus of the heart, the latter to paralysis of the cardiac nerves arising from the cardiac ganglia. Upon man he did not find caffeine act very powerfully; a dose of 7.8 grains produced only confusion in the head, tremor of the hands, and quickening of the pulse. Other writers, however, have observed much more formidable symptoms, and there is little reason to doubt that, in sufficient quantity, caffeine would prove as directly poisonous to man as to animals.

THERAPEUTIC ACTION.—Coffee in strong decoction is very well known as an antidote to various conditions of nervous depression.

Opium-poisoning affords perhaps the most familiar field for the reviving powers of coffee; it is a common and successful practice to ply the patient with repeated small cups of the strongest black coffee, very hot. In this way the increasing torpor of the nervous centres is arrested, and the patient's attention is kept sufficiently roused to enable him to co-operate with the physician in those unremitting muscular exercises which the latter urges him to keep up. The great object is no doubt to maintain the activity of respiration and of the eliminative processes during a certain period, after which safety will have been reached. The coffee effects this in a twofold manner: partly by direct stimulation of the brain centres, and partly by the general augmentation of reflex excitability. In this action of recovering patients from opium-stupor, it does not appear that caffeine can replace coffee. It was proposed by Dr. Campbell, of America, for this purpose, but it did not succeed in his hands; and in a subsequent case, recorded by Dr. Anstie, it had only a temporary influence.[1] Either the methylamine or the aromatic principle of roasted coffee must exert some powerful influence of a stimulating kind.

In various adynamic Fevers coffee has often been used with much success; and there are some physicians who almost always employ it instead of alcohol for this purpose. Here perhaps the whole effect is not due to its mere action on the nervous system; for it has been stated by

---

[1] Medical Times and Gazette, 1862.

some observers that coffee has the power to diminish the febrile waste of nitrogenous tissues, as displayed in the large discharge of urea.[1]

Even as a true Antiseptic and Antiperiodic, coffee is adopted by many physicians; the practice comes from the Philippine Islands, and has been introduced into Europe by Dutch physicians, and the opinion seems quite as strong in Holland as in the East that coffee will cure intermittents as effectually as quinine itself. Upon this subject there is no English experience, so far as I am aware.

In **spasmodic Asthma** coffee is a remedy of old standing. It was especially recommended by the late Dr. Salter, who, in his excellent treatise on that disease, speaks of *café noir* as one of the most generally useful of all remedies for asthma. It is to be taken very hot, in small quantities of the most concentrated infusion, and upon an empty stomach. Asthmatics should not use coffee as a daily beverage, but reserve it for the time of an attack.

In **Hay Fever** black coffee has been well spoken of by Mr. Worthington.

In **Whooping-cough** it has been recommended.

In **Headaches dependent on the nervous system**, rather than the digestive organs, coffee is a most valuable remedy; in fact, from the simplest "nervous headache" up to the most typical migraine, coffee, with few exceptions, affords either cure or palliation.

In **the typical Migraine**, however, we shall do well to throw aside the use of coffee, and employ only the hypodermic injection of caffeine. I believe that Eulenburg was the first to speak of the value of caffeine in this complaint; he employed for injection a solution in water with a little spirit, in which some twenty-five or thirty minims contained the dose (0.06 gramme, or about one grain), and found great benefit from its use.

In **other Neuralgias** caffeine has also been found very useful by some authors. Dr. Anstie employed it successfully in intercostal neuralgia accompanying shingles. A solution in glycerine has been recommended by Lorent—one grain in two and a half minims; but this gives a deposit, and requires using at the moment of injection.

For migraine, caffeine has also been extensively used in the form of the so-called citrate of caffeine; but this appears to be a salt of indefinite composition, and much less trustworthy than caffeine hypodermically injected.

In **Diarrhœa**, especially when occurring in children, coffee is useful by virtue of its astringent properties, which depend upon the peculiar coffee-tannin. But this property is very unequally exerted upon different persons; and there are people with whom, for unknown reasons, coffee always acts rather as an aperient than otherwise.

In **Delirium Tremens** coffee is often exceedingly useful; and a case is recorded in which the hypodermic injection of caffeine (one-grain dose) proved useful in the insomnia of chronic alcoholism. It must be borne in mind that there is a restlessness and insomnia connected with great lowering of general nervous power, which is quite as marked as the similar condition which may be produced by exalted reflex irritability.

---

[1] Dr. Radcliffe informs me that, in the wards of Westminster Hospital, he for many years employed coffee to the absolute exclusion of alcohol in typhus and typhoid fever, and with the most satisfactory results.

PREPARATIONS AND DOSE.—These are sufficiently discussed above.

Tea, the product of several different species of Thea, a genus which belongs to the order *Camelliaceæ*, is remarkable in a pharmacological point of view from containing in its leaves an alkaloid, Theine, generally believed to be not merely isomeric but identical with caffeine. It is needless to say that if this be the case the popular idea is amply confirmed which attributes many of the qualities and virtues of tea to the subtler aromatic ingredients, seeing that there is an unmistakable difference between the general effects of the two beverages. At present, however, we know nothing of the action of these additional ingredients. Concerning the general differences between the effects of tea and coffee, we might say that tea is more refreshing and stimulating in states of fatigue of brain or muscle, and that when taken in excess (more particularly in chronic excess), it more powerfully affects the stability of the motor-system, causing a tremulousness of the limbs which is less frequently produced by excesses in coffee; on the other hand, coffee is much more prone to produce distressing palpitations of the heart, and, in some persons of weak digestive powers, coffee taken with substantial food will altogether arrest the digestion of the latter.

Paraguay tea, or Maté, the great national drink of the inhabitants of a very large area of South America, is made from the leaves of the *Ilex Paraguayensis*, a member of the order *Aquifoliaceæ*. It contains caffeine in large proportion, and is by many supposed to be intermediate in its effects between tea and coffee. The infusion is drunk without milk or sugar, and is taken extremely hot; it is sucked through a tube.

Guarana, the last on our list, is prepared from the *Paullinia sorbilis*, being in fact a paste made from the fruit of that plant, which belongs to the order *Sapindaceæ*. The alkaloid, at first called guaranine, has long been known to be really identical with caffeine; hence guarana-powder is only a somewhat uncertain, because varying preparation of caffeine. Nevertheless it is a somewhat convenient form for the internal administration of this drug. It has been used for many years in France, and some two or three years ago Dr. Wilks introduced it into this country as a remedy for migraine. It is sold in boxes of powders, or in small bottles, the stopper being hollowed out, so as to contain a measure of twenty grains. It is undoubtedly a very effective palliative of sick-headache. One dose of twenty grains may be taken every half hour, infused in a teacup of boiling water until about two or three have been taken; more than this tends in some subjects to produce distressing palpitation.

OLEACEÆ.

THE OLIVE (OLEA EUROPÆA).

ACTIVE INGREDIENTS.—Pure olive oil is unctuous, pale yellow, or yellow with a greenish tinge, very slightly odorous, and possessed of a bland and slightly sweetish flavor. The sp. gr. is 0.92. At a temperature of 38° F. it begins to congeal, and at freezing point a portion of it becomes solid. Ether dissolves half its own volume of the oil, but in alcohol it is only partially soluble, unless the alcohol be in very large proportion. The constituents, according to Braconnet, are olein, about seventy-two per cent.; and palmatine, or margarine, about twenty-eight per cent. Treated with nitrous acid, or with nitrate of mercury, it becomes a peculiar fatty solid, called elaïdin. With alkalies and certain other bases, it unites so as to form a soap. Exposed to the atmosphere, it is apt to become rancid, acquiring an unpleasant taste and smell, a deeper color, and a more dense consistence, which changes are quickened by heat.

PHYSIOLOGICAL ACTION.—Taken into the system, olive oil is mildly laxative, emollient, and in some degree nutritive. Externally applied, it relaxes the cuticular tissues.

THERAPEUTIC ACTION.—As a very gentle laxative, in cases where there is much intestinal irritation, forbidding the use of powerful medicines, olive oil may be given with the best effect. It is especially valuable, in the form of enema, for infants and young children.

In various forms of irritant poisoning, if olive oil be promptly administered it mechanically entangles much of the noxious matter, and interposes itself as a shield to the mucous surface.

It has been recommended as a vermifuge, but this can only be the case in so far as its not very powerful purgative qualities extend.

As a mechanical defence from the atmosphere for diseased cutaneous surfaces, olive oil is often very useful. The special value of olive oil in pharmacy, however, is its convenience for use in the making of liniments, plasters, ointments, and other preparations for external use.

PREPARATIONS AND DOSE.—Oleum Olivæ, ℨj.— ℥j. (4.—30.). It is an ingredient of: Liniment. Ammoniæ; Liniment. Camphoræ; Liniment. Chloroformi; Liniment. Plumbi Subacetatis; Emplast. Ammoniaci cum Hydrargyro; Emplast. Hydrargyri; Emplast. Plumbi; Ceratum Cetacei; Ceratum Plumbi Subacetatis; Ceratum Saponis; Charta Cantharidis.

SOAP.

The soaps prepared with olive oil are of two distinct kinds: *Sapo durus,* or hard soap, in which the other chief ingredient is soda; and *Sapo mollis,* or soft soap, in the manufacture of which potash is employed. Hard soap is of a grayish-white color, corneous when cold and dry, and then reducible to powder, though, if warmed, readily yielding to pressure. Water dissolves it, and, if the solution be treated with lime or lead, it precipitates. Rectified spirit dissolves it entirely. Incineration leaves a residue of non-deliquescent ash.

Soft soap, instead of being solid, in consistence resembles honey or jelly. It is yellowish, translucent, and scentless, often speckled with white, owing to the presence of minute crystals, and is soluble in rectified spirit. The residue, after incineration, is an ash, which rapidly deliquesces.

These two descriptions of soap, though differing much in physical qualities, are both composed of oleates and palmates. Hard soap consists of oleate and palmate of soda; soft soap of oleate and palmate of potash. Neither of them, when pure, should impart an oily stain to paper.

Soap, whether hard or soft, is in pharmacy useful for little besides external purposes. The *Sapo durus* used to be given as an antacid, in doses of five to twenty grains; it is employed in the preparation of the Pilula saponis composita, and of various other pills, not so much, however, for any service it may render as a medicine, as for its suitableness as a vehicle or adjunct; it is valuable, also, in the preparation of certain plasters, such as the Emplastrum saponis, and of the Linimentum saponis. The *Sapo mollis* has likewise been administered as an antacid, in doses of five to twenty grains.

PREPARATIONS AND DOSE.—Pilula Saponis Co., gr. v. (.30); Ceratum Saponis; Emplast. Saponis; Linimentum Saponis.

GLYCERINE.

Glycerine, discovered in 1789 by Scheele, who gave it the name of the "sweet principle of oils," is a thick and syrupy liquid produced by the saponification of the fats and fixed oils. It is either colorless, or of a pale amber tint, unctuous, inodorous when pure, and very sweet to the taste, whence the current appellation. The sp. gr. is 1.260; the composition is $C^3H^4O^3$. It contains a small percentage of water.

[From a chemical standpoint, glycerine belongs to the alcohol series, and experimentally has been found to produce symptoms resembling those of alcohol. In animals intoxication follows the ingestion of large doses, and, if they be excessive, death, preceded by coma, is the result.]

Glycerine does not become rancid: it does not evaporate, but with heat it decomposes, with evolution of intensely irritating vapors, and at a full red heat ignites and burns with a blue flame. When submitted to strong heat in a capsule, it leaves no residue. With water and with alcohol it mixes readily, dissolving in each of these in any proportions, but in ether it is not soluble. It dissolves, in turn, a great variety of different substances, such as iodine, common salt, the fixed alkalies, the vegetable acids, tannic acid in particular, and many of the salts of the vegetable alkalies. Hence it becomes an excellent recipient of many medicinal matters. To lotions it is a very useful adjunct, by reason of its soft and unchanging oiliness: it is excellent also as a simple lubricant for dry, chapped, or slightly abrased skin, and in cases of deafness, either alone or in combination with olive oil. It has been tried as a substitute for cod-liver oil, but not satisfactorily.

[Its internal use has been recommended on insufficient theoretical grounds in some forms of acne, and from reports appears to have been somewhat useful.]

PREPARATIONS AND DOSE.—Glycerina, ℨss.—j. (2.50—5.).

[NITRO-GLYCERINE (Glonoinum).

When glycerine is subjected to the action of nitric and sulphuric acids, a body known as nitro-glycerine or glonoine, discovered by Sobrero,[1] is formed. Its composition is $C^3H^5N^3O^9$. It has an oily consistence, but is heavier than water (sp. gr., 1.60), in which it is insoluble. It is very soluble in alcohol.

PHYSIOLOGICAL ACTION.— Its physiological effects were first studied by Dr. C. Hering (homœopath), of Philadelphia, in 1848, since which time it has been experimented with by many, who, without exception, confirm the more prominent phenomena described by him. Taken in doses of one-tenth grain, it produces rapid increase of pulse, with congestion of the head (internal and external), accompanied with severe bursting headache, and sometimes with nausea and vomiting. In doses of $\frac{1}{100}$ grain it produces the same kind of phenomena, but less in degree, and in susceptible persons even much smaller quantities are capable of producing appreciable effects, as we know by personal experience. These effects, after moderate doses, pass off after a few hours. In many respects it greatly resembles the nitrite of amyl.

THERAPEUTIC ACTION.—Dr. Murrell,[2] of England, while working with nitro-glycerine, was struck with the resemblance referred to, and employed it with satisfaction in three cases of angina pectoris in doses of $\frac{1}{100}$ minim. Drs. Hammond, Hamilton, and others of this city have used it for some time in epilepsy and other anæmic conditions of the brain. It is somewhat slower in its action than amyl nitrite, but its effects are more enduring.

PREPARATIONS AND DOSE.— None officinal. The most convenient form for its administration is a one per cent. solution in alcohol—dose, ♏j. ±.]

MANNA (Fraxinus Ornus).

ACTIVE INGREDIENTS.—Manna is of the consistence of cheese, yellowish-white, faintly odorous, and of a sweetish but sickly flavor. About sixty or eighty per cent. consists of mannite: the other constituents are sugar, of which there are two kinds, one crystallizable, the other not so; a small quantity of gum, some resin, and a little nauseous extractive matter. When pure, it is soluble in three to five parts of cold, and in its own weight of boiling water; and is also soluble in rectified spirit. It melts with heat, and burns with a bluish flame.

Mannite, $C^6H^7O^6$, may be extracted from the crude manna by means

---

[1] Comptes Rendus, 1847.
[2] Ringer : 7th Ed., 1878, p. 372.

of boiling alcohol, from which, on cooling, it separates in shining, whitish, four-sided, and acicular crystals. It is odorless, sweet, soluble in boiling, less so in cold, alcohol, and is incapable of fermenting with yeast.

PHYSIOLOGICAL ACTION.—Manna is mildly laxative, though it sometimes causes flatulence and griping.

THERAPEUTIC ACTION.—Manna is well adapted for administration, as a gentle laxative, to infants and young children; also for introduction into purgative mixtures, as an adjuvant.

PREPARATIONS AND DOSE.—Manna, ȝj.—ij.(4.—8.), for children.

THYMELACEÆ.

MEZEREON (DAPHNE MEZEREUM).

ACTIVE INGREDIENTS.—Mezereon bark contains daphnine, a resin, and a volatile oil. The daphnine is a neutral crystalline principle, bitter, and slightly astringent, but inert; the resin is dark green, acrid, and soluble in alcohol and in ether; the volatile oil is acrid and highly irritant. Upon this, in conjunction with the resin, depend the active properties. When mezereon root is boiled in water an acrid vapor is given off.

PHYSIOLOGICAL ACTION.—That the mezereon is a poisonous plant appears incontrovertible. The berries have long been known to be highly dangerous, and, in a case recorded by Linnæus, are said to have caused death. The bark of the root (and of the stem also, but in a less degree) is excessively caustic. Chewed, it gives at first an impression of sweetness, but an acrid and burning sensation soon arises in the mouth and fauces, lasting for several hours; and if the impregnated saliva be swallowed, or any portion of the bark itself, the burning extends to the throat and stomach. Pereira says that "all parts" of the plant, "swallowed in large quantities," . . . "prove poisonous." The topical action of the bark is that of an irritant, and, when applied to the skin, that of a vesicant. A decoction, taken internally, appears to promote the action of the skin and of the kidneys; but Dr. Alex. Russell disputes these effects. In large doses it causes irritation both of the kidneys and of the alimentary canal.

Decoction of mezereon bark often proves laxative. In some constitutions it increases the flow of saliva, and causes pain in the stomach and bowels; and sometimes vomiting and purging.

THERAPEUTIC ACTION.—Mezereon bark has been very strongly recommended in rheumatic and scrofulous diseases; also in chronic cutaneous disorders, and in cases of syphilis.

There would, however, seem to be some uncertainty in its action; or perhaps certain constitutions only are open to its effects. Mr. Pearson absolutely denies the utility of mezereon in relation to syphilitic disease in any form or in any stage.

It has also been resorted to with good effect as a cure for toothache, and as a masticatory in cases of paralysis of the tongue.

PREPARATIONS AND DOSE.—Extract. Mezerei Fluid., ♏ v.—xxx. (.30—2.); Ungt. Mezerei; also an ingredient of Decoct. Sarsap. Co. and Syr. Sarsap. Co.

ZINGIBERACEÆ.

GINGER (ZINGIBER OFFICINALE).

ACTIVE INGREDIENTS.—The taste of the rhizome is hot and biting, with a little acridity superadded: these characters it retains when dried. The odor, although in recent examples faint, then becomes pleasantly aromatic and pungent. These qualities depend chiefly upon a volatile oil and a resinous matter: the sp. gr. of the former is 0.893. The rhizome also contains a considerable quantity of starch, with other substances ordinarily found in roots.

PHYSIOLOGICAL ACTION.—Ginger, in its action on the system, is stimulant. If a piece of the rhizome be chewed, it causes a considerable flow of saliva; and the powder, if passed into the nostrils, excites sneezing. The application of the powder, mingled with water, to the skin produces intense heat and tingling. On being swallowed, its stimulating effects are perceived first in the alimentary canal and in the organs of respiration. Administered in moderation, it is a valuable stomachic, and, if the body is relaxed and enfeebled, especially in old age, or if gout be present, it promotes digestion. Compared with pepper, it is less acrid.

THERAPEUTIC ACTION.—Ginger is advantageously given in dyspepsia and in flatulency. It also constitutes an excellent adjuvant, especially to tonic medicines, to which it communicates cordial and carminative qualities; and in connection with drastic purgatives it is useful to check or prevent nausea and griping. For these purposes it may be employed either in substance or in infusion; and similarly in regard to flatulent colic and tympanitis. In India, Europeans of delicate constitution are accustomed to use an infusion of ginger instead of tea.

Ginger is useful in cases of relaxed uvula, and in slight paralytic affections of the tongue. It is serviceable also in toothache, and, from its action on the skin, may be resorted to for the relief of headache. To this end a paste formed of it should be spread upon paper and applied to the forehead.

PREPARATIONS AND DOSE.—Zingiber, gr. v.—xv. (.30—1.); Oleoresina Zingiberi, gr. ss —ij. (.03—.12); Extr. Zingib. Fl., ♏ v.—xv. (.30—1.); Tinct. Zingib., ♏ xx.— ʒ j. (1.20.—4.); Syr. Zingib., ʒ j.—ij. (5.—10.); Infus. Zingib., ℥ ss.—ij. (15.—60.); Trochisci Zingib., No. 1—3, p. r. n.

CARDAMOMS (Elettaria Cardamomum).

Active Ingredients.—Cardamom seeds are fragrant; the taste is warm, pungent, and highly aromatic. Their principles are extracted by water, and still more easily by alcohol. The important one is a volatile oil of sp. gr. 0.945, which is present in the proportion of about 4.5 per cent., and rises with the water on distillation. It is colorless, possessed of an agreeable and penetrating odor, and of a strong, burning, camphoraceous, and slightly bitter taste; this, however, as well as the scent, is lost by exposure to the atmosphere. In addition to the volatile oil, cardamom seeds contain a fixed oil, some coloring matter, salts, and other unimportant ingredients.

Physiological Action.—Cardamoms well illustrate the aromatic qualities of the seeds of the Zingiberaceæ. They are gratefully fragrant and pungent, carminative and stomachic, and, compared with many other substances of their class, less heating and stimulating.

Therapeutic Action.—Combined with bitters, cardamoms are serviceable in dyspeptic cases; and combined with purgatives, they check flatulence and griping, for the prevention of which last they are particularly useful. The tincture is advantageously combined with stomachic infusions; and in cases of flatulent colic, and in gouty and spasmodic affections of the stomach, it may be usefully employed as an adjunct to ether, opium, or other antispasmodics. For rendering mineral waters and saline solutions easy and agreeable to the stomach there is probably no more suitable or appropriate drug.

Preparations and Dose.—Tinct. Cardamomi; Tinct. Card. Co., of either ℨ j.—ij. (4.—8.).

IRIDACEÆ.

[BLUE FLAG (Iris Versicolor).

Active Ingredients.—The chemistry of this plant has not been thoroughly studied. No attempts, so far as we are aware, have been made to isolate its active principles and to determine their ultimate composition; from it, however, is obtained an article, used chiefly by eclectics, called irisin or iridin. It is said by Coe to consist of a resin, a resinoid, alkaloid, and a neutral principle.

Physiological Action.—In the fresh state it possesses acrid properties, lost in part by drying. According to Bigelow,[1] "water distilled from the root has a highly nauseous taste and odor," showing that part of its principles are volatile. When taken in sufficient dose, twenty to sixty grains of the fresh root, it produces violent vomiting with purga-

---

[1] Am. Med. Botany, vol. i., p. 158, Boston, 1818.

tion and great depression. In smaller doses it is diuretic and cholagogue, and in very small ones it produces annoying constipation from inactivity of the rectum. Fæces accumulate in the lower bowel, but the rectum seems to have lost all power to expel them. This we know by personal experience and complaints of patients to whom we have given it in doses too small to produce full cholagogue effects. Preparations from the dried plant we have seen produce headache and cerebral disturbance. The cholagogue action of iridin was the subject of experiment by Rutherford and Vignal, who write (*Brit. Med. Journ.*, May 5, 1877): "1. Five grains of iridin, mixed with a little bile and water, and placed in the duodenum, very powerfully stimulate the liver. It is not so powerful as large doses (four grains) of podophyllin; but it is more powerful than euonymin, as is shown by the amount of bile secreted per kilogramme of dog; the fractions for the two euonymin experiments being 0.4789 c. c., and 0.4678, whereas in the iridin experiments they are 0.537 c. c., and 0.638 c. c. The high fraction in the second iridin experiment probably resulted from a much smaller dog getting the same dose as in the first experiment, the smaller liver being thereby stimulated to a proportionably greater amount of work. 2. Iridin is also a decided stimulant of the intestinal glands. Judging from these experiments, its irritant effects on the intestinal mucous membrane are decidedly less than those of podophyllin, while the purgative effects are greater than in the case of euonymin. The statement of the writer in the *Lancet* above quoted" (*vide infra*) "that in man it is gentler in its action than podophyllin, is fully supported by these experiments, and there seems every reason why this substance should be removed from its present obscurity and placed in a prominent position in practical medicine."

THERAPEUTIC ACTION.—Blue flag was held in high esteem by the aborigines of this country, and by many tribes was specially cultivated as a medicinal plant. Schoepf (1787) writes: "*Mirifice valet in ulceribus tibialibus et vulneribus.*" Iris versicolor cannot be recommended as a simple cathartic, its action being too distressing; but Bigelow speaks highly of its diuretic properties. The writer in the *Lancet*, above referred to, says: "In our hands iridin has produced effects similar to those occasioned by a combination of blue-pill, rhubarb, and aloes. It seldom fails to produce a mild catharsis with bilious evacuations; it appears to possess the advantages of (1) not requiring the addition of a mercurial; (2) not irritating the rectum, as aloes is apt to do; and (3) it has no astringency, and therefore does not produce subsequent costiveness like rhubarb when given alone. In a sluggish state of the bowels, arising from torpidity of the liver, or when the stools are pale, particularly as we find them in the intervals of overt attacks in gouty persons, we have found the iridin one of the best aperients, much gentler than podophyllin, and more reliable when a slight cholagogue action is required to be maintained for a lengthened period."

Personally we have never employed iris versicolor for cathartic or diuretic purposes. But as an hepatic stimulant in those gouty conditions which so frequently give rise to chronic eczema it is invaluable. In these cases we usually prescribe from five to ten minims of the tincture from the fresh root twice daily. If the smaller dose constipates, it is increased or a laxative dose of leptandra is combined with it. In one other distressing condition we have found iris of the utmost service, namely, a blinding headache, the pain occupying the region of the right supra-orbital and accompanied with nausea or vomiting. This form of headache is usually the result of hepatic derangement, and when thus caused may usually be promptly relieved by iris. The dose under these circumstances should not exceed one minim of the tincture, to be repeated every twenty minutes or half hour till three doses are taken. As a rule, the headache will be greatly relieved and perhaps entirely gone before the third dose is taken. If the three doses do not relieve, it is not worth while to continue it longer.

PREPARATIONS AND DOSE.—Although contained in the secondary list of the Pharmacopœia, no officinal preparations of iris are given. We have employed fluid extracts, "concentrated tinctures," and irisin made from the dry root, but have given them up in favor of the tincture (one part root, two parts alcohol) made from the fresh root. The dry root, as found in the shops, varies so much in quality and activity that tinctures and fluid extracts made from it cannot be thoroughly depended on. If the only variation were in the strength of the preparation, this could be easily rectified by proper adjustment of the dose, but, after a large comparative experience, we are satisfied that the *quality* and character of the action is not identical in the fresh and dry root. Wet or dry seasons, etc., may influence the *proportional amount* of active constituents in the plant, but they will probably still preserve their own peculiarities as regards composition and relative proportion. On the other hand, after a plant is gathered, its juices sometimes undergo decomposition and chemical change, and some of them, as we well know, become practically inert; in others there is a change in the kind of effect.

LABIATÆ.

LAVENDER (Lavandula Vera).

ACTIVE INGREDIENTS.—The flower spikes, which alone are officinal, contain a considerable quantity of oil, combined with which is a bitter principle. The former is an oxidized volatile oil or hydrocarbon, and contains in solution a camphor-like substance called stearoptine. In color it is pale yellow; the sp. gr. somewhat varies, but is usually about 0.95, the lightest oil being the purest. It is fragrant, of a pungent and aromatic taste, soluble in rectified spirit, and in two parts of proof spirit.

PHYSIOLOGICAL ACTION.—Lavender oil is, in large doses, a narcotic poison, causing death preceded by convulsions. Doses of one drachm are sufficient to produce these effects in rabbits. It is also a very effective poison to skin parasites.

THERAPEUTIC ACTION.—Occasionally administered in cases of hysteria, and for nervous headache, hypochondriasis, and flatulence. The oil is employed for scenting evaporating lotions, ointment, and liniments. It is used in the preparation of the compound liniment of camphor, and the compound tincture is a constituent of the *Liquor Potassæ Arsenitis.* The leaves and flowers have been employed in sternutatories.

PREPARATIONS AND DOSE.—Oleum Lavendulæ, ♏j.—v. (.05—.25); Spts. Lavendul., ʒ ss.—j. (2.—4.); Spts. Lavendul. Co., ʒ j.—ij. (2.—4.).

SPEARMINT (Mentha Viridis).

ACTIVE INGREDIENTS.—Spearmint, cultivated from time immemorial in every kitchen-garden, owes its general properties to a volatile oil, believed to be a hydrocarbon, and containing a peculiar camphor in solution, as also a substance isomeric with carvol. This oil is lighter than water, the sp. gr. being 0.914; the odor is strong and pungent; the taste is the same as that of the living plant; in color it is pale yellow, but with age it becomes reddish. At 320° F. it boils.

PHYSIOLOGICAL ACTION.—Stimulant and carminative.

THERAPEUTIC ACTION.—Like lavender, spearmint is usefully employed in flatulence and to relieve the pains of colic; it is said to be a good adjunct to purgative medicines.

PREPARATIONS AND DOSE.—Aqua Menthæ Viridis, ʒ j.—ij. (30.—60.); Oleum Menth. Virid., ♏j.—v. (.05—.25); Spts. Menth. Virid., ♏v.—xxx. (.30—1.20).

PEPPERMINT (Mentha Piperita).

ACTIVE INGREDIENTS.—In flavor peppermint is the most agreeable of its genus. The taste and odor, as in the other species, are given by an oil which, on being distilled from the recent plant while in bloom, possesses in high degree a peculiar and penetrating fragrance, with a warm and aromatic flavor, followed by a singular sense of coldness in the mouth. The sp. gr. is 0.92, the boiling point 365° F. While new, this oil is pale greenish yellow, but with keeping the color deepens. Two isomeric constituents are found in it—one liquid, the other solid. The solid one makes its appearance when the oil is reduced to a temperatre of 12° or thereabouts, and receives the name of "peppermint-camphor." It presents itself in the form of acicular white crystals; retains the taste and odor of the plant; is soluble in alcohol and in ether, almost insoluble in water, and fusible at 92° F. The composition is $C^{10}H^{19}HO$. By the action of chloride of zinc, peppermint-camphor is decomposed, with production of Menthine, $C^{10}H^{18}$, a transparent and mobile liquid of sp. gr. 0.85.

PHYSIOLOGICAL ACTION.—Except that the action corresponds with that of spearmint, nothing is exactly known on this subject.

THERAPEUTIC ACTION.—Peppermint is administered for the relief of nausea and of griping pains in the alimentary canal; also to expel flatus, and sometimes as an antispasmodic. It is convenient also as a cover for the taste of other medicines and as an adjunct to purgatives.

PREPARATIONS AND DOSE.—Aqua Menthæ Piperitæ, ʒ j.—ij. (30.—60.); Ol. Menth. Pip., ♏j.—v. (.05—.25); Spts. Menth. Pip., ♏v.—xx. (.30—1.20); Trochisci Menth. Pip., No. 2—5, p. r. n.

PENNYROYAL (Mentha Pulegium).

Active Ingredients.—In composition, oil of pennyroyal is believed to resemble oil of spearmint, *i. e.*, to be an oxidized hydrocarbon, containing camphor in solution. The color is pale yellow; the odor corresponds with that of the recent plant, which is peculiar and agreeable; the taste is warm, aromatic, and somewhat bitter. The sp. gr. is 0.95; the boiling point, 395° F. Pennyroyal contains tannin in small quantity, with other ingredients apparently of little or no importance.

Physiological Action.—The operation of pennyroyal corresponds pretty nearly with that of the larger species of Mentha. Some persons consider it to be very specially antispasmodic and emmenagogue.

Therapeutic Action.—The chief medicinal value of this little herb is found in the directions just named, viz., as an antispasmodic and emmenagogue. It is also esteemed in whooping-cough and hysterical complaints.

Preparations and Dose.—None officinal. Preparations analogous to those of the other mints may be made and administered in corresponding doses.

ROSEMARY (Rosmarinus Officinalis).

Active Ingredients.—The oil upon which depends the well-known scent of this plant is colorless; it is apparently of the same composition as the Labiate oils already described, and retains both the odor and the taste of the plant. Sp. gr., 0.888. Boiling point, 365° F.

Physiological Action.—Resembles that of the other aromatic Labiates.

Therapeutic Action.—Nervous headache and hysteria are said to be relieved by the smell of rosemary, and the powers of the mind have been said to be quickened., It is celebrated also for its property of encouraging the growth of the hair and mitigating baldness. The green hue of the pomatums employed for these purposes is chiefly, if not entirely, owing to the presence of rosemary. It likewise is supposed to have the singular power of preventing the hair from uncurling when exposed to a damp atmosphere, whence probably the very ancient fame of rosemary for the toilet. Rosemary is used also as an emmenagogue.

Preparations and Dose.—Oleum Rosmarini, ♏j.—v. (.05—.25); Spts. Rosmarini (B. Ph.), ♏xv.— ʒj. (1.—4.).

PIPERACEÆ.

BLACK PEPPER (Piper Nigrum).

Active Ingredients.—Pepper is hot, pungent, and aromatic, and along with other ordinary constituents of small berries, contains the above-named Piperine, a resin, and a volatile oil. Piperine, $C^{17}H^{19}NO^{3}$, occurs, when pure, in white, crystalline, rhomboidal prisms, which are odorless and almost tasteless. In cold water it is insoluble, in boiling water it is scarcely soluble; but in alcohol and acetic acid it dissolves readily and, in a somewhat less degree, also in ether. It is volatile, and fuses at 212° F.

The resin, which exists in large quantities, is soft and acrid, solid at 32° F., extremely pungent, soluble in alcohol and ether. With fatty bodies it unites readily.

The volatile oil, $C^{10}H^{16}$, when pure, is colorless, and has the taste and odor of pepper. Sp. gr., 0.9932.

Physiological Action.—Pulverized pepper, allowed to remain in contact with the skin, induces redness and pain. Taken into the mouth, or if the corns be chewed, it excites sensations of intense burning; and, if absorbed by the nostrils, burning and severe sneezing. Swallowed in moderate quantities, as a condiment, it stimulates the stomach, exciting a sense of warmth, and slightly accelerating the pulse. Digestion is assisted—this is often seen when substances which alone do not admit of being readily assimilated have been associated with pepper as medicine—diaphoresis is promoted, and the mucous surfaces become excited. Taken in excess, pepper induces intestinal inflammation, with violent burning pain in the epigastric region, accompanied by great thirst. In several cases that have been recorded, the immoderate use of pepper has been followed even by rigors, convulsions, and delirium; in one instance vomiting ensued. Upon the whole, when used legitimately, pepper may be described as a warm carminative stimulant, capable of producing systemic excitement, and of directly acting upon the mucous membrane of the alimentary canal. The action is not diffusible, but local; and hence, in assisting feeble digestion, it checks tendency to flatulence, or, if flatus be present, the escape of it is promoted. This last effect comes of the local action of pepper upon the rectum. The urethra is likewise subject to its operation.

Therapeutic Action.—Pepper is of very old celebrity, and whether or not cared for as a condiment, was certainly employed as a medicine in remote ages. The ancient Greek physicians recommended it with warm water, for the relief of "cold fit." In dyspepsia it has been found useful as a gastric stimulant; it is given also to check vomiting, to abate nausea, and to stop singultus. In cases of relaxed uvula it may be advantageously resorted to for use in the form of a gargle; or it may be employed as a masticatory, which method of securing the action is useful also in cases of paralysis of the tongue, and affections of the mouth and throat which demand the employment of a powerful local remedy. As a febrifuge, pepper has been found serviceable by the French and German physicians; and in our own country it has long been a popular medicine in intermittents, being administered in spirit-and-water when the paroxysms are about to commence.

In the form of confection, pepper is well adapted to weak and leucophlegmatic habits, but the use of it in this way requires patience and perseverance; and there is always the objection that it is liable to accumulate in the colon, so that a laxative is required for the discharge of it. Sir Benjamin Brodie says that the accumulation of the confection-paste in the colon has sometimes caused the disappearance of severe hæmorrhoids, the local effect being energetic in the extreme. The direct intent of the confectio is for diseases connected with the rectum, such as fistula and ulcers, with hæmorrhoids, of course, the affected parts being gently stimulated.

The alkaloid, piperine, is said to possess febrifugal properties that allow of its comparison with the alkaloids of the cinchonas. But while in no way superior as a drug, the cost alone would render it unpopular.

PREPARATIONS AND DOSE.—Piper, gr. ij.—x. (.13—.65); Oleoresina Piperis, gr. ss.—ij. (.03—.13).

CUBEBS (CUBEBA OFFICINALIS).

ACTIVE INGREDIENTS.—The odor of cubebs is peculiar, camphoraceous, and aromatic; the taste is hot and peppery. They contain a volatile oil, a resin, and a substance called cubebine.

The volatile oil, *Oleum cubebæ*, $C^{15}H^{24}$, is colorless, or pale greenish yellow, and, like that of black pepper, lighter than water, the sp. gr. being 0.929. It yields the odor and possesses the flavor of cubebs, and evaporates on exposure to the atmosphere. Upon this volatile oil the active properties mainly depend; cubebs, therefore, should not be pulverized for medicinal use until actually wanted. The powder is dark in color and presents an oily appearance.

The resin is soft and acrid; the cubebine, $C^{33}H^{34}O^{12}$, is neutral, and analogous to piperine. It crystallizes in small needles.

PHYSIOLOGICAL ACTION.—Cubebs, though possessing all the ordinary characteristics of common black pepper, are devoid of the agreeable flavor which recommends the last named for use as a condiment. Their properties have relation more especially to the mucous membrane of the genito-urinary apparatus; they have others which entitle them to be classed with acrid substances. Experiments upon rabbits show that with these animals the oil acts as a poison, the symptoms induced being particularly like those which ensue upon the administration of copaiva, but manifested yet more strongly. In man, the operation of the oil is shown especially in the region indicated. It quickens the action of the kidneys, causes vascular injection of those organs, and gives an albuminous character to the urine, which also acquires the peculiar odor of cubebs.

In addition to these more striking effects, under the influence of cubebs the peritoneum becomes affected; the gastric mucous membrane becomes injected; and, if the dose be large, there are nausea and vomiting. Tried in small or experimental quantities, the oil induces thirst and heat in the throat, which, on increasing the dose, are followed by feverishness, nausea, eructation, and headache. Taken when the stomach is in an irritated or inflammatory condition, the results are nausea, vomiting, burning pain, griping, and even diarrhœa. In some instances the use of cubebs has caused an eruption upon the skin resembling urticaria.

Compared with the operation of copaiva, it is to be observed that continued resort to cubebs in small doses is unattended by the hurtful effects in regard to the digestive functions and to the appetite which ensue upon the use of the former. The tendency is rather to energize them, so that cubebs have decidedly the advantage when copaiva, for any reason, may be objectionable.

THERAPEUTIC ACTION.—At what period cubebs were first employed is uncertain, but that they were known in England at least five hundred years ago appears to be well established.

Practically, at the present day, the principal use made of this drug is in the treatment of gonorrhœa, for which disorder cubebs were recommended in 1818, and still more urgently during the two or three following years. Though unquestionably efficacious, it must be confessed that they have sometimes induced conditions very distressing to the patient. Hæmorrhoids, hæmaturia, and deep-seated headaches result in certain constitutions from the use of cubebs, while in others there is concomitant diarrhœa. The nausea is occasionally so severe as to forbid perseverance with the medicine. The stimulant operation of cubebs upon the bladder is well illustrated in a case described by Sir Benjamin Brodie.[1] They are good for catarrh of this organ, and for leucorrhœa.

The utility of cubebs is not limited to their action upon the mucous membrane of the genito-urinary organs. They have proved useful in catarrhal affections of the air-passages, especially when the secretion is copious and the system relaxed. They are valuable in all cases of inflammation of the mucous membrane lining the bronchial tubes and the intestinal canal. When so employed, it has been recommended by some that oxide of bismuth should be conjoined. Originally they seem to have been resorted to as a gastric stimulant and carminative in atonic dyspepsia. In India they are still accounted useful as a stomachic. Rheumatism has likewise been treated with cubebs, but the success has been very questionable.

PREPARATIONS AND DOSE.—Cubebæ, ʒ ss.—ij. (2.—8.); Extr. Cubebæ Fl., ʒ ss.—j. (2.—4.); Tinct. Cubebæ, ʒ j.—ij. (4.—8.); Oleum Cubebæ, ℳ v.—xx. (.25—1.); Oleoresina Cubebæ, gr. v.—xx. (.30—1.20).

MATICO (ARTANTHE ELONGATA).

ACTIVE INGREDIENTS.—Matico leaves contain a volatile oil, a dark green resin, a bitter principle called maticine, traces of tannic acid, and another acid readily crystallizable, termed the artanthic. The oil is of a light green color, and on keeping becomes thick and crystalline. The maticine is soluble in alcohol and in water, but insoluble in ether.

PHYSIOLOGICAL ACTION.—Very little is yet known beyond the general fact that matico is aromatic, stimulant, and laxative.

---

[1] London Medical Gazette, i. 300.

THERAPEUTIC ACTION.—Matico is useful as a substitute for cubebs, in cases of discharge from mucous surfaces. When introduced to this country in 1839, it was with the reputation, however, of being an internal or chemical styptic. Externally it may be employed for stanching the flow of blood from trivial cuts and wounds, such as leech-bites, on which it acts mechanically.

PREPARATIONS AND DOSE.—Extr. Matico Fl., ℨ ss.—j. (2.—4.); Infus. Matico (B. Ph.), ℥j.—ij. (30.—60.).

LAURACEÆ.

CINNAMON (Cinnamomum Zeylanicum).

ACTIVE INGREDIENTS.—The dried inner bark of the cinnamon tree possesses a warm, sweet, and aromatic taste, and evolves a corresponding odor. These qualities are referable to a volatile oil, $C^9H'OH$, which, when separated, varies a little in color, when good is bright yellow, but by keeping becomes red. The odor and taste are pleasant, and resemble the corresponding qualities of the bark; it is heavier than water, and on exposure to the atmosphere absorbs oxygen, and forms cinnamic acid, two resins, and water. Cinnamon bark also contains tannic acid in considerable quantity, a resin, and an acid called cinnamic acid.

PHYSIOLOGICAL ACTION.—Cinnamon is warm and cordial to the stomach, astringent, and carminative. The essential oil is devoid of the astringency found in the bark, and taken in moderate doses becomes an agreeable stimulant, producing a sensation of warmth in the epigastrium, and promoting assimilation. The too constant use of cinnamon has a tendency, however, to produce costiveness.

In full doses cinnamon acts as a general stimulant to the nervous and vascular systems. Some practitioners believe that it exerts a specific influence over the uterus.

THERAPEUTIC ACTION.—Being cordial, stimulant, and tonic, cinnamon is useful in many cases characterized by feebleness and atony. As an astringent it checks diarrhœa, for which purpose it is best combined with chalk, opium, or some vegetable infusion. In the later stages of low fever it is also employed with advantage; and in flatulent and spasmodic affections of the alimentary canal it may often be used to good purpose. Nausea and vomiting are checked by the administration of cinnamon. It has also been employed with excellent effect in uterine hæmorrhage, and is an ingredient in various powders, infusions, and tinctures. The oil is sometimes employed as a powerful stimulant in cases of paralysis of the tongue, in syncope, and in cramp of the stomach. The principal value, after all, as with the bark, is to modify the flavor of bitter infusions, and, when combined with purgatives, to check their griping action.

PREPARATIONS AND DOSE.—Ol. Cinnamomi, ℳj.—v. (.05—.25); Tinct. Cinnamomi, ℨj.—ij. (4.—8.); Spts. Cinnam., ℨj.—ij. (4.—8.).

CAMPHOR (Camphora Officinarum).

Active Ingredients.—Camphor, when extracted and purified, is white, concrete, crystalline, semi-translucent, somewhat granular, tough, and difficult to pulverize. The odor is strong, penetrating, and characteristic; the taste is pungent, bitter, and aromatic. It is lighter than water, the sp. gr. being 0.98; at ordinary temperatures it is solid, but, owing to its volatility, slowly evaporates. Under the influence of heat it rapidly and entirely sublimes; it melts at 288°, and boils at 399°, and when ignited burns with a bright flame and copious production of smoke. In water it is very sparingly soluble, so that if water be added to the alcoholic solution, which is readily made, the camphor immediately precipitates. It is soluble also in ether, and in the volatile and fixed oils. The characters of camphor are thus pretty much those of a concrete volatile oil. By constant heating with nitric acid it becomes oxidized, and converted into camphoric acid.

Physiological Action.—It is singular how little is yet known with accuracy on this point, notwithstanding the large number of strong statements that have been made by various writers respecting the action of camphor.

It is certainly known, however, that camphor will, in very large doses, cause death, preceded by delirium, coma, and convulsions. There are certain concentrated spirituous preparations of camphor, originally intended to be taken in very small quantities, which the folly of patients not unfrequently induces them to prescribe for themselves in large quantities: in this way many serious consequences have occurred. In man I am not aware of camphor having proved fatal except in one case. But there are numerous instances upon record in which it placed the recipients in imminent danger; and one of the earliest cases that were recorded with exactness still remains as instructive as any, viz., that of Mr. Alexander.[1] That gentleman took forty grains of camphor, mixed with syrup of roses, for experimental purposes. In twenty minutes he was languid and listless; in an hour he became giddy, confused, and forgetful. All objects quivered before his eyes, and a tumult of undigested ideas floated through his mind. At length he passed into total unconsciousness, and in that state was attacked with strong convulsions and maniacal frenzy. The nature of the accident having been suspected, an emetic was given, with the effect of expelling almost the whole dose of camphor that had been swallowed. The graver symptoms then disappeared, but a variety of singular mental affections continued for some time. Considering how small a proportion of the camphor had been dissolved at the time when the emetic was so happily administered, one cannot doubt that much less than the forty grains, if fully taken into the blood, would have proved fatal to this particular person. There is no doubt, however, that much larger doses than this have been taken without inducing more than temporary and not very severe effects. How much of this variability of effect is really due to idiosyncrasy, and how much to mere difference in the rate of absorption must, however, remain at present undecided.

Upon animals many persons have experimented with large and fatal doses; and the post-mortem appearances have proved that, besides its

---

[1] Quoted in "Christison on Poisons."

action upon the nervous centres, camphor has the power of a direct irritant to the alimentary canal, and also to the mucous tract of the genito-urinary organs.

One of the phenomena which have been often, though not always, noted in severe camphor-poisoning, is dilatation of the superficial vessels, especially of the head and face; this is usually accompanied by delirium.

Upon the skin camphor is well known to act as an irritant. A concentrated solution rubbed in soon causes heat and bright redness; and if it be applied to a raw surface there are intolerable burning and consecutive inflammation.

The recent researches of Grisar (upon valerian oil and other ethereal oils) also included experiments with camphor, and it was found that in certain doses camphor decidedly exhibits a power of reducing exalted reflex irritability. It is not improbable that to this function we must attribute its distinct, though not well understood, capacity of subduing certain forms of diarrhœa.

On the whole, while our knowledge of the physiological action of camphor must be allowed to be at present very partial and confused, it is probable that Nothnagel is right in his statement that the apparently conflicting attributes of sedation and stimulation which have been ascribed to this drug are both correct; the difference being in fact a matter of dose and occasion.

THERAPEUTIC ACTION.—There are at least two ways in which camphor often proves distinctly beneficial.

1. **As a remedy for Functional Nervous disorders** not of the severest type, it is often efficacious. In that type of **nervous headache** which, without attaining the character of well-marked rhythmical migraine, very closely borders upon it as regards the character of the pain, and is most commonly seen in hysterical females, camphor is very useful. From three to five grains of camphor rubbed down with a little spirit, and suspended in water by the aid of some tragacanth, and a twenty- or thirty-grain dose of carbonate of magnesia, is an excellent though apparently rather clumsy form. The magnesia is not superfluous, for it often assists the cure by correcting gastric acidity which is present at the time.

In **Hysteric Excitement**, and also in **Chorea**, when these are not of the graver types, camphor is often of considerable use; and it has even been seen by Van der Kolk to prove useful in a case of acute mania in which other very powerful remedies had failed. In **Delirium tremens** and in **spasmodic Asthma** it has also been strongly recommended by various authors. But with regard to all these maladies the same remark must be made about camphor as was made by Van der Kolk respecting its use in insanity, viz., that its effects are widely different in different individuals, and that we never can tell beforehand whether it will act well or not.

2. **In many forms of Diarrhœa** camphor proves extremely useful, and it is of much importance to define accurately the cases in which it may be expected to prove efficacious. They are evidently those in the production of which nervous irritation has a large share. Among the cases of diarrhœa that occur in high summer heat, there are many distinguished by clear red tongue, in which the mingled exhaustion and irritation of the nervous system are for the most part the true sources of the flux, and here camphor will do very much good; indeed, a few doses of three or four grains or drops will often completely check the disorder.

Camphor is less useful in proportion as the diarrhœa is more dependent on local irritation from unwholesome food, etc. And as to the assertions that have been made with regard to the power of this drug to arrest **Asiatic cholera**, I have never witnessed such effects, except in the first (or doubtful) stage, and I cannot but suppose that some error in diagnosis has been made in most instances of supposed cure.

If we carefully consider all the best ascertained therapeutic effects of camphor, there appears a high probability that the secret of its beneficial action under very varying superficial conditions depends upon the power to subdue reflex excitability, which is indicated in the experiments of Grisar made with it and with valerian oil, etc. It is difficult otherwise to explain its ancient high reputation as an anaphrodisiac, and the strong modern recommendations of it in all kinds of irritable conditions of the external and internal sexual organs. No one who tries it in an unprejudiced manner can help observing that it fails quite as often as it succeeds in any one of these disorders, and in fact it is impossible to discern any proof of a direct action either upon the sexual organs or upon the nervous centres that govern them, except so far as a general state of exalted reflex irritability may exist, and may be calmed by the remedy. It is even probable also that a large portion of the so-called stimulant effects for which it has been so warmly praised by Copland and others in the treatment of adynamic fevers, etc., is really only a tranquillization of the reflex apparatus which can be obtained with less of general depression than would be likely to follow the use of other narcotics.

As an **External remedy** for pains, camphor is a good deal employed in the form of liniment, and is generally credited with good effects, but its action is very uncertain.

Finally, some persons recommend its use as a **diaphoretic**. It has been already mentioned that in many cases of camphor-poisoning an immense dilatation of the superficial blood-vessels has occurred. This dilatation of peripheral vessels, attended by copious sweat, can be produced by the use of camphor vapor, from the heating of camphor on a plate over which the patient sits with a blanket pinned around his neck, so that the fumes do not enter the throat.

PREPARATIONS AND DOSE.—Camphora, gr. iij.—x. (.20—.65); Oleum Camphoræ, ℥j.—iij. (.05—.15); Aqua Camphoræ, ʒij.— ℥j. (8.—30.); Spts. Camph., ℥x.— ʒj. (.65—4.); Linimentum Camphoræ.

SASSAFRAS (Sassafras Officinale).

ACTIVE INGREDIENTS.—Sassafras wood has a warm and aromatic taste, and a peculiar fragrance. It contains a volatile oil, fatty matter, resin, a principle called sassafrine, tannin in small quantity, and some other less important ingredients. The volatile oil is light yellow; it has a pungent taste like the odor of the wood, a sp. gr. of 1.094, and is itself composed of two other oils, one of which sinks in water, while the other swims. The properties are yielded to hot water and to alcohol. It is to be observed, however, that the wood of the tree contains a less amount of the active principles than exists in the bark and in the roots.

PHYSIOLOGICAL ACTION.—Taken in infusion, both the wood and the bark of the root are reputed stimulant and sudorific, but little is accurately known on this subject. The sudorific action is much assisted by the use of warm clothing and tepid drinks. Employed as an adjuvant to other medicines, sassafras mitigates disagreeable flavors, and renders them more tolerable to the stomach.

THERAPEUTIC ACTION.—On account of its sudorific and alterative properties, sassafras is advantageously employed in cutaneous disorders, also in rheumatic and syphilitic complaints. It is seldom used alone; more usually in combination with sarsaparilla and guaiacum. When the system presents febrile and inflammatory conditions, it should, on account of its stimulating nature, not be resorted to.

PREPARATIONS AND DOSE.—Oleum Sassafras, ♏ij.—x. (.10—.50). Sassafras is an ingredient of the Comp. Syrup and Comp. Decoct. of Sarsaparilla.

BEBEERU, OR BIBIRU (NECTANDRA RODIÆI).

ACTIVE INGREDIENTS.—Bibiru bark contains an alkaloid called Beberia or Biberine, $C^{19}H^{21}NO^3$, a yellow resinous-looking body, which does not crystallize, is soluble in alcohol, slightly so in ether, and sparingly in water. In acids it dissolves, neutralizing them, and forming amorphous yellow salts which are uncrystallizable. There is also a certain quantity of tannic acid, which gives a green color with a salt of iron. The Beberiæ sulphas of the Brit. Pharm. ($C^{35}H^{20}NO^4HOSO^3$) presents itself in the form of thin brown scales, which are translucent, becoming yellow when pulverized, very bitter in taste, and soluble in alcohol and in water. The bark itself is hard, heavy, and brittle, cinnamon-colored within, and very bitter and astringent.

PHYSIOLOGICAL ACTION.—The properties of Bibiru bark are tonic, anti-periodic, and febrifugal. The physiological action of biberine has already been described under the subject of Buxin.

THERAPEUTIC ACTION.—Although noticed as far back as 1769 by Bancroft, in his "Natural History of Guiana," it was only in 1834 that the value of Bibiru bark as a medicine was made public. In that year Dr. Rodie discovered that the alkaloid contained in the bark and also in the fruit could be used as a substitute for cinchona, and employed it successfully in intermittents. His discoveries were announced in the *Edinburgh Medical and Surgical Journal* for October, 1835.

In 1843 Dr. Douglas Maclagan published, in the "Transactions of the Royal Society of Edinburgh," a full account of the composition and properties of the bark, confirming all that had been stated by Dr. Rodie; and at the close of the following year we have in Hooker's *London Journal of Botany* the full botanical account of the tree itself.

In practice, the peptic and tonic effects of Biberine appear to be produced without causing at the same time the headache, giddiness, and ringing in the ears, which often accompany the exhibition of quinia. Hence, with patients with whom quinia disagrees, it becomes an exceedingly useful substitute.

In febrifugal and in anti-periodic properties Bibiru bark appears, however, to be inferior to the Peruvian drug.

It is distinctly stated at the same time by Dr. Rodie that biberine, when properly administered, generally cures intermittents which quinine has failed to remove, and that it appears not to affect the head, and not to produce its effects by counter-morbid action, in the way that the alkaloids derived from the cinchonas usually do, or are supposed to do.

Both bibiru bark and the sulphate of biberine have been exhibited as a peptic in anorexia and in dyspepsia; also as a general tonic where the constitution is debilitated; as in protracted phthisis, and strumous affections. As a febrifuge, bibiru has been employed both in intermittent and remittent complaints. As an anti-periodic it has also been resorted to in headaches requiring such a medicine, and also in intermittent neuralgias.

PREPARATIONS AND DOSE.—The dose of the sulphate of biberine is from one to three grains, when employed as a tonic; and from five to twenty grains, when given as a febrifuge. In substance, bibiru is given in pills, prepared with conserve of roses.

THE SWEET BAY (Laurus Nobilis).

ACTIVE INGREDIENTS.—In England the bay tree is immediately distinguished from all other out-door evergreens by the cinnamon odor of the bruised leaves. This proceeds from a volatile oil, contained also in the fruit. The latter holds in addition a fixed fatty oil, and other ingredients. The "oil-of-bays" of commerce, oleum lauri expressum, which is a compound of the volatile oil and of the fixed fatty oil, is obtained from the drupes, whether fresh or dried, by means of heat and pressure.

PHYSIOLOGICAL ACTION.—The volatile oil obtained by distillation of the berries with water is aromatic, stimulant, and narcotic.
In large doses the leaves are emetic.

THERAPEUTIC ACTION.—Both the leaves and the fruit of the bay tree have been employed to strengthen the digestive organs; also to prevent flatus. They are also said to act as an emmenagogue.

The oil obtained from the fruit is employed externally as a stimulating liniment in cases of sprain and bruises. It has likewise been given in paralysis, and has been employed for the relief of colic, and to mitigate deafness; but these uses are now obsolete.

The fruits enter into the composition of the emplastrum cumini.

ARISTOLOCHIACEÆ.

SERPENTARIA (Aristolochia Serpentaria).

ACTIVE INGREDIENTS.—The root of this plant contains a volatile oil, a resin, and a bitter and acrid extractive matter, which last is soluble both in water and in spirit. The volatile oil is yellowish when newly prepared, but with age it becomes brownish; the odor is aromatic, the taste warm, bitter, and camphoraceous. All the properties of serpentary are yielded to water, as well as to proof spirit and to alcohol. By distillation it furnishes a considerable amount of camphor.

PHYSIOLOGICAL ACTION.—Taken internally in small doses, serpentaria promotes the appetite. In larger doses, it causes nausea, flatulence, uneasy sensations in the stomach, and frequent, though not watery stools. When the absorption is completed, the pulse increases both in frequency and fulness; the surface-heat of the blood is heightened, and secretion and exhalation generally become augmented. A certain degree of analogy with camphor seems to be manifested by disturbance of the cerebral functions, serpentaria being apt to induce headache and a feeling of oppression, and at night to disturb the sleep. Emphatically, it is stimulant, tonic, and diaphoretic.

According to Jörg, after employment of the powder and the infusion of serpentary, in a large number of experiments, the conclusions come to were the following: "Serpentaria occasions nausea, eructation, vomiting, constriction and pain in the stomach, borborygmi, colic in the small intestine, discharge of flatus, and a disposition to go to stool, but without evacuation, or the evacuation of consistent fæces only. The appetite is sometimes impaired, and sometimes increased; the stomach and bowels often become distended with flatus; itching about the anus, and even hæmorrhoids, are produced occasionally. Hence it would appear that the medicine acts upon the alimentary canal as an irritant, producing a secretion, not of liquid, but of gas.[1]

THERAPEUTIC ACTION.—As is implied in the name, the virtues of this plant were originally believed to render it an antidote to serpent-bites.

The conditions of the system in which the value of serpentaria is best declared are those of atony and torpor. In the low stages of typhus, it is by some persons considered a valuable remedy, in combination with sesqui-carbonate of ammonia, and administered when the tongue is dry, and brown or black, the pulse, at the same time, being feeble. Sydenham recommended a scruple of serpentary, taken in three ounces of wine, as a remedy for tertians.

In **Dyspepsia, chronic Rheumatism, and gouty Rheumatism,** serpentary is given. In the chronic forms of the last named it is recommended by Garrod, who considers it a remedy of some power. In certain **exanthematous** diseases it has been administered with advantage, either alone or in combination with other tonics. In **adynamic Fevers,** the low delirium, watchfulness, and other irregular actions of the nervous system which often ensue, appear to be amenable to the powers of serpentary.

In **remittent Fever,** especially when the remission is obscure, serpentaria is by some thought preferable to cinchona, being seldom offensive to the stomach, and quite free from proneness to do mischief. Sydenham remarks that in all cases where it is expedient to combine wine with cinchona the effects are much improved by the addition of serpentaria. It also enables the stomach to retain the cinchona with more comfort.

In **bilious Vomiting,** serpentaria is found in America to be effica-

---

[1] Stillé, "Therapeutics," vol. ii., p. 533.

cious in checking the nausea and tranquillizing the stomach. For this purpose it is given in decoction, in doses of a tablespoonful frequently repeated.

In **Cynanche maligna**, serpentaria is employed externally and as a gargle.

PREPARATIONS AND DOSE.—Extract. Serpentariæ Fluid., ʒ ss—j. (2.—4.); Tinct. Serpentariæ, ʒ ij.—iv. (8.—15.); Infus. Serpentariæ, ℥ j.—ij. (30.—60.).

URTICACEÆ.

THE HOP (HUMULUS LUPULUS).

ACTIVE INGREDIENTS.—The active principles of the hop reside chiefly, though not exclusively, in the lupuline, which possesses the peculiar flavor of the hop, and determines, by its abundance or otherwise, the value of any particular sample. According to Payen, it contains two per cent. of volatile oil, about eleven per cent. of bitter principle, and fifty to fifty-five per cent. of resin. Tannin is present to the extent of about five per cent., also a nitrogenous substance and a gummy one.

The bitter principle, called lupulite or humuline, is yellowish-white, uncrystallizable, neutral and very bitter. It is soluble in alcohol, sparingly so in water, and insoluble in ether: it contains no nitrogen, and is destitute of narcotic properties.

The volatile oil is obtained by submitting lupuline to distillation with water: it is said to consist of an oxygenated oil and a hydrocarbon, $C^{10}H^{16}$, isomeric with oil of turpentine. In color it is pale greenish-yellow, though, when re-distilled, it becomes colorless. The taste is acrid; the odor resembles that of hops; it is very soluble in alcohol and ether, less so in water, and has a sp. gr. of 0.910. By keeping and exposure to the atmosphere it becomes a resinous mass.

The resin is bright-yellow, changing to orange color on exposure to the atmosphere. It is soluble in alcohol and in ether.

PHYSIOLOGICAL ACTION.—In regard to this very little can be said with any confidence. The general belief is that hops possess tonic and moderately narcotic properties; and that pain and nervous irritation may be allayed by the use of them, since they often assist sleep, if they do not actually induce it. The volatile oil has been stated to be narcotic; but this is denied by Wagner.

THERAPEUTIC ACTION.—Lupuline is of value as a medicine for the relief of chordee, and for checking involuntary seminal emissions. It is said to be of use in nocturnal incontinence of urine, and in cases of irritable bladder.

Lupuline has been resorted to as an anodyne in rheumatism, and as a tonic in dyspepsia, and has been found serviceable in gouty spasm of the stomach.

With patients who suffer from insomnia, nervousness, or over-excitement, the hop-pillow is deserving of trial; it relieves pain and allays nervous irritation. It would seem as if the volatile oil must be the source of the influence.

PREPARATIONS AND DOSE.—Tinct. Humuli, ʒij.—iv. (8.—15.); Infus. Humuli, ℥ij.—iv. (60.—1.20); Lupulina, gr. v.—xv. (.30—1.); Tinct. Lupulinæ, ʒj.—ij. (4.—8.); Fl. Ext. Lupulinæ, ʒss.—j. (2.—4.); Oleoresina Lupulinæ, gr. j.—v. (.06—.30).

HEMP (Cannabis Sativa).

ACTIVE INGREDIENTS.—The vegetative portions of the hemp plant exhale a powerful and rather repelling odor when fresh, but the scent is much diminished or lost altogether by drying. The taste is feebly bitter. These qualities are referable to the active principle of the plant, called cannabine, a resinous body possessed of a peculiar fragrance, and a warm, bitter and acrid taste. It is soluble in alcohol and in ether, but not soluble in water, which throws it down from the alcoholic solution as a white precipitate. Hemp likewise contains a small proportion of volatile oil, with some lignin, albumen, gum, and other less important ingredients. The seeds do not share in the composition of the stems and leaves, and of course not in their properties.

PHYSIOLOGICAL ACTION.—Observations and experiments upon animals with regard to the influence of hemp appear to indicate that carnivorous creatures suffer a kind of intoxication. Herbivorous ones, on the contrary, seem to be unaffected by it, although the dose administered has been considerable.

In man, when subjected to the operation of this drug, the pulse in some instances becomes quickened, though not remarkably so, and the respiration somewhat slower; in other cases no change is observable. At the same time there is a sensation of warmth in every part of the body excepting the feet, which, ordinarily, seem cold by contrast. It is said also to increase the appetite for food. The secretions do not appear to be materially affected. It does not cause dryness of the tongue, nor is there any constipation of the bowels.

Should the individual taking cannabis be of a temperament favorable to its peculiar action, there is very generally an agreeable exhilaration of the spirits, followed, more or less, by stupor and sleep. At times there is a delightful state of reverie. Having myself experimented with this drug, I can state from my own personal knowledge that the reverie, when it does come on, is agreeable beyond description. But the next day there are sensations of dulness and heaviness, which are anything but enviable. Because of the exhilarating effects, the natives of India are in some parts addicted to the use of it as a pleasant inebriant. It is certain, however, that in other countries the peculiar energy of the plant is not manifested in the same degree, nor to the same extent, nor always in the same manner; whence it is justly inferred that climatic and ethnological considerations have both to be taken into account in estimating the absolute powers of cannabis. Instead of agreeable sensations, it would seem that with certain individuals the effects are sometimes nausea, vomiting, and oppressive thirst. Cases are known where it has induced a kind of catalepsy. Indian hemp is said likewise to act as an aphrodisiac.

The medicinal properties of cannabis, strictly so called, are well summed up by Dr. Glendinning, who employed this drug extensively. "It acts as a soporific or hypnotic in causing sleep; as an anodyne in lulling irritation; as an antispasmodic in checking cough and cramp; as a nervine stimulant in removing languor and anxiety, also in raising the pulse and enlivening the spirits, without any drawback or deduction of indirect or incidental inconvenience; conciliating tranquil repose without causing nausea, constipation, or other signs or effect of indigestion, without headache or stupor." Notwithstanding these praises, it must be acknowledged that much has yet to be learned in regard to the physiological action of cannabis, though in all likelihood what more is learnt will be in its favor. As an anodyne it is decidedly inferior to opium. [In doses of ¼ grain of the extract, three times a day, and continued for several days, we have seen C. Indica produce involuntary twitchings and jerkings of the lower extremities.]

Therapeutic Action.—Indian hemp is employed in medicine as an antispasmodic and anodyne in **Traumatic Tetanus.** Dr. O'Shaughnessy, and the late Prof. Miller, of Edinburgh, gave it in cases of traumatic tetanus, and state that the beneficial effects which followed its use were very marked. They record several severe cases which were treated with the extract, and all of which were cured. Dr. O'Shaughnessy administered as much as two to three grains of the resin, once, twice, or even thrice in the twenty-four hours.

Uterine Hæmorrhage, such as occurs at the period of the cessation of the menses, is often moderately arrested or abated by the use of cannabis. Dr. Churchill, of Dublin, gave for this purpose five to fifteen or twenty drops of the tincture three times in the day, and with excellent results.

Uterine contraction during labor.—Dr. Alexander Christison claims for Cannabis Indica a power of promoting contraction equal to that which is possessed by ergot. He states that the effects are perceived more quickly than those of ergot, and that they are more energetic but of shorter duration. He adds that these effects are far from certain to be experienced. The observations of Christison have not been confirmed by other authorities.

Gonorrhœa.—Cannabis Indica has been recommended in gonorrhœa, and there can be no doubt that the employment of it causes a diminution of the discharge, and relieves the violent burning pain in the urethra which comes on during and after micturition. [In gonorrhœa we have had considerable and very favorable experience with a tincture made from fresh native hemp. (C. sativa, var. Americana.) It should be given in doses of a few drops three or four times a day, but should not be given until the more acute symptoms have commenced to subside. It is fully as effectual as copaiva or sandal, and is infinitely pleasanter to take. It is referred to by Schoepf as being in use in this country for this purpose prior to 1787. It is omitted by Thacher, Bigelow, and Barton, but is noticed by Rafinesque.]

Neuralgia, and **Neuralgic Headache** more particularly, has been palliated and even cured by one-quarter to one-half grain doses of the extract, given two or three times a day. Cannabis also relieves dysuria, accompanied by bloody urine.

Chorea and **Delirium Tremens** are likewise said to have yielded to it; in the latter, if hypnotics must be employed at all, Indian hemp is one of the least dangerous and most useful. Half to one grain of the extract should be given as a dose.

[There is a good deal of confusion on the subject of *hemp* in this country, as there are at least three substances that pass under this name. These are, first, the resin obtained from the C. sativa grown in Asia, and distinguished in the Pharmacopœia as the C. Indica; second, the product of the C. sativa of this country, and called (in the Pharmacopœia) C. Americana; third, the indigenous "Indian hemp" or Apocynum cannabinum. The narcotic properties of the C. Americana are much feebler than those of the C. Indica. The Apocynum is an active diuretic.]

PREPARATIONS AND DOSE.—Tinctura Cannabis (Indicæ), ℥ v.—xx. (.30—1.20); Extract. Cannabis Indicæ, gr. ⅙—⅓ (.01—.03); Extract. Cannabis Americanæ, gr. ½—ij. (.03—.12). The tincture of C. Americana that we have used in gonorrhœa is made from the fresh leaves and young twigs, one part; alcohol, two parts.

CORYLACEÆ.

THE OAK (QUERCUS ALBA).

ACTIVE INGREDIENTS.—The principal active ingredients contained in oak bark are tannic acid, gallic acid, and pectin, all of which are yielded both to water and to spirit. The amount of tannic acid varies with the age of the branch from which the bark is taken, also with the age of the bark itself, and with the season of the year at which it is removed from the tree. For pharmaceutical purposes these two acids are prepared from the galls of the *Quercus infectoria*, under which head it will be more convenient to speak of their composition.

PHYSIOLOGICAL ACTION.—Oak bark is powerfully astringent, and similar in its operation to other vegetable astringents containing tannin.

THERAPEUTIC ACTION.—Powdered oak bark is valuable in cases of passive hæmorrhage, and in diarrhœa. The infusion or decoction may be employed with advantage as an astringent gargle. Injections and fomentations of a powerfully astringent character may also be conveniently prepared from oak bark.

PREPARATIONS AND DOSE.—Decoctum: Quercûs Albæ, ℥ ss.—ij. (15.—60.).

GALLS (QUERCUS INFECTORIA).

ACTIVE INGREDIENTS.—Galls contain about thirty-five to forty per cent. of tannic acid, and about six per cent. of gallic acid; also a substance called ellagic acid, discovered by Chevreul and Braconnot, with some sugar, starch, gum, extractive, saline matter, and other subordinate ingredients. The soluble principles are taken up by forty times their weight of boiling water; the residue is tasteless. The light brown color of the infusion is deepened by ammonia, and changed to orange by nitric acid. Bichloride of mercury renders it milky, but with not one of these reagents is there any precipitate, nor yet with litmus, which the infusion reddens.

Tannic acid, $C^{27}H^{22}O^{17}$, is well described in the British Pharmacopœia, 1867, as occurring, when separated, in "pale yellow vesicular masses, or thin and glistening scales, with a strongly astringent taste, and an acid reaction; readily soluble in water and rectified spirit, very sparingly soluble in ether. The aqueous solution precipitates solution of gelatine yellowish white, and the persalts of iron of a bluish-black color. When burned with free access of air it leaves no residue."

Gallic acid, $H^3C^7H^2O^4.H^2O$, presents itself, when prepared, in the shape of crystalline, acicular prisms, or silky needles, sometimes nearly white, but generally of a pale fawn color. It requires about one hundred parts of cold water for solution, but dissolves in three parts of boiling water. It is soluble also in rectified spirit. The aqueous solution gives no precipitate with solution of isinglass, but with a persalt of iron it gives a black one. When dried at 212° the crystalline acid loses 9.5 per cent. of its weight. Burned with free access of air, it leaves no residue.[1]

Ellagic acid, $C^{14}H^4O^8$, appears in the separated state as a white crystalline powder, and differs from tannin and gallic acid in being almost insoluble alike in alcohol, ether, and water. According to M. Pelouze,[2] it does not exist ready formed in galls, but originates in the action of atmospheric oxygen upon their tannin; and the same, this author informs us, is the origin also of gallic acid. The quantity of ellagic acid contained in galls is only trifling, but the intestinal concretions called bezoars are said to be composed of it almost exclusively.

PHYSIOLOGICAL ACTION.—1. Of Tannic Acid. Tannic acid, taken into the mouth, acts as a powerful astringent upon the mucous membrane of all the parts it comes in contact with, acting also upon the fauces.

Tannin scarcely deserves the name of a poison; its effects, even when given in the largest doses, being entirely local and not severe. Upon the mucous membrane of the stomach and intestinal canal these very large doses are capable of producing a peculiar effect by virtue of its power to combine with albuminoid matters in a manner similar to that which constitutes the essence of the tanning process in the curing of skins for leather. The consequences of this, in cases where large quantities have been taken, have been obstinate constipation and vomiting, thirst, and, in a few cases, cutting pains in the abdomen.

It is by virtue of its tendency to combine with protein tissues that tannic acid possesses a decidedly superior power as a local astringent over gallic acid.

Tannic acid, when absorbed into the blood, appears to suffer decomposition, the products of which are gallic acid, pyrogallic acid, and a humin-like substance. The principal fact in confirmation of this belief is that ordinary tannic acid is not found in the urine,[3] but only the decomposition products already named, or some of them. Gallic acid is nearly always

---

[1] Brit. Pharm., 1867.
[2] United States Dispensatory, 12th Ed., p. 404.
[3] Catechu-tannin and several other varieties do pass off in the urine unchanged.

present, and can be recognized by its characteristic effect on persalts of iron; but Bartels has shown that occasionally only the humin-like body passes over, and sometimes even this trace of the tannin is lacking. Neither liver, pancreas, skin, nor pulmonary mucous membrane appears to excrete tannin unchanged; but the sputa, like the urine, have often been observed to be colored almost black.

The intensely powerful coagulating action of tannin upon blood, when mixed with it out of the body, might give rise to a belief that similar changes occur within the blood-vessels. But there is no proof of this, and there is much reason to think that very little tannin *as such* is absorbed into the blood. Tannin, in the form of albuminate, has been found in plenty in the fæces; but it is not likely that much passes through the gastro-intestinal mucous membrane without previous decomposition. The remote effects of tannin are therefore with great probability to be ascribed to its secondary products.

2. **Of Gallic Acid** the local action, as already said, is far less powerful than that of tannin, but the former passes rapidly through the body, appearing unchanged in the urine, and affecting various organs and functions in a manner which becomes important from a therapeutic point of view.

THERAPEUTIC ACTION.—1. **Of Tannic Acid.** This is now understood to be mainly local; but even with this limited range, the drug is highly useful in a variety of cases.

As a local Astringent, tannic acid fulfils many functions. Thrown (in the form of a fine powder, or in the shape of gargle, or of spray for inhalation) upon the mucous membrane of the fauces and air-passages, it checks congestion and inflammatory swelling, reduces excessive secretion, and arrests the tendency to capillary hæmorrhage. Hence it is useful in quinsy, catarrhal affections, chronic bronchitis, laryngeal phthisis, and hæmoptysis. Applied to a bleeding surface anywhere, if no large vessel be engaged, tannic acid is prompt in its styptic action, and among other uses may be applied to modify the fungating and easily bleeding surfaces of ulcers. But perhaps the most decided instances of its local utility are to be found in its application in the shape of the glycerine of tannin to eczematous skin (Ringer); blenorrhœal and leucorrhœal states of the urethra and vagina; sore throat (painted on with a brush); intertrigo, rectal fissures, and prolapsus ani.

As an Antidote to alkaloid Poisons, tannin holds a high place. The observation was first made in regard to emetin-poisoning (or rather ipecacuanha-poisoning), sufficient to produce excessive vomiting; but the observation has been extended to opium and its salts, digitaline, nicotine, and other alkaloids. **In metallic poisoning**, particularly that with antimony, tannin has gained some repute; but it is not as effective as albuminous matter (white of eggs).

2. **Of Gallic Acid**, while the local effects are not particularly worth mentioning, since those of tannin are so superior, the remote effects are so remarkable as to render it in every way one of our best remedies. As **a remote astringent**, gallic acid enjoys a reputation which is doubtless somewhat exaggerated, but is in large part deserved. There are two principal forms in which this action can be brought into play.

In Hæmorrhage of the most various kinds, but especially in hæmoptysis, and in hæmaturia which depends upon morbid conditions of the mucous membrane of the bladder, gallic acid is often exceedingly effective.

Other remedies, and particularly ergot of rye, may perhaps supersede it to a considerable extent, but gallic acid remains one of the most useful agents of this class. "Ruspini's styptic" was mainly composed of it.

In Dropsy and Albuminuria, however, the astringent power of gallic acid is probably even more useful than in hæmorrhagic conditions. There is a very general agreement among those who have practically tried it, that gallic acid can often do great service in checking the waste of valuable albuminoid matters which the system would otherwise have to support. It is remarkable that while albuminuria is strikingly diminished, there is usually an increase in the flow of urine. The action of gallic acid upon the tracts, pulmonary and urinary, through which it is eliminated, is used as a strong argument by the school of therapeutists which teaches that for one of the most promising modes of therapeutic influence we ought to look to the curative effects of drugs on organs by which they are cast out of the body.

PREPARATIONS AND DOSE.—Tinct. Gallæ, ℨj.—ij. (4.—8.); Ungt. Gallæ; Acidum Gallicum, gr. v.—xx. (.30—1.20); Glyceritum Acidi Gallic., ♏ xx.— ℨj. (1.20—4.); Acidum Tannicum, gr. v.—xx. (.30—1.20); Glycerit. Acid. Tannic., ♏ xx.— ℨj. (1.20—4.); Suppositoria Acid. Tannic.; Ungt. Acid. Tannic.

MELANTHACEÆ.

COLCHICUM (Colchicum Autumnale).

ACTIVE INGREDIENTS.—The corm and the seeds both contain a crystalline principle, which may be separated from a solution in dilute spirit in needles and prisms, and is called Colchicine, $C^{17}H^{19}NO^5$. Pelletier and Caventou believed it to be veratria in combination with gallic acid, but Geiger and Hesse proved this opinion to be erroneous. Colchicine is a powerful poison; the taste is bitter, but not so burning as that of veratria; it is destitute of odor, does not excite sneezing, and is soluble in water and in alcohol. With nitric acid it exhibits a play of colors commencing with violet. In the corm there are also contained fatty matter, yellow coloring matter, starch, gum, lignin, and a peculiar acid called cevadic acid. The active properties are partially taken up by water, readily so by alcohol, by dilute spirit, and by vinegar.

PHYSIOLOGICAL ACTION.—Colchicum is acrid and sedative, and to graminivorous animals generally poisonous. Hence they never eat the herbage unless it be inadvertently mingled with their fodder and unperceived by them, in which case it is apt to cause inflammation of the intestines and bloody evacuations, sometimes ending fatally. Störck, who, in 1763, published a pamphlet upon colchicum, administered it to dogs; the result being shown in vomiting, tremor of the limbs, and convulsive movements of the belly. The animals howled, passed urine in large quantity, had bloody evacuations and prolapsus of the rectum, and finally died. In some cases the dogs lay prostrate, languid and perfectly still, with feeble pulse, slow and often irregular respiration, refusing food, and having their eyes glazed and swollen. So peculiarly severe are the effects of colchicum upon dogs, that in France the medicine bears the name of *mort au chien*, and from my own experiments I can testify to its appropriateness. Cows, when they have eaten colchicum, lose all desire for food and water; they cease to ruminate; they suffer from running at the eyes and nostrils, and from a peculiar kind of diarrhœa; instead of diuresis, as in dogs, there is a tendency to diminution of the urine; the belly becomes distended, and the animal sinks into a half stupefied state.

Upon certain other creatures colchicum appears to produce but little effect. Experiments made with rabbits had scarcely any result beyond exciting some diuresis; upon frogs the effects were trifling and very temporary. In all observations made upon the action of colchicum, whether in regard to animals or to the human subject, it must be remembered that the drug varies in energy according to the period of the year at which the corms are collected.

In man, colchicum, administered in small and repeated doses, promotes the action of the secreting organs and quickens especially that of the intestinal mucous membrane. The skin, the liver, and the kidneys are affected by it much less obviously, and the nervous system, as a rule, seems quite as little apt to receive its influence. Headache and vertigo are stated to have sometimes occurred among the effects; and even results such as ensue from the use of narcotics have been ascribed to colchicum. But the symptoms in question, when manifested, were probably referable not so much to the action of the drug *per se* as indirectly to the exhaustion which follows the employment of excessive doses. For, if taken in excess, colchicum produces nausea, vomiting, colic in some degree, failure of appetite, coating of the tongue, borborygmi, and diarrhœa. Still larger doses intensify the emesis and bring on purging; the tormina and tenesmus become severe; and the stools resemble those of patients suffering from dysentery, indicating much intestinal inflammation. This last is a very special symptom of overdosing with colchicum. Finally, in doses beyond medicinal ones, colchicum becomes, to man as well as to dogs, a powerful poison, death being ushered in by acute pains in the bowels, incessant vomiting, and renewed purging and tenesmus, accompanied by great prostration and clammy sweats.

The effects of colchicine have been tried upon a variety of animals. As with colchicum itself, the principal and most characteristic phenomena are those of its drastic operation on the alimentary canal; whether given by the mouth or applied to a wound, the alkaloid constantly produces these effects, while convulsions are a much more occasional result. Geiger destroyed kittens in twelve hours with doses of six-tenths grain, the symptoms being those of gastro-enteritis; and Bley found one-fourth of this quantity fatal to a kitten of three months; vomiting, purging, and convulsions preceded death.

Various experimenters have obtained substantially similar results in operating upon rabbits, pigeons, and dogs. The researches of Schroff and of his pupil Heinrich showed that the fatality in animals was by no means constantly proportioned to the dose. Heinrich, however, himself took successively .156 and .312 grain of colchicine; the smaller dose, taken pure, caused an acrid and burning taste, nausea, retching, and an increased flow of saliva, lasting for several hours. The pulse was lowered about eleven beats during the first two hours. The larger dose, taken eight days later, produced similar symptoms during the first four hours, followed, however, by symptoms of severe gastro-enteritis, which did not disappear until the fifth day. Heinrich also describes a case of accidental poisoning, in which a girl, aged twenty, took less than one-fourteenth grain of colchicine; this caused pain in the stomach, repeated vomiting of green matter, liquid diarrhœa, drowsiness and collapse, with a pulse of ninety-six. There was also extreme dilatation of the pupils, with occasional convulsive twitchings of the right hand, and streaming perspiration of the face. The epigastrium was extremely tender, and there was bloody vomiting for eight or ten days. It appears that colchicine represents nearly, if not quite accurately, the total power of the fresh corm, and that it is eighty to one hundred times more active than the latter.

The general results of inquiries into the physiological action of colchicum are probably summed up with accuracy in the words of Gubler: " Colchicum is a drastic purgative, accidentally an emetic, and indirectly a sedative of the circulation; a depressant diaphoretic, sialagogue, and a diuretic." The slowing effect on the pulse seems always to be produced where doses have been given just so large, or so often repeated, as to approach but not to reach the line at which decided gastro-enteritis sets in. The latter condition is attended by increased rapidity of the pulse.

THERAPEUTIC ACTION.—At what date colchicum was first employed in medicine and for what diseases is not known, but its use reaches back at least to the sixteenth century. Störck, as above stated, published a pamphlet concerning it in 1763, which may be considered as the first attempt to obtain exact knowledge respecting its action.

As a specific for Gout, it was, however, that colchicum first obtained wide celebrity, and principally under the form of the *Eau médicinale* which became fashionable in France during the reign of Louis XV. The popularity of this preparation was immense at the time; in subsequent periods the reputation of colchicum has fluctuated, at times reaching an extraordinary height, at others sinking very low. This periodical discredit of the drug has perhaps been due to its varying strength, attributable to collection of the corms at improper seasons.

At present, and especially since the publication of Dr. Garrod's masterly treatise on gout, scarcely any one disputes the fact that, in the acute stages of the disease, colchicum has the power to produce a remarkable amendment of the symptoms.

Upon the inflammation and pain of acute gout it acts very directly, and Garrod observes that this action is independent of any evident purgation, sweating, or diuresis, though it is well known that large doses will produce those effects. Nevertheless there is no doubt that, in the treatment of many cases of gout, the increase of secretions is an important matter; and, instead of saying that colchicum is a specific remedy for gout, we may content ourselves with the assertion that it more powerfully affects the gouty than any other known form of inflammation. It is also certain that colchicum is not a lasting and final remedy for gout; it does not prevent relapses, and its power becomes weaker on successive occasions, till at last it will not check pain and inflammation unless given in dangerously large doses. As regards what was at one time considered an unfailing action of colchicum, namely, an augmentation of the elimination of uric acid, Garrod has shown that this is at least very doubtful, and thus its claims for being regarded as a rational specific for gout are further weakened.

In **Acute Rheumatism** there are surprisingly different opinions as to the efficacy of colchicum. Dr. Garrod thinks that in this disease colchicum does good, if at all, by controlling the heart's action. Many authorities describe it as *not less* remedial in rheumatism than in gout; Gubler speaks of it in the same terms for both diseases.

On the other hand, it is certainly the case that of late years colchicum has been losing its prestige in London hospital practice as a remedy for acute rheumatism, and is even considered by many physicians either useless or altogether noxious, as being apt to seriously embarrass the circulation.

It may, at any rate, be stated that the long-continued use of colchicum, or its employment in large doses, is not to be thought of in rheumatism. The peculiar anæmia which attends this disease produces an irritable weakness of the heart, which renders it liable to suffer alarmingly from the effects.

In certain **Acute Inflammations** we are often convinced by experience that colchicum exercises a happy influence, and yet it may be very difficult to say what is the *modus operandi*. In **bronchitis**, for example, we are often not at all sure whether the remarkably fortunate results of colchicum treatment are not really due to the fact that the chest affection is but a part of the gouty diathesis. The same remark applies to its use in **ophthalmia**.

In **Fevers** of a zymotic origin no one any longer thinks of recommending colchicum. It is far too treacherous a drug, looking to its power of suddenly depressing the circulation, to be worth resorting to on the chance of its producing some mitigation of the febrile symptoms.

In **Nervous Affections**, such as hysteria, chorea, etc., for which it was formerly recommended, colchicum is no longer considered of any value. To produce decided impressions on the nervous system we must give such doses as would always be more or less dangerous.

Among the miscellaneous uses for which it may be employed, colchicum is deservedly in favor as an addition to other aperients, when a slight increase of their activity is desirable. Half a grain to one grain of the acetic extract is thus often added to a small dose of colocynth or of aloes. Indeed, we may sometimes get a very effective combination by adding such a dose of colchicum to such a medicine as nux vomica, which otherwise might be quite unable to act as an aperient at all.

As a **Diuretic** colchicum can be no longer recommended. Its action is most uncertain, and, now that we possess such a diuretic as digitalis, there is no occasion to employ it.

PREPARATIONS AND DOSE.—Extr. Colchici Aceticum, gr. j.—ij. (.06—.12); Extr. Colch. Radicis Fluid., ♏ ij.—x. (.12—.60); Vinum Colchici Radicis, ♏ v.—xxx. (.30—2.); Extr. Colchici Seminis Fluid., ♏ ij.—x. (.12—.60); Tinct. Colchici, ♏ v.—xxx. (.30—2.); Vinum Colchici Seminis, ♏ v.—xxx. (.30—2.).

AMERICAN HELLEBORE (Veratrum Viride).

Active Ingredients.—The researches of Dr. Eugene Peugnet[1] have satisfactorily settled many vexed questions in regard to this subject. He shows that the principal alkaloid is veratrine,[2] identical with that which is obtained from the *album*, but different from the veratrine in cevadilla seeds. The second alkaloid (called viridia by Bullock) proves to be identical with Simon's jervine. There is no irritant poisonous resin corresponding to that which is found in *V. album*. Peugnet considers that the combination of the two alkaloids which is present in *V. viride* forms the best representative of the medicinal virtues of the genus veratrum.

Physiological Action.—Having regard to the distinction, rendered probable by Peugnet, between veratrine (procured commercially, as at present, from cevadilla) and the veratroida which is the true therapeutic ingredient of veratrum, we shall have to describe first the properties of veratrine and then those of the veratrum viride.

(*a*) The action of veratrine has received a great deal of investigation of late years. It has been shown to possess in high degree those irritant properties which are characteristic of all the species which contain it. Upon the skin it acts in very dilute dose somewhat as an anæsthetic; but in concentrated form it is highly irritant, producing erythema, or even pustular eruptions. Applied to the true skin, or to the mucous membranes, it produces a violent sense of heat. Taken into the stomach, or injected subcutaneously, it is absorbed with great rapidity, and produces spasmodic action of the intestines, mucous outpour from their glands, and salivation. At the same time the patient experiences curious sensations of heat and cold in various parts of the body. Many of the occasional effects are very uncertain and irregular in their occurrence; thus on one occasion there may be much diuresis, on another much sweating, on a third much sneezing and lachrymation. Purging, again, which is sometimes a marked symptom, is absent in the great majority of instances. One of the most constant phenomena is the steady descent of the pulse, which often reaches a very low level.

Given in fatal doses, veratrine produces violent vomiting and collapse, intense depression of the pulse, and a kind of tetanic spasms which usher in asphyxia and death. The spasmodic condition of the muscles has been the subject of much discussion. Gubler, without any sufficient reason, as it appears to me, regards this phenomenon as the reflex effect of the irritation of the intestinal canal. The excellent researches of Prevost, of Geneva, demonstrate, I think conclusively, that the muscular spasm is due to direct irritation of the muscles by the alkaloid: veratrine is a heart poison because it is a muscular poison. Other views of the cause of death are, that the heart spasm is due to stimulation of the vagus, or even that death is not dependent on the heart at all, but on spinal paralysis. It seems very unlikely that the latter is the case, or that the central nervous system is affected at all. The convulsions which occasionally occur, and also the alterations in sensibility, can at present be only imperfectly explained. Nor is it possible to say why vomiting or diarrhœa should occur,

---

[1] Medical Record, 1872.
[2] Peugnet calls it "Veratroida."

as it does in a considerable number of cases. At any rate, it is not the result of inflammatory irritation, for none such takes place. No signs of inflammation are found in the alimentary canal after death: the later researches have proved this conclusively, although a contrary belief formerly prevailed.

(*b*) Veratrum viride has been reported on by several observers. Dr. Hadlock [1] says that in large doses it causes nausea and vomiting, complete relaxation of the system, copious sweat, and a pale, cool skin. The pallor is apt to become extreme, and to be accompanied by syncope, especially if the patient rises suddenly from the recumbent position. There was no evidence of any primary action on the nervous centres; no delirium, coma, or stupor; nor was there any effect on the kidneys or bowels. Opium and morphia act very happily as antidotes to the severer symptoms. Hadlock considers that the primary action of V. viride is on the heart.

Oulmont,[2] who used a resinous extract of V. viride, found that in man, dogs, rabbits, and frogs, it affected chiefly the respiratory, circulatory, and digestive systems, rapidly causing nausea, vomiting, diarrhœa, and almost immediate slowing of the pulse and breathing. Even in cases where the vomiting and diarrhœa were excessive, there were no inflammatory lesions to be found after death.

Oulmont further observed a remarkable reduction of temperature: sometimes this amounted to two, three, or five degrees in the course of an hour and a half or two hours, and this low temperature would be maintained during as much as twenty hours.

This latter effect was not observed by Dr. Squarey [3] in some very interesting experiments which he made on patients in University College Hospital. Squarey used a tincture of V. viride; the patients, four in number, were suffering from slight maladies. In no case was there any notable depression of temperature, which fact was the more striking because the pulse was decidedly reduced. In all other respects Squarey's observations agreed with Oulmont's. Nausea and vomiting were produced in every case where more than twenty minims were given at once, and in one case there was a little diarrhœa. The respirations were also reduced in correspondence with the pulse, and in all cases there was a sense of great general loss of power when under the influence of the drug. Squarey concludes that V. viride acts specially on the heart, and that it is very analogous in its effects to digitalis. Ringer, on the other hand, considers its action more closely allied to that of aconite than to that of digitalis.

THERAPEUTIC ACTION.—Both veratrine and veratrum viride have attracted much attention as remedies during the last fifteen years.

(*a*) Veratrine has been largely used in France and Germany; to a much less extent in this country.

1. **The external application** may first be mentioned, as this mode of employment produces results concerning which there is no dispute.

In **Neuralgias,** and other severe localized pains, the ointment of veratrine is frequently applied with much benefit. In average cases the pharmacopœial [B. Ph.] unguentum (eight grains to the ounce) answers

---

[1] Med. and Surg. Reporter, 1870.
[2] Gaz. Hebdom., iii. 7, 1868.
[3] Practitioner, 1870, vol. i., p. 211 et seq.

well; it produces a slight pricking numb sensation, and speedily relieves pain.¹ It seems, however, to be frequently forgotten that in high concentration veratrine is intensely irritant to the skin, and that on susceptible skins even the pharmacopœial ointment will produce inflammation, and sometimes pustulation. The stronger ointments which are sometimes carelessly recommended, are still more likely to produce such effects, and in a very severe and unpleasant degree.

2. **The internal uses** of veratrine are much more varied, and are also much more the subject of difference of opinion.

In acute Inflammations, veratrine has been largely employed of late years.

In Pneumonia, more especially, it has been considered by many one of the most valuable additions to our resources that have ever been made. One of the earliest authorities on this subject is Vogt,² of Berne, who gave it in doses of $\frac{1}{13}$ of a grain (five milligrammes) every two or three hours, till it produced vomiting or retardation of the pulse, and found it very effective. If the stomach proved too irritable the dose was reduced, and the veratrine was given either with opium or in an effervescing draught. The mortality was eight per cent.

Aran³ highly recommended veratrine in pneumonia, and this recommendation was afterwards indorsed by Trousseau. Aran gave it in pills, with small doses of opium, which possibly mitigated the local effects.

Biermer⁴ tried veratrine extensively in pneumonia, and highly recommends it. Doses of $\frac{1}{20}$ grain, in pill, were employed, and produced decided effects in reducing pulse and temperature. But he remarks that its effects are very variable, and that it is often apt to cause a troublesome amount of nausea and vomiting, and sometimes diarrhœa.

The collective results of foreign experience seem to agree with my own, to the effect that veratrine may produce striking alleviation in pneumonia, but is on the whole so unmanageable a remedy as to be of doubtful applicability. I shall have to speak rather more favorably of veratrum viride.

In acute Rheumatism there is much evidence in favor of veratrine. It was highly recommended for this purpose by Trousseau, Alies, Bouchut, Léon, and others. Yet, after all that has been said upon the subject, it does not appear to be proved that veratrine acts in any other way than in reducing the febrile phenomena.

It is as an **antipyretic**, in fact (or this together with some special influence in reducing pain), that veratrine effects, or seems to effect, some good in the treatment, not only of acute rheumatism, but of a number of pyrexial diseases for which it has been recommended. Some, indeed, of the maladies for which it has been specially prescribed are very unfit subjects for its operation; thus it was once particularly praised in the treatment of typhoid fever, but careful investigation of the recorded results of its use in this disease gives no encouragement; and, on the other hand, the treacherous qualities of the alkaloid as a depressor of the heart's action have led to many disastrous consequences.

(*b*) Veratrum viride no doubt stands in a somewhat more favorable position than veratrine, yet it is necessary to be cautious in estimating the worth of the various statements that have been made in its favor.

---

¹ [The *Unguentum Veratriæ* of the U. S. Ph. contains twenty grains to the ounce.]
² Bull. Gén. de Thérap., 1860.
³ Ibid., xiv., p. 385.
⁴ Quoted by Niemeyer.

The American authors, in particular, are not always free from suspicion of exaggeration in this matter, although, as I have shown, some of the most trustworthy information which we possess concerning the veratrums was first published in an American journal by Dr. Peugnet.

As an Antipyretic in general, V. viride certainly possesses considerable power. I may say of it (as I might have said of veratrine) that, however doubtful may be its efficacy to lower temperature in the healthy body, there can be no doubt at all of its potency in reducing febrile heat, together with abnormal rapidity of pulse and breathing. Nor are these results to be despised, even though the course of the disease were not shortened, if we can but satisfy ourselves that the remedy is otherwise safe; since the mere reduction of fever is in itself a source of great comfort to the patient. There is much evidence to show that veratrum viride is more manageable than veratrine.

In acute Rheumatism it was strongly recommended by Dr. Osgood, and his commendations of it for this purpose are probably better justified than some of his statements as to its effects in some other acute diseases. It should be given in small doses, and may often be advantageously combined with opium.

In Pneumonia a strong case seems to have been made out for the use of veratrum viride. The experience of Dr. Kiemann[1] forms perhaps one of the most favorable testimonies; he gave V. viride in forty cases, and, although five of these proved fatal (making a mortality of 12.5 per cent.), he says that they were in a desperate condition from the first, and that the results were otherwise decidedly good. The reduction of fever was marked and constant, and although Kiemann acknowledges that V. viride is a powerful remedy, never to be intrusted to any one but a skilled physician, he states that he has not observed it produce *collapse*, as some authors have reported.

By Drasche,[2] also, the antifebrile power of V. viride in pneumonia is fully acknowledged. He observes, however, that if the use of the medicine be ceased at all before the moment of defervescence, the pulse at once rises again. And he states that, so far from forwarding the process of subsequent resolution, V. viride rather delays it.

Perhaps the testimony of Oulmont is the most important that we possess to the fact that in pneumonia, and indeed in pyrexial diseases generally, V. viride is more effective than veratrine.

In chronic Rheumatic and Neuralgic affections modern experience does not appear to encourage the expectations that were formerly entertained of the usefulness of V. viride, any more than of the internal employment of veratrine.

PREPARATIONS AND DOSE.—Extr. Veratri. Viridis Fluid., ♏ i.—v. (.06—.30); Tinct. Veratri. Viridis, ♏ ij.—x. (.12—.60); Ungt. Veratriæ.

---

[1] Prag. Vierteljahrssch., Bd. iii., 1868.
[2] Quoted in Ringer, "Handbook," p. 349.

SALICACEÆ.

THE CRACK WILLOW (Salix Fragilis).

Active Ingredients.—Willow bark is bitter and astringent. These qualities are referable partly to a certain proportion of tannin, and more particularly to the presence of a crystalline neutral principle called salicine, the composition of which is $C^{13}H^{18}O^7$. Besides these there are gum, extractive, and other constituents ordinarily occurring in the bark of trees. Salicine, when separated, presents itself in the form of white and silky acicular crystals and laminæ; it is bitter and inodorous, and soluble in water and in alcohol.

Physiological Action.—The action of salicine is very slight, if measured by any sensible disturbance which it can cause in the human organism. Various experimenters have taken enormous doses for days and weeks together, without suffering any disturbance of digestion, and with no other noticeable effect than slight disturbance of vision and persistent noises in the ears. One or two cases are related in which the visual disturbances (clouds and sparks of fire) were troublesome even for some days after the medicine was left off. But the most important fact is one that has been fully established by recent experiments, namely, that salicine does not possess the power of arresting putrefaction, and consequently lacks that peculiar property which belongs in pre-eminence to the cinchona alkaloids. Salicine further differs from the cinchona alkaloids in being rapidly decomposed within the organism.

Therapeutic Action.—At the present day salicine is much discredited as a remedy, especially in England. As an antiperiodic, Garrod denies its power, or rates it very low; and with this agrees the general feeling. It is nevertheless somewhat puzzling that salicine, if it be really inert, should have acquired so extensive a reputation for curing intermittents; and it is well not to forget the fact that so good an observer as Küchenmeister declared that salicine somewhat reduces the size of the spleen in the healthy subject. Some of the favorable American reports are especially hard to be explained away, unless we suppose gross carelessness of observation. Yet it cannot be denied that the best authorities have everywhere given up the hopes which were at first excited, of salicine taking the place of quinine. At most it can only be allowed that in very mild cases of intermittents salicine may apparently be used with success. If employed for such a purpose, doses of from ten to sixty grains must be administered between the fits of fever.

Salicine also possesses a doubtful reputation as a repressor of **passive hæmorrhages and chronic mucous discharges.**

Probably it is only as a mild **stomachic tonic,** in cases of atonic dyspepsia, that salicine will continue to be used. It may be specially convenient when we need a change from other remedies which have begun to lose their power.

Preparations and Dose.—None officinal. Salicine, gr. v.—ᴣj. (.30—4.)

ULMACEÆ.

SLIPPERY-ELM (Ulmus Campestris).

Active Ingredients.—Elm bark contains about three per cent. of tannin, and a peculiar mucilaginous and gummy principle, not met with elsewhere, called ulmine. Ulmine is dark brown, insoluble in cold water, slightly soluble in boiling water, very soluble in alkaline solutions. Richter classes elm bark with the mucilaginous astringents.

Physiological Action.— Elm bark is demulcent and diuretic. Being slightly astringent, it is also in some degree tonic.

Therapeutic Action.—Decoction of elm bark has been recommended in cases of cutaneous eruption, such as psoriasis, perhaps as a cheap substitute for sarsaparilla.

Preparations and Dose.—Mucilago Ulmi, *ad libitum;* Pulvis Ulmi, in poultice.

FUNGI.

ERGOT (Secale Cornutum; Cordyceps Purpurea).

Active Ingredients.—Ergot contains about thirty-five per cent. of a fixed oil, and about fifteen per cent. of a peculiar reddish-brown substance called ergotine, $C^{50}H^{59}N^2O^3$. The fixed oil is lighter than water, and soluble in alcohol, and (at all events when obtained by expression) is inactive as a medicine. The ergotine, popularly understood to be the active ingredient, has an acrid, bitter, and disagreeable taste; it is soluble in spirit, supposed to be insoluble in ether, and is certainly insoluble in water, which it simply colors red. The coloring matter is supposed to resemble the hæmatine of the blood. Being insoluble in water, it is scarcely likely, as remarked by Ingenhohl, to be the principle which renders ergot capable of contracting the uterus, since aqueous solutions are prescribed with the greatest advantage when uterine contraction is required.

From Winckler's analysis, it appears that having separated the fixed oil by means of ether, he found in the residue not merely ergotine, but another principle—a volatile basic substance analogous to propylamine, and to which he gave the name of Secalia. Ergotine, which appears to possess the properties of an acid, combining with the last named, gives Ergotate of Secalia, a product insoluble in ether, but soluble in alcohol and in water, and which Winckler believed to be the true representative of the active properties of the drug.

Ergot, from which the inactive fixed oil had been removed by means of ether, was employed by Kilian, of Bonn, with the best results, for the purpose of causing contraction of the uterus.[1]

In addition to the other principles, trimethylamine, C^3H^9N, is believed to have been obtained from ergot by distillation with potash.[2] This substance, as stated under Arnica, may be recognized by its offensive and peculiar odor.

PHYSIOLOGICAL ACTION.—The action of ergot is remarkable, not only in the human subject, but upon animals. Dogs, pigs, and other creatures are affected by it in a very obvious and decided manner, and results of the same character are produced in all. Tessier says that dogs, rather than voluntarily swallow ergot, will die of hunger. The principal symptoms produced in them by the administration are dilatation of the pupils, with subsequently injected conjunctivæ; rapid respiration, quickened pulse, tremors, paraplegia, profuse flow of saliva, thirst, vomiting, and frequently diarrhœa, attended by prostration and convulsions, and often followed by death. Tessier's experiments on pigs showed that the first effects of ergot in the animals were redness of the eyes and ears, with coldness of the last, and also of the limbs, accompanied by swelling of the joints. Later on, the ears, limbs, and tail became gangrenous, and the animals expired in convulsions. Post-mortem examination of the intestines showed these parts likewise to be gangrenous. Bonjean administered three drachms of the oil of ergot to a rabbit; after a period of suffering it became convulsed and died of opisthotonos. Tried upon the gravid uterus of animals the operation of ergot is by no means uniform. In some cases the result is abortion; sometimes the contents of the uterus are simply destroyed, and, after the lapse of a time, expelled; in other cases, again, there are no obvious effects.

In the healthy human subject the symptoms most commonly induced by ergot are colic, pains and spasms in the stomach, salivation, nausea, vomiting, a sensation of fulness in the head, painful diarrhœa, occasionally with great prostration, and attended by mucous discharge; a tendency to suppression of the urine or to difficulty of micturition, and dilatation of the pupils, though the eyesight is not materially affected. In addition to these effects there are headache, which often lasts for several days, noise in the ears, flushed face, vertigo, stupefaction, sense of weight and uneasiness in the limbs, often palpitation of the heart, with dyspnœa, oppression of the chest, and unsteady gait, resembling that which is associated with drunkenness. The pulse at the same time becomes depressed, and frequently intermittent, and the beats are often diminished, ten, twenty, or more per minute.

Administered to pregnant women, ergot excites uterine action in a manner truly remarkable, facilitating parturition, or bringing on abortion. There is this great difference, however, in regard to the labor pains: natural pains are intermittent, the expectant mother experiencing intervals of rest ; whereas the pains induced by ergot are continuous, and attended by a persistent sense of bearing-down. If there be any mechanical obstruction to delivery, the continuous character of the pains is apt to prove highly dangerous, through causing rupture of the uterus or of other delicate structures. Ergot, in a word, is eminently parturifacient, not only inducing uterine pains to liberate the fœtus, but subsequently promoting

---

[1] Boyle and Headlam, "Mat. Med.," p. 707.
[2] Garrod, p. 351.

the expulsion of the placenta, and putting an end to or checking uterine hæmorrhage. In certain cases of the administration of ergot to parturient women, the death of the child is said to have been caused by it. Experience proves that the fault has not lain with the medicine, but with the unwise administration; the usual causes of failure and injurious results being either malformation of the pelvic structures, or premature resort to ergot, which is hurtful if given too soon, or while the uterus is not sufficiently dilated.

Among the effects produced by ergot, it is important to mention those which constitute the frightful condition called "ergotism," and which are induced either by the long-continued administration of the drug in medicinal doses, or by eating bread made from corn which has been infected by it while growing. This dreadful disorder is manifested under two distinct forms: the convulsive and the gangrenous—the former not terminating in gangrene, whereas the latter is always distinguished by gangrene. The convulsive form commences with *malaise*, irritation of the whole surface of the body, with formication, numbness and coldness of the extremities, often accompanied by cramps and by pains in the head and loins. Some time later the digestive organs become affected. There is a sense of tightness and oppression about the epigastrium, heart-burn, a feeling of lightness in the head, difficulty in hearing, with faintness, abnormal twitchings of the muscles of the face, often attended by strabismus and irregular contractions of the joints. These symptoms are generally accompanied by delirium bordering upon mania, with cold sweats; and the whole body is pervaded by a sense of great heat.

Convulsive ergotism, as above described, often carries with it a ravenous desire for food. A vesicular eruption on the skin, often with petechiæ, is also a very frequent addition. In continental Europe it has often come on in the shape of an epidemic, the period during which the disorder has prevailed extending to three or four weeks, and life being extensively destroyed.

The gangrenous form of ergotism commences with a sense of pain and weariness in the limbs; the countenance becomes heavy-looking and stupid, and the skin acquires a leaden or cadaverous hue. Afterwards there is *malaise*, with exhaustion, coldness of the whole body, and numbness of the hands and arms, while formication is perceived in every part of the surface. The lower extremities become similarly affected with the upper ones; later on, the abdominal muscles are spasmodically contracted; and, about the sixth day from the commencement, nausea, vomiting, and diarrhœa make their appearance, with severe colicky pains in the bowels and bladder.

In the case upon which these details are founded, upon the fourteenth day two of the children of the family lay as if stupefied. If they were disturbed they raved wildly, and complained of pains in the head and limbs. At the same time a pruriginous eruption appeared upon the skin, the unfortunate sufferers expiring on the twenty-first day in violent convulsions.

Bonjean relates a fatal case in which there was pain in the left groin, coldness and pain in the legs, a dark spot upon each calf, formed by an eruption of vesicles, the itching of which was violent, and gangrene of the lower third of each leg. The feet were black and dry; the upper portions of the legs became affected with humid gangrene; in three weeks the sphacelated parts began to separate, and soon afterwards it became necessary to amputate each limb below the knee. Very little blood flowed while the operation was in course of being performed.

The epidemics of which ergotism has constituted so frightful and by no means rare a feature have always been of one or other of these two descriptions. In some of these epidemics the convulsive form has prevailed; in others it has assumed the gangrenous form. It is upon the distinction in the character of the epidemics that the disorder itself has been resolved into the dual form described. In almost every instance the individuals who suffered and survived became for a time either imbecile or idiotic. A favorable crisis in the progress of the disorder was invariably supposed to be indicated by swelling of the feet, bleeding ulcers, diarrhœa, and eruptions upon the skin.[1]

Therapeutic Action.—In **Parturition** the therapeutic value of ergot is unquestionably its most important feature as a medicinal agent. The nature of this action has been spoken of in the previous section, and to repeat the facts is unnecessary. There are other uses of ergot of very considerable importance. Uterine hæmorrhage, for instance, accompanied by spasmodic contraction of that organ, and by distressing pains of a bearing-down description, is speedily checked by ergot. So, too, is hæmorrhage accompanied by atony of the uterus. In uterine hæmorrhage of any kind, whether or not connected with pregnancy, ergot, in a word, is generally of great service. It is recommended also for the expulsion from the cavity of the uterus of clots of blood, polypi, and hydatids. In cases of threatened miscarriage, when slight hæmorrhage has commenced, accompanied by bearing-down pains, I have often employed this medicine in small doses, or such as are sufficient to partially brace the uterus without inducing the physiological result of uterine pains, and this with excellent results.

In **Amenorrhœa.**—Some practitioners frequently prescribe ergot with a view to the restoration of the monthly flow after a suppression of considerable duration, and where preparations of iron have failed of success. Neligan states that he has employed it in chlorotic amenorrhœa, also with the most beneficial results. Leucorrhœa, occurring in the intervals of menorrhagia, is frequently arrested by a decoction of ergot, given by vaginal injection. This also conduces to the arrest of the tendency to menorrhagia at the following monthly period.

As a general Styptic, ergot is believed to possess the property of causing contraction of the capillaries in general. At any rate, it is most useful in removing purpura and other capillary hæmorrhages. In dysentery and hæmorrhages arising from enteric fever, and in hæmorrhages arising from inflammation of other mucous surfaces, it is unquestionably of great service.

Epistaxis, Hæmoptysis, and **Hæmatemesis**, I have often seen stopped by resort to ergot, and this when most other means of cure have appeared to fail.

Paralysis, whether accompanied or not by contractions of the limbs, is often treated beneficially with ergot; paralytic dysuria also, particularly when the patient has a very distinct and continuous sensation of the bladder being only very partially relieved of its contents.

When this medicine proves inefficacious, the failure is often to be

---

[1] I am indebted to Stillé for many valuable particulars in regard to the physiological action of ergot; but to many of the less urgent symptoms it is capable of producing I can speak from personal knowledge, having frequently administered it as a medicine, and taken it experimentally myself.

ascribed either to peculiarity of constitution in the individual, or to inferiority in the quality of the ergot, a condition of very frequent occurrence.

In **Neuralgias** ergot has been much employed of late years. It is chiefly in visceral neuralgias (gastralgia especially) that this remedy seems likely to prove useful.

Ergotine.—A special section must be devoted to the therapeutic use of this substance, which has rapidly assumed great importance of late years. It must, however, be understood that the ergotine which has been employed in actual medicine is not the pure alkaloid, but an impure preparation—either that of Wiggers or that of Bonjean. The latter is decidedly the most convenient from its complete solubility in water and in alcohol. Bonjean's was called *extrait hæmostatique*, and especially recommended for uterine and pulmonary hæmorrhages. It is a reddish-brown substance, with a roast-meat-like smell, and a pungent bitter taste.

Aneurism, however, was the disease, the treatment of which with ergotine fully aroused attention to the importance of this remedy. In 1869 Langenbeck published a remarkable paper, in which he gave an account of two cases of aneurism (one probably aortic) successfully treated by the subcutaneous injection of ergotine under the skin covering the tumor, repeated for many days together. Similar results were afterwards obtained by Dutoit, of Berne, and others. The treatment has since been repeated in various countries, and, on the whole, with an encouraging amount of success.

In **Hæmoptysis**, since the renewed attention to it which was caused by Langenbeck's observations, ergotine has proved a very valuable remedy. A very interesting paper by Dr. Currie Ritchie[1] relates a number of cases in which the results were striking. Ergotine, from T. and H. Smith, of Edinburgh (five grains in ten minims of distilled water), was subcutaneously injected in seven cases; in an eighth this was found too irritant, and Langenbeck's formula was successfully followed, viz., three grains of ergotine in equal parts of glycerine and rectified spirit. Either one or two injections of five grains of ergotine seemed always sufficient to completely arrest the bleeding.

In **Menorrhagia**, even when connected with fibroid tumors, and other formidable organic affections of the uterus, there can be no doubt of the efficacy of ergotine. In these affections, as also indeed in hæmoptysis, I have had large personal experience of its efficacy, particularly when subcutaneously injected. Dr. Meadows deserves the credit of having pointed out, as long ago as 1868,[2] that sub-involution, chronic subacute metritis, and hypertrophy of the uterus "are alike in this respect, that they are invariably associated with increased vascularity of the organ, though this is mostly of the passive or congestive kind; they are consequently liable to excessive discharges either of mucus or of blood; and they are further characterized by increased bulk of tissue." Dr. M. brings forward much valuable evidence to show that these maladies may be successfully dealt with by means of ergot of rye.

I have personally used both the ethereal extract (which Meadows employed), and, with preference, the subcutaneous injection of ergotine. I wish to give a caution against the employment of the large doses of the latter that have been recommended by several authors. Instead of four or

---

[1] Practitioner, vol. vi., 1871.
[2] Ibid., vol. i., Sept., 1868.

five grains, which seems to be the quantity ordinarily used, I inject either one or two grains in glycerine and water, and never inject more than one grain in the same place, much preferring to make a second puncture if necessary. I find the smaller doses sufficient for the production of the characteristic action of the drug upon blood-vessels, and upon unstriped muscular fibre generally, and the local effects of the injection are much less severe than with larger doses. It does not seem to be generally known that ergotine is very irritant to the subcutaneous tissue, and that quite a common result of a five grain injection is the formation of a large, livid, tumor-like swelling, with the appearance of a black cicatrix in the centre. Fortunately this kind of swelling seldom or never suppurates, but it is a very awkward occurrence, and is apt to annoy and alarm the patient, besides giving a great deal of pain in some cases.

I have on three different occasions injected one to two grains of ergotine into patients suffering from hæmorrhage of the bowels as a consequence of enteric fever. I have also used it in a similar manner with several patients having severe epistaxis; and in both sets of cases with the best result.

Hildebrandt treated nine cases of fibroid tumors of the uterus by the subcutaneous injection of ergotine. "In four," he says, "the diminution of the tumor was free from doubt; in the others troublesome symptoms subsided."

PREPARATIONS AND DOSE.—Ergota, ʒ ss.—j. (2.—4.); Extr. Ergotæ Fluid., ʒ ss.—j. (2.—4.); Vinum Ergotæ, ʒ ij.—iv. (8.—15.). "Ergotine" and Extr. Ergotæ (Squibb) are not officinal: dose of either, gr. v.—xv. (.30—1.).

[CORN SMUT (Ustilago Maidis).

So far as studied, its effects are similar to those of ergot. It may be used in the same preparations and doses. It is readily obtained at the proper season, and good ustilago is better than bad ergot].

FILICES.

MALE FERN (Lastræa Filix-mas).

ACTIVE INGREDIENTS.—The particular ingredients of the rhizome of the Filix-mas are a volatile oil, a fatty fixed oil, and a resin, besides which there are tannin, filicic acid, coloring matter, some starch, gum, uncrystallizable sugar, and other subordinate substances. Bock represents the proportion of the fixed oil to be that of 60.0 in 1000 of the dried rhizome, while the volatile oil exists to the extent only of 0.4. The active principles of the plant, which are probably identified with the oils, or with one of them, are separable by means of ether, in which they are soluble. The ethereal extract, termed "oil of male-fern," contains also the resin;

it is blackened by the presence of coloring matter, and retains the peculiar odor of the rhizome.

PHYSIOLOGICAL ACTION.—Filix-mas is of very ancient repute as a vermicide. It appears to act as a direct poison to the worms, though the expulsion of the dead creatures has generally to be accomplished through some other agency, the natural action of the intestines fulfilling its own part. Besides the essentially vermicidal action, there is a slightly purgative one.

THERAPEUTIC ACTION.—The chief success of the oil of male-fern appears to have been found in connection with tape-worm, and most especially in cases of the presence of *Bothriocephalus latus*. Its value in such cases can hardly be over-estimated, and an additional good feature is the smallness of the dose required. The patient should fast before receiving the medicine, or, at all events, take it upon an empty stomach; and after an interval, say from five to six hours, it should be followed by a gentle laxative. If the alimentary canal be occupied by food, the medicine is too much diluted to be operative, concentration being always very important in regard to vermifuges. A dose of castor oil the night before taking the medicine is often preferable to a laxative subsequently.

PREPARATIONS AND DOSE.—Oleoresina Filicis, gr. x.—xxx. (.60—2.); Extract. Filicis Liquidum (B. Ph.), xv.—xxx. (1.—2.).

LICHENES.

ICELAND MOSS (Cetraria Islandica).

ACTIVE INGREDIENTS.—Iceland moss is devoid of odor, but possessed of a bitter taste, and in substance is mucilaginous. It contains about eighty per cent. of farinaceous matter, which is resolvable into lichenine, or lichen-starch, and the starch-like body called inuline. The former is insoluble in alcohol, ether, and cold water, in which last it swells up, but is soluble in boiling water. Iodine turns it blue, whereas inuline remains unaffected.

The bitter principle resides in the cortical portion, and is called cetrario acid. When pure it possesses the form of acicular white crystals, which are intensely bitter, and almost insoluble in water, though when boiled in water the crystals communicate to it a bitter taste. It is sparingly soluble in alcohol and ether, and readily soluble in alkaline solutions, forming soluble compounds. The proportion of acid present is about three per cent.; there are found also a little gum, extractive, uncrystallizable sugar, and other ingredients of minor importance.

PHYSIOLOGICAL ACTION.—Iceland moss, when deprived of its bitterness by prolonged maceration in water, becomes a nutritious food, and is employed as such by the Icelanders.

Therapeutic Action.—Iceland moss, having no properties that can be strictly called medicinal, takes its place simply with the light and farinaceous foods that often become useful for invalids, the bitterness being first removed by steeping the plant in some alkaline solution.

It has been recommended for disorders of the stomach, such as dyspepsia and chronic dysentery; also in affections of the respiratory organs, as phthisis and chronic catarrh.

Preparations and Dose.—Decoctum Cetrariæ, ℥ ss.—ij. (15.—60.)

INDEX.

INDEX.

MEDICINES.

Abies Balsamea, 109
  excelsa, 108
Absinthe, 164
Absinthic acid, 164
Absinthine, 164
Aconite, 1
Aconitia, 1
Ægle Marmelos, 97
Akazga, 135
Akazgia, 135
Alder Buckthorn, 105
Almond, 205
Aloe, Socotrina, 202
  vulgaris, 202
Aloes, 202
Aloetic acid, 203
Aloin, 203
Althæa officinalis, 93
American Hellebore, 208
Amygdaline, 206
Amygdalus communis, 205
Anacardiaceæ, 158
Anacardium occidentale, 163
Anacardium orientale, 163
Anacyclus Pyrethrum, 164
Anamirta Cocculus, 59
Anemone pratensis, 10
Anemonine, 10
Anemonic acid, 10
Anethol, 224
Anethum graveolens, 224
Angustura, 138
Angusturine, 138
Anise, 224
Anisi Camphor, 55
Anithol, 55
Anthemis nobilis, 168
Apiaceæ, 387
Apocynum Cannabinum, 291
Apomorphia, 78
Aporetine, 99
Arbutine, 228
Arbor Vitæ, 111
Arctostaphylos Uva-ursi, 228

Aricine, 239
Aristolochia Serpentaria, 286
Aristolocheaceæ, 286
Arnica montana, 170
Arnicine, 170
Artanthe elongata, 280
Artemisia Absinthium, 164
  Santonica, 166
Asafœtida, 220
Asclepiadaceæ, 233
Asparagine, 93
Atropia, 25
Atropa Belladonna, 25
Aurantiaceæ, 95

Baël, 97
Balsam of Peru, 194
  of Tolu, 194
Balsamodendron Myrrha, 162
Barosma crenulata, 138
Barosmine, 139
Bay, Sweet, 286
Bebeerine, 285
Bebeeru, 285
Beberia, 285
Belladonna, 25
Belladonnine, 25
Benzoic acid, 231
Benzoin, 230
Berberia, 17, 56
Bibiru bark, 285
Bittersweet, 48
Blue Flag, 273
Brassic acid, 84
Brayera anthelmintica, 208
Broom, 192
Brucia, 133
Bryonine, 121
Bryonia, 121
  alba, 122
Bryonia dioica, 121
Bryoretine, 121
Buckthorn, Purging, 104
Buchu, 138
Burgundy Pitch, 108

Buxine, 58
Byttneriaceæ, 103

Caffeine, 264
Cajeput-oil, 182
Cajeputine, 182
Calabar Bean, 187
Calumba, 56
Calumbic acid, 56
Calumbine, 56
Cambogia, 112
Camphor, 282
Camphor officinarum, 282
Canada Balsam, 100
Canellaceæ, 108
Canella alba, 108
Cannabine, 280
Cannabis Americana, 200
Cannabis Indica, 280
  sativa, 280
Capsicine, 48
Capsicum annuum, 48
Caraway, 223
Cardamoms, 273
Cardol pruriens, 163
Cardol vesicans, 163
Carolina Pink-root, 130
Carum Carui, 223
Carvol, 223
Caryophylline, 180
Caryophyllus aromaticus, 180
Cascarilla bark, 115
Cascarilline, 115
Cashew-nut, 163
Cassia obovata, 191
Castor-oil, 113
Catechine, 103
Catechu-pallidum, 258
Cathartic acid, 191
Cathartiginic acid, 191
Celandine, 83
Cephælis Ipecacuanha, 250
Cerasus Lauro-cerasus, 200
Cetraria Islandica, 300
Chamomile, 168
Chamomilline, 168

INDEX.

Chelerythrine, 79, 83
Chelidonine, 79
Chelidonium majus, 83
Cherry-Laurel, 206
Chinaphiline, 229
Chimaphila umbellata, 229
Chirata, chiretta, 175
Chirettine, 175
Chrysophane, 98
Chrysophanic acid, 98
Cicutine, 214
Cimicifuga Racemosa, 21
Cinchonaceæ, 238
Cinchona, 238
Cinchonidine, 239, 245
Cinchonine 239, 244
Cinnamic acid, 281
Cinnamon, 281
Cinnamomum Zeylanicum, 281
Cissampelos Pareira, 58
Citric acid, 95, 96
Citrullus Colocynthis, 119
Citrus Aurantium, 95
   Limonum, 96
   vulgaris, 96
Clove, 180
Cocculine, 59
Cocculus Indicus, 59
   palmatus, 56
Cochlearia Armoracia, 86
Cocoa, 103
Codeia, 63, 66, 75
Coffee, 264
Cohosh, 21
Colchicine, 294
Colchicum autumnale, 294
Collodium, 94
Colocynth, 119
Colocyntheine, 119
Colocynthine, 119
Colocynthitine, 119
Coltsfoot, 173
Columbo, 56
Compositæ, 163
Conia, 214
Coniic acid, 214
Coniferæ, 105
Conin, coniine, 214
Conium maculatum, 214
Convolvulaceæ, 175
Convolvuline, 177
Convolvulin oil, 178
Convolvulinic acid, 178
Convolvulus scammonia, 175
Copaifera multijuga, 195
Copaiva, 195
Copaivic acid, 195
Coriander, 223
Coriandrum sativum, 223
Cordyceps purpurea, 303
Corn-smut, 303
Corylaceæ, 291
Cotton, 93

Cotton-wool, 94
Crack Willow, 302
Crotonic acid, 114
Croton Eleuteria, 115
Croton-oil, 114
Crotonol, 114
Croton Tiglium, 114
Cruciferæ, 84
Cubeba, officinalis, 279
Cubebine, 279
Cubebs, 279
Cucurbitaceæ, 116
Cumin, 222
Cuminol, 222
Cuminum Cyminum, 222
Curara, 131
Curarine, 131
Cusparia bark, 138
Cusparine, 138
Cymol, 222

Dandelion, 160
Daphne Mezereum, 271
Daphnine, 271
Datura Stramonium, 43
Daturia, 43
Delphinine, 20
Delphinium Staphisagria, 20
Demerara Pink-root, 130
Digitalis purpurea, 142
Digitaline, 142
Digitaleine, 142
Digitalose, 143
Digitoxin, 143
Dill, 224
Diosmine, 139
Drymis Winteri, 55
Dulcamara, 45
Dulcamarine, 45

Ecbalium Agreste, 116
Elaterine, 117
Elaterium, 116
Elecampane, 163
Elettaria cardamomum, 273
Ellagic acid, 291
Elm-bark, 540
Emetine, 87, 259
Emulsine, 206
Ergot, 303
Ergotine, 303
Ericaceæ, 228
Erucic acid, 84
Eserine, 187
Eugenic acid, 180, 198
Eugenine, 180
Euphorbiaceæ, 113
Euphrasia officinalis, 156
Euphrastic acid, 156
Euryangium Sumbul, 224
Eye-bright, 156

Fabaceæ, 187
Fennel, 222

Filices, 308
Filix-mas, 308
Flax, Common, 91
   Purging, 92
Fœniculum dulce, 222
Flexible collodion, 174
Foxglove, 142
Frangulin, 105
Fraxinus Ornus, 270
Fungi, 303

Galipea Cusparia, 138
Galls, 291
Gallic acid, 291
Gamboge, 112
Gambogic acid, 112
Gelsemin, 157
Gelsemium sempervirens, 157
Gentianaceæ, 174
Gentian, 174
Gentiana lutea, 174
Gentianic acid, 174
Gentianine, 174
Gentio-picrine, 174
Ginger, 272
Githagine, 163
Glonoinum, 270
Glycerine, 269
Gossypium herbaceum, 93
Granataceæ, 141
Granatine, 141
Grape, 198
Guaiacum officinale, 179
Guaiaconic acid, 179
Guaiaretic acid, 179
Guarana, 267
Gun-cotton, 94
Guttiferæ, 112

Hæmatin, 195
Hæmatoxylin, 195
Hæmatoxylon Campechianum, 195
Hamamelaceæ, 211
Hamamelis Virginica, 211
Helenine, 163
Hellebore, American, 298
Helleboretin, 14
Helleborein, 12
Helleborin, 12
Helleborus niger, 12
   viridis, 14
Hemidesmic acid, 233
Hemidesmine, 233
Hemidesmus Indicus, 233
Hemlock, 214
Hemp, 289
Henbane, 39
Hop, 288
Horse-radish, 86
Humuline, 288
Humulus Lupulus, 288
Hydrastia, 17
Hydrastin, 17

INDEX.

Hydrastis Canadensis, 17
Hydrocyanic acid, 206
Hyoscyamia, 39
Hyoscyamus, 30

ICELAND Moss, 309
Igasuria, 123
Ignatia, 123
Ilex Paraguayensis, 267
Illicium anisatum, 55
Indian Baël, 97
   Hemp, 289
Inula Helenium, 163
Inuline, 163, 309
Ipecacuanha, 259
Iridaceæ, 273
Irisin, 273
Iris versicolor, 273
Ipomœa Jalapa, 177

JABORANDI, 139
Jalap, 177
Jalapine, 175
Juniper, 108
Juniperus communis, 108
   Sabina, 108

KAMALA, 116
Kinic acid, 239
Kino, Indian, 193
Kino-tannic acid, 239
Kinovic acid, 239
Kinovine, 239
Kinovi-tannic acid, 239
Krameria triandra, 90
Krameric acid, 90
Kousso, 208
Koussine, 208

LABIATÆ, 275
Lastræa Filix-mas, 308
Lauraceæ, 281
Laurus nobilis, 286
Lavandula vera, 275
Lavender, 275
Lemon, 96
Lichenes, 309
Lichenine, 309
Liliaceæ, 201
Linaceæ, 91
Linine, 92
Linseed, 91
Linum catharticum, 92
   usitatissimum, 91
Lobeliaceæ, 225
Lobelia inflata, 225
Lobelic acid, 225
Lobeline, 225
Logwood, 195
Lupuline, 288
Lupulite, 288

MACE, 232
Macrotin, 21
Magnoliaceæ, 55

Magnolia acuminata, 56
   glauca, 55
   tripetala, 55
Male Fern, 308
Malvaceæ, 93
Manna, 270
Mannite, 270
Marsh-mallow, 93
Maté, 267
Maticine, 280
Matico, 280
May Apple, 15
Meconic acid, 62
Meconine, 64, 76
Malaleuca Cajeputi, 182
Melanthaceæ, 294
Menispermaceæ, 56
Menispermum Cocculus, 108
Mentha piperita, 276
   pulegium, 277
   viridis, 276
Methylcaprinol, 251
Methylpelargonylketon, 251
Mezereon, 271
Mimotannic acid, 193
Monesine, 88
Monninine, 88
Morphia, 62, 65, 73
Mustard, 84
Myristicaceæ, 232
Myristic acid, 232
Myristica fragrans, 232
Myronic acid, 84, 86
Myrosin, 84, 86
Myrospermum Peruvianum, 194
   Toluiferum, 194
Myroxylon Toluiferum, 194
Myrrh, 162
Myrrhic acid, 162
Myrrhus, 162
Myrtaceæ, 180

NAPELLINE, 1
Narceia, 63, 66, 75
Narcotine, 63, 67
Narthex Asafœtida, 220
Nauclea Gambir, 258
Nectrandra Rodiœi, 281
Nicotiana Tabacum, 50
Nicotine, 50
Nitro-glycerine, 270
Nutmeg, 232
Nux-vomica, 122

OAK, 291
Olea Europœa, 268
Oleaceæ, 268
Olive, 268
Ophelia Chirata, 175
Ophelic acid, 175
Opianine, 64
Opianyl, 64, 76
Opine, 64
Opium, 62, 68

Papaveraceæ, 62
Papaver somniferum, 62
Papaverine, 63, 67, 76
Paraguay Tea, 267
Pareira Brava, 58
Paullinia sorbilis, 267
Pennyroyal, 277
Pepper, 278
Peppermint, 276
Peru, Balsam of, 194
Physostigma venenosum, 187
Physostigmine, 187
Picræna excelsa, 186
Picrotoxine, 59
Pilocarpus pinnatifolius, 139
Pilocarpia, 139
Pimenta vulgaris, 181
Pimento, 181
Pimpinella Anisum, 221
Pinus sylvestris, 108
Piperaceæ, 278
Piperine, 278
Piper nigrum, 278
Pipsissewa, 229
Pitch, Burgundy, 108
Pix liquida, 108
Plum, 208
Podophyllin, 15
Podophyllum peltatum, 15
Polygalaceæ, 88
Polygaline, 88
Polygala Senega, 88
Polygonaceæ, 98
Pomegranate, 141
Prunus domestica, 208
   Virginiana, 208
Prussic acid, 206
Pseudo-aconitia, 1
Pterocarpus Marsupium, 193
   santalinus, 211
Puccine, 79
Pulsatilla nigricans, 10
   Nuttalliana, 12
Punica Granatum, 141
Punicine, 141
Purging Buckthorn, 104
   Flax, 92
Pyridine, 50
Pyrethrum, 164
Pyrethrine, 164
Pyro'aceæ, 229
Pyroxylon, 94

QUASSIA, 186
Quassine, 186
Quassite, 186
Quercus alba, 291
   infectoria, 291
Quillagine, 88
Quinic acid, 239
Quinidine, 239, 245
Quinine, 238, 242

RANUNCULACEÆ, 1
Rhamnaceæ, 104
Rhamnine, 104
Rhamnocathartine, 104
Rhamnus catharticus, 104
  Frangula, 105
Rhatanine, 90
Rhatany, 90
Rheum, 98
Rhubarb, 98
  Tannin, 98
Rhus diversiloba, 161
  toxicodendron, 158
  venenata, 161
Ricinic acid, 113
Ricinate of soda, 113
Ricinus communis, 113
Rosaceæ, 205
Rosemary, 277
Rosmarinus officinalis, 277
Rottlera tinctoria, 116
Rottlerine, 116
Rutaceæ, 136
Rue, 136
Ruta graveolens, 136
Rutine, 136
Rye, Ergot of, 303

ST. IGNATIUS' BEAN, 128
Salicaceæ, 302
Salicine, 302
Salix fragilis, 302
Sandal-wood, 211
Sanguinaria Canadensis, 79
Sanguinarine, 79
Santalaceæ, 211
Santalum album, 211
Santonica, 166
Santonine, 166
Saponine, 88
Sarothamnus Scoparius, 192
Sarsaparilla, 182
Sassafras officinale, 284
Sassafrine, 284
Savin, 108
Scammonine, 175
Scammony, 175
Scillitine, 201
Scoparine, 192
Scrophulariaceæ, 142
Sculline, 201
Secalia, 303
Secale Cornutum, 303
Senega, 88

Senegine, 88
Senna, 191
Serpentaria, 286
Simarubaceæ, 186
Sinapine, 84
Sinapis, 84
Smilaceæ, 182
Slippery Elm, 303
Smilacine, 182
Smilax officinalis, 182
Soap, 268
Solanaceæ, 25
Solanine, 45
Solanum Dulcamara, 45
South American Arrow-poison, 131
Sparteine, 192
Spearmint, 276
Spigeliaceæ, 122
Spigelia Anthelmia, 130
  Marilandica, 130
Squill, 201
Squirting Cucumber, 116
Staphisagrine, 20
Star-anise, 55
Stavesacre, 20
Stramonium, 43
Struthiine, 88
Strychnia, 122
Strychnos Ignatia, 128
  Nux-vomica, 122
Styracaceæ, 230
Styracine, 230
Styrax Benzoin, 230
  officinale, 230
Styrole, 230
Sumbul, 224
Sweet Bay, 296
  Orange, 95
  Violet, 87

TANNIC ACID, 291
Tar, 108
Taraxacum, 169
Tea, 267
Terebinthina, 105
Thebaine, 64, 68
Theobroma Cacao, 103
Theobromine, 103
Thuja occidentalis, 111
Thujin, 111
Thujigenin, 111
Thymelaceæ, 271
Tobacco, 50

Tolu, Balsam of, 194
Toxicodendric acid, 158
Trimethylamine, 170
Turpeutine, 105
Tussilago Farfara, 173

ULMINE, 303
Ulmaceæ, 303
Ulmus campestris, 303
Umbelliferæ, 214
Umbelliferine, 224
Urginea Scilla, 201
Urticaceæ, 288
Ustilago Maidis, 308
Uva-ursi, 228

VALERIAN, 234
Valerianaceæ, 234
Valeriana officinalis, 234
Valerian-camphor, 234
Valerianate of Ammonia, 238
  of Atropine, 237
  of Iron, 238
  of Quinine, 237
  of Soda, 237
  of Zinc, 237
Valerianic acid, 237
Veratrine, 298
Veratrum viride, 298
Violaceæ, 87
Viola canina, 87
  odorata, 87
  tricolor, 88
Violine, 87
Vitaceæ, 198
Vitis vinifera, 198

WEST AFRICAN ORDEAL POISON, 135
Wild Cherry, 209
Willow, Crack, 302
Winter's Bark, 55
Wild Pansy, 88
Witch-hazel, 201
Wormgrass, 130
Wormwood, 104

XANTHOPUCCINIA, 17

YELLOW JASMINE, 157

ZINGIBERACEÆ, 272
Zingiber officinale, 272
Zygophyllaceæ, 177

INDEX.

DISEASES.

ABDOMINAL CRAMPS (See COLIC).
ABDOMINAL PLETHORA: Grape-cure.
ABSCESSES: Belladonna, Sarsaparilla.
AGUE (See INTERMITTENT FEVER).
ALBUMINURIA: Cannabis Indica, Chimaphila, Gallic Acid, Turpentine.
ALCOHOLISM: Cannabis Indica, Quinine, Strychnia.
AMAUROSIS: Arnica.
AMENORRHŒA: Aloes, Cimicifuga, Colocynth, Ergot, Guaiacum, Myrrh, Pulsatilla, Rue, Sanguinaria, Savin, Senega, Stavesacre, Storax.
AMPUTATIONS: Arnica.
ANAL FISTULA (See FISTULA).
ANASARCA (See DROPSY).
ANEURISM: Ergotine.
ANGINA PECTORIS: Aconite, Chamomile, Cherry-laurel, Glonoinum, Opium, Strychnia.
ANTIMONY POISONING: Tannin.
ANOREXIA (See DYSPEPSIA).
ANUS, FISSURE OF THE (See FISSURE).
ANUS, SPASMS OF THE SPHINCTER OF THE: Belladonna.
APHRODISIAC DISORDERS: Camphor.
APHONIA: Ignatia.
APPETITE, DISEASED: Ignatia.
APOPLEXY: Aconite, Cocculus Indicus, Colocynth, Elaterium.

ARTICULAR AFFECTIONS: Burgundy Pitch.
ASCARIDES (See WORMS).
ASTHMA: Aconite, Asafœtida, Cimicifuga, Balsams of Peru and Tolu, Belladonna, Camphor, Chamomile, Coffee, Dulcamara, Indian Bael, Ipecacuanha, Lobelia, Mace, Morphia, Nutmeg, Nux-vomica, Opium, Pulsatilla, Quinine, Senega, Stramonium, Strychnia, Sumbul, Tobacco.
ATACTIC MOVEMENTS: Nux-vomica, Strychnia.
ATONIC CONDITIONS OF THE BOWELS (See CONSTIPATION).
ATONIC DYSPEPSIA: Calumba, Chamomile, Cubebs, Quinine, Salicine.

BALDNESS: Rosemary.
BELLADONNA-POISONING: Opium.
BILIOUSNESS: Angustura, Bryony, Calumba, Podophyllum, Serpentaria, Taraxacum.
BLADDER, SPASMS OF THE: Belladonna.
BLADDER, CATARRH OF THE: Belladonna, Benzoin, Buchu, Castor-oil, Chimaphila, Cocculus Indicus, Copaiva, Cubebs, Gallic Acid, Grape Cure, Lupuline, Pareira, Tar, Uva-ursi.
BLENORRHŒA: Tannic Acid, Turpentine, Sandal Wood Oil.
BOILS: Collodion, Belladonna.

BONE DISEASE: Sarsaparilla.
BOWELS, ULCERATION OF: Turpentine.
BRAIN, INFLAMMATORY AFFECTIONS OF THE: Aconite, Belladonna, Colocynth, Jalap, Opium.
BREASTS, DISTENDED: Belladonna, Hyoscyamus, Stramonium.
BREASTS, INDURATED AND INFLAMED: Belladonna.
BRIGHT'S DISEASE (See ALBUMINURIA).
BRONCHITIS: Anise oil, Asafœtida, Balsams of Peru and Tolu, Benzoin, Cascarilla Bark, Cherry-laurel, Codeia, Colchicum, Coniine vapor, Cubebs, Ipecacuanha, Mustard, Myrrh, Nux-vomica, Opium, Sanguinaria, Senega, Squill, Storax, Sumbul, Tannic acid, Tar, Turpentine.
BRUISES: Arnica, Oil of Bay.
BURNS AND SCALDS: Collodion, Cotton wool, Linseed oil, Oil of Turpentine.

CALCULUS: Castor-oil.
CANCER: Belladonna, Codeia, Conium, Hydrastis, Hyoscyamus, Opium.
CARBUNCLES: Belladonna, Collodion.
CATAMENIA, ARRESTED (See MENSTRUATION).
CATARRH: Aconite, Asafœtida, Cimicifuga, Canada Balsam, Chamomile, Cherry-laurel, Copaiva, Cubebs, Elecampane,

Euphrasia, Grape-cure, Guaiacum, Iceland moss, Ipecacuanha, Myrrh, Mustard, Nux-vomica, Storax, Tannic acid, Tar, Uva-ursi.

CEREBRAL DISEASES: Colocynth, Conium.

CHAPPED SKIN (See CRACKED SKIN).

CHLOROSIS: Benzoin, Cocculus Indicus, Ergot.

CHOLERA, ASIATIC: Camphor, Opium.

CHOLERA, ENGLISH (See also DIARRHŒA): Calumba, Camphor, Mustard, Opium, Sumbul.

CHOLERAIC AFFECTIONS: Cajeput Oil.

CHORDEE: Aconite, Cannabis Indica, Hyoscyamus, Lupuline, Sandal Wood Oil.

CHOREA: Cimicifuga, Calabar-bean, Camphor, Cocculus Indicus, Colchicum, Conium, Curara, Indian Hemp, Picrotoxine, Quinine, Strychnia, Valerian, Valerianate of Zinc.

CLAVUS HYSTERICUS: St. Ignatius' Bean.

COLIC (See GRIPING PAINS): Anise oil, Cajeput oil, Caraway oil, Cardamoms, Chamomile, Cloves, Cocculus Indicus, 'Dill, Marsh Mallow, Oil of Bay, Opium, Rue, Spearmint, Strychnia.

CONDYLOMATA: Arbor Vitæ (*Thuja occidentalis*).

CONFINED BOWELS (See also CONSTIPATION): Almond, Buckthorn, Linseed Oil, Manna, Olive Oil, Prunes, Purging Flax, Rhubarb, Senna, Turpentine.

CONSTIPATION (See also CONFINED BOWELS): Aloes, Belladonna, Castor-oil, Cocculus Indicus, Colocynth, Croton Oil, Elaterium, Hydrastis, Hyoscyamus, Indian Bael, Jalap, Nux-vomica, Opium, Podophyllum, Scammony, Senna, Strychnia, Tar, Turpentine.

CONSUMPTION (See PHTHISIS, BRONCHITIS, COUGH, &c.): Nutmeg, Wild Cherry.

CONVULSIONS: Aconite, Asafœtida, Belladonna, Hemlock, Hyoscyamus, Opium, Rue, St. Ignatius' Bean, Valerian.

CORYZA: Aconite, Camphor, Cimicifuga, Hydrastis, Nux-vomica, Pulsatilla, Strychnia.

COUGH (See BRONCHITIS, &c.): Aconite, Almond, Asafœtida, Capsicum, Celandine, Chamomile, Codeia, Coltsfoot, Hyoscyamus, Lobelia, Marsh Mallow, Narceia, Pulsatilla, Sanguinaria, Senega, Tar, Valerian, Wild Cherry.

CRACKED NIPPLE OR SKIN: Benzoin, Collodion, Hydrastis.

CROUP: Aconite, Belladonna, Ipecacuanha, Sanguinaria, Senega.

CUTANEOUS DISORDERS: Buchu, Dulcamara, Elm Bark, Hellebore, Iris-Versicolor, Thuja, Laurel-Water, Mezereon, Olive Oil, Quinine, Sarsaparilla, Rhus, Sassafras, Viola Canina, Viola Tricolor.

CUTS: Arnica, Collodion, Matico.

CYSTITIS: Belladonna, Benzoin, Buchu, Cubebs, Opium, Pareira brava, Turpentine.

CYSTORRHŒA: Canada Balsam.

DEAFNESS: Glycerine, Oil of Bays.

DEBILITY: Angustura, Cascarilla, Myrrh, Quinine, Sanguinaria, Sarsaparilla.

DELIRIUM: Digitalis, Hyoscyamus.

DELIRIUM TREMENS: Belladonna, Camphor, Calabar-bean, Cayenne-pepper, Cimicifuga, Coffee, Conium, Digitalis, Hyoscyamus, Indian Hemp, Opium, Sumbul.

DENTITION: Belladonna, Calumba, Dulcamara, Hyoscyamus.

DIABETES: Almond Bread, Laurel water, Opium, Rhatany.

DIARRHŒA: Aconite, Aloes, Angustura, Arnica, Calumba, Camphor, Capsicum, Cascarilla Bark, Castor-oil, Catechu, Chamomile oil, Cinnamon, Coffee, Dulcamara, Hydrastis, Indian Bael, Ipecacuanha, Kino, Kinovate of Lime, Kinovic acid, Logwood, Mastic, Nutmeg, Nux-vomica, Oak bark, Opium, Podophyllum, Pomegranate Bark, Pulsatilla, Rhubarb, Sumbul, Tar, Veratrum.

DIPHTHERIA: Sanguinaria.

DISCHARGES: Canada Balsam, Matico, Pareira, Salicine, Turpentine.

DROPSY: Broom, Bryonia, Buckthorn, Buchu, Cajeput oil, Capsicum, Chimaphila, Cimicifuga, Colocynth, Digitalis, Elaterium, Gallic Acid, Gamboge, Hellebore, Horseradish, Hydrastis, Jalap, Mustard, Rhatany, Scammony, Senega, Squill, Turpentine.

DYSCRASIC MALADIES: Grape cure.

DYSENTERY: Aconite, Angustura, Calumba, Cascarilla Bark, Castor-oil, Catechu, Cotton seeds, Ergot, Gamboge, Grape-cure, Iceland Moss, Indian Bael, Ipecacuanha, Kinovate of Lime, Logwood, Mastic, Nutmeg, Nux-vomica, Opium, Sumbul.

DYSMENORRHŒA: Belladonna, Cimicifuga, Guaiacum, Pulsatilla, Rue, Sumbul.

DYSPEPSIA: Aloes, Angustura, Anise oil, Bibiru Buchu, Capsicum, Cardamoms, Cascarilla, Catechu, Chamomile, Cherry-laurel, Chiretta, Cloves, Cocculus Indicus, Cubebs, Elecampane, Gentian, Ginger, Hemidesmus, Horse-radish, Hydrastis, Iceland Moss, Ipecacuanha, Lupuline, Mustard, Myrrh, Nux-vomica, Oil of Bay, Opium, Orange-peel, Pepper, Pimento, Podophyllum, Pulsatilla, Quassia, Quinine, Rhatany, Rhubarb, St. Ignatius' Bean, Salicine, Sanguinaria, Ser-

pentaria, Sumbul, Tar, Taraxacum, Valerian, Wild Cherry, Winter's Bark, Wormwood.
DYSPHAGIA : Cajeput Oil.
DYSPNŒA OF CARDIAC DISEASE: Cimicifuga, Spigelia, Valeria, Wild Cherry.
DYSURIA : Chimaphila, Ergot, Indian Hemp.

EAR-ACHE : Digitalis, Pulsatilla.
ECCHYMOSIS: Arnica.
ECTHYMATA: Grape-cure.
ECZEMA : Celandine, Grape-cure, Hamamelis, Iris Versicolor, Rhus, Tannic acid, Tar, Viola tricolor.
ELEPHANTIASIS : Sarsaparilla.
EMPHYSEMA : Senega.
ENCEPHALITIS: Belladonna.
ENDOCARDITIS : Bryonia, Spigelia Anthelmia.
ENURESIS, NOCTURNAL : Belladonna.
EPIDIDYMITIS : Pulsatilla, Tobacco.
EPILEPSY: Belladonna, Bryonia, Calabar-bean, Cocculus Indicus, Conium, Curara, Glonoinum, Indigo, Opium, Quinine, Rue, Stramonium, Sumbul, Valerian, Valerianate of Atropine, Valerianate of Zinc.
EPISTAXIS: Aconite, Arnica, Belladonna, Catechu, Digitalis, Ergot, Ergotine, Hamamelis, Rue.
ERYSIPELAE: Aconite, Belladonna, Collodion, Cotton wool, Digitalis, Quinine, Rhus.
EXANTHEMATA : Elecampane, Serpentaria.
EXOSTOSIS: Aconite.
EYE AFFECTIONS : Euphrasia, Physostigma, Rue.

FACIAL SPASM : Curara.
FEBRILE DISORDERS : Orange-juice.
FEVERS : Aconite, Angustura, Arnica, Belladonna, Bibiru Bark, Calumba, Camphor, Capsicum, Cascarilla Bark, Cinnamon, Cimicifuga, Cocculus Indicus, Coffee, Colchicum, Digitalis, Hydrastis, Hyoscyamus, Kinovic Acid, Lemon juice, Mustard, Morphia, Nutmeg, Pepper, Piperine, Podophyllum, Pomegranate juice, Quinine, Rhatany, Salicine, Sarsaparilla, Sumbul, Turpentine, Valerian, Veratrum.
FISSURE OF THE ANUS : Belladonna, Collodion, Ratanhia, Sanguinaria.
FISTULA : Pepper, Sanguinaria.
FISTULAR ULCERS : Capsicum.
FLATULENCE (See also DYSPEPSIA): Aloes, Anise oil, Asafœtida, Columba, Cardamoms, Caraway oil, Chiretta, Cinnamon, Cocculus Indicus, Dill, Fennel oil, Ginger, Lavender oil, Marsh Mallow, Oil of Bay, Orange-peel, Pennyroyal, Peppermint, Pimento, Spearmint, Valerian.
FRECKLES: Benzoin.
FLUOR ALBUS : Copaiva, Rhatany, Tannic Acid.
FUNGOUS ULCERATIONS: Stavesacre.

GANGRENE: Cinchona, Oil of Turpentine, Quinine, Sanguinaria, Turpentine.
GASTRALGIA : Cherry Laurel, Ergot, Nux Vomica, Opium, Pulsatilla, Quinine, Strychnia, Sumbul.
GLANDULAR SWELLINGS: Ammoniacum, Belladonna, Hydrastis, Valerian.
GLEET : Balsams of Peru and Tolu, Buchu, Canada Balsam, Copaiva, Mastic, Tannin.
GLOBUS HYSTERICUS: Asafœtida, Ignatia, Valerian.
GONORRHŒA : Aconite, Balsams of Peru and Tolu, Buchu, Canada Balsam, Cannabis Americana, Cannabis Indica, Castor-oil, Chimaphila, Copaiva, Cubebs, Hydrastis, Matico, Pareira, Pulsatilla, Sandal Wood Oil, Tannin, Uva-ursi.
GOUT : Aconite, Belladonna, Buckthorn, Capsicum, Chimaphila, Cloves, Codeia, Colchicum, Cotton wool, Dulcamara, Elaterium, Guaiacum, Hyoscyamus, Lemon juice, Savin, Wormwood.
GOUT OF THE STOMACH ; Belladonna.
GRAVEL : Cotton root.
GRIPING PAINS (See Colic) : Caraway, Cardamoms, Coriander, Fennel, Ginger, Orange Peel, Pennyroyal, Pimento.
GUMS, DISORDERS OF : Mastich, Myrrh, Pomegranate Bark.

HÆMATEMESIS: Ergot, Ipecacuanha.
HÆMATURIA : Chimaphila, Gallic Acid, Quinine, Turpentine.
HÆMOPTYSIS : Aconite, Arnica, Digitalis, Ergot, Ergotine, Hamamelis, Ipecacuanha, Tannic Acid, Turpentine.
HÆMORRHAGE: Arnica, Belladonna, Cimicifuga, Cinnamon, Digitalis, Ergot, Ergotine, Gallic Acid, Grape-cure, Ipecacuanha, Logwood, Matico, Oak Bark, Rhatany, Rose Water, Salicine, Savin, Tannic Acid, Turpentine.
HÆMORRHAGE, UTERINE : Cannabis, Cinnamon, Digitalis, Ergot, Ergotine.
HÆMORRHOIDS, Castor-oil, Celandine, Copaiva, Hamamelis, Hellebore, Hydrastis, Hyoscyamus, Nux-vomica, Opium, Pepper, Rhubarb, Stramonium, Tannin, Wild Cherry.
HAY FEVER : Aconite, Coffee, Ipecacuanha, Quinine.
HEADACHE : Balsams of Peru and Tolu, Belladonna, Bibiru Bark, Cajeput-oil, Camphor, Cannabis Indica, Cimicifuga, Coffee, Colocynth, Ginger, Guarana, Hydrastis, Indian Hemp, Lavender oil, Nux-vomica, Rosemary, Valerian.
HEART DISEASE: Cimicifuga, Digitalis, Hellebore, Indian Bael, Mustard, Strychnia, Wild Cherry.
HEARTBURN (See also DYSPEPSIA) : Almonds, Capsicum, Podophyllum, Strychnia.

HECTIC FEVER: Calumba, Cinchona, Wild Cherry, Quinine.
HEPATIC DISEASES: Bryonia, Celandine. Nux-vomica, Podophyllum, Strychnia, Sanguinaria, Taraxacum.
HERNIA: Lobelia, Tobacco enemata.
HERPETIC DISEASES: Celandine. Dulcamara, Kamala, Rhus, Tar.
HICCOUGH (See also DYSPEPSIA): Cherry-Laurel, Nux-vomica, Pepper, Strychnia.
HOARSENESS: Horseradish.
HOOPING-COUGH: Asafœtida, Belladonna, Cannabis Indica, Chamomile, Coffee, Ergot, Ipecacuanha, Laurel-water, Lobelia, Morphia, Pennyroyal, Pulsatilla, Quinine, Sanguinaria, Senega, Squill, Syrup of Violets, Tannin.
HOSPITAL GANGRENE: Sanguinaria.
HYDROCELE: Digitalis.
HYDROCYANIC ACID POISONING: Atropia.
HYDROTHORAX: Digitalis, Elaterium.
HYPERÆMIA: Belladonna, Digitalis, Hyoscyamus.
HYPERTROPHY: Aconite, Digitalis.
HYPOCHONDRIASIS: Cimicifuga, Hyoscyamus, Indian Bael, Lavender oil, St. Ignatius' Bean, Scammony, Sumbul, Valerian.
HYSTERIA: Asafœtida, Bryonia, Cajeput oil, Camphor, Cannabis Indica, Chamomile oil, Cocculus Indicus, Colchicum, Dulcamara, Gentian, Hyoscyamus, Iguatia, Lavender oil, Orange-peel oil, Pennyroyal, Quassia, Quassia Wood, Rosemary, Rue, St. Ignatius' Bean, Sanguinaria, Strychnia, Sumbul, Syrup of Violets, Valerian.
HYSTERICAL CHOREA: Cimicifuga.
IMPETIGO: Grape-cure, Laurel-water.
INCONTINENCE OF URINE: Belladonna, Buchu, Ergot, Lupuline, Santonine, Strychnia.

INFLAMMATION: Aconite, Arnica, Belladonna, Bryony, Buchu. Castor-oil, Colchicum, Collodion, Copaiva, Cotton-wool, Hydrastis, Lemon juice, Marsh Mallow, Morphia, Mustard, Oil of Turpentine. Opium, Orange-juice, Pulsatilla, Quinine. Trimethylamine, Veratrine.
INSANITY: Arnica, Conium, Hellebore, Opium, Quinine.
INSOMNIA: Codeia, Coffee, Hop-pillow, Laurel-water, Morphia, Narcein, Sumbul.
INTERMITTENT FEVERS: Bitter Almond, Capsicum, Cascarilla, Cloves, Digitalis, Elm Bark, Gentian, Horse-radish, Hydrastis, Kino, Pepper, Quinine, Salicine, Sarsaparilla, Strychnia, Wild Cherry, Wormwood.
INTERTRIGO: Tannic Acid.
INTESTINAL CATARRH: Opium.
IPECACUANHA POISONING: Tannin.
IRITIS: Belladonna, Turpentine.
IRRITATION OF MUCOUS MEMBRANES: Indian Bael, Linseed, Marsh Mallow, Opium.
ISCHURIA: Chimaphila.
ITCH: Anise Oil, Kamala.
ITCHING: Almond, Laurel-water.

JAUNDICE: Aloes, Celandine, Dulcamara, Gentian.

KIDNEY DISEASE (See also ALBUMINURIA): Buchu, Calumba, Castor-oil, Digitalis, Hemidesmus, Hyoscyamus, Senega.

LABOR: Cimicifuga, Belladonna, Cotton-root, Ergot, Indian Hemp, Marsh Mallow.
LARYNGEAL IRRITATION: Marsh Mallow.
LARYNGISMUS STRIDULUS: Quinine.
LARYNGITIS: Sanguinaria.
LEPRA: Dulcamara, Elm bark.

LEPROSY: Oil of Cashew.
LEUCORRHŒA: Cimicifuga, Balsams of Peru and Tolu, Canada Balsam, Cocculus Indicus, Copaiva, Cubebs, Ergot, Kino, Logwood, Myrrh, Pareira, Pulsatilla, Savin, St. Ignatius' Bean, Sumbul, Tannic Acid, Uva-ursi.
LITHIASIS: Buchu.
LUMBAGO: Cimicifuga, Belladonna, Burgundy Pitch, Canada Balsam.

MANIA: Arnica, Bryonia, Camphor, Colocynth, Conium, Dulcamara, Opium, Scammony, Stramonium.
MEASLES: Aconite, Pulsatilla.
MELANCHOLIA: Indian Bael.
MELANISMUS: Turpentine.
MENINGITIS: Belladonna, Hyoscyamus.
MENORRHAGIA: Digitalis, Ergot, Ergotine, Hamamelis, Ipecacuanha, Pomegranate bark, Savin.
MENSTRUAL DISORDERS (See AMENORRHŒA and DYSMENORRHŒA).
MENSTRUATION: Aconite, Cimicifuga, Aloes, Cocculus Indicus, Ignatia, Mustard, Myrrh, Pulsatilla, Rue, Sanguinaria, Savin.
MENTAL DISEASES (See INSANITY and MANIA).
MIGRAINE: Cimicifuga, Camphor, Caffeine, Coffee, Guarana, Iris versicolor, Sanguinaria.
MISCARRIAGE, THREATENED: Ergot.
MONOMANIA: Hyoscyamus.
MUCOUS DISCHARGES (See also LEUCORRHŒA): Canada Balsam, Matico, Storax, Turpentine.

NARCOTIC POISONINGS: Lemon-juice, Mustard.
NASAL ULCERS: Hydrastis.
NASAL INFLAMMATION: Pulsatilla.
NAUSEA: Calumba, Cinnamon, Cloves. Cocculus Indicus, Ginger.
NAUSEA: Nutmeg, Pepper, Peppermint, Pimento.
NEPHRITIS: Belladonna, Chimaphila, Juniper, Uva-ursi.

INDEX. 321

NERVOUS AFFECTIONS: Asafœtida, Camphor, Cocculus Indicus, Coffee, Digitalis, Ignatia, Orange-peel Oil, Quinine, Senegr, Sumbul, Theobromine.

NEURALGIA: Aconite, Belladonna, Bibiru Bark, Celandine, Codein, Coffee, Digitalis, Ergot, Indian Hemp, Morphia, Mustard, Oil of Turpentine, Opium, Pyrethrum, Quinine, St. Ignatius' Bean, Spigelia Anthelmia, Stavesacre, Strychnia, Sumbul, Theobromine, Turpentine, Valerian, Veratrine.

NIPPLES, SORE: Balsams of Peru and Tolu, Benzoin, Collodion.

NOCTURNAL EMISSIONS: Cimicifuga, Digitalis, Lupuline.

NOCTURNAL ENURESIS: Belladonna.

NOSE, BLEEDING OF THE (See EPISTAXIS).

NYMPHOMANIA: Stramonium.

OBSTRUCTION OF THE BOWELS (See CONSTIPATION).

OPHTHALMIA: Colchicum, Hydrastis, Physostigma, Pulsatilla, Rue, Spigelia Anthelmia, Stavesacre.

OPIUM POISONING: Belladonna, Coffee, Lemon-juice.

ORCHITIS: Digitalis.

OTITIS: Pulsatilla.

PALPITATION: Aconite, Cherry-Laurel, Hyoscyamus, Spigelia Anthelmia, Valerian.

PARALYSIS: Aconite, Arnica, Capsicum, Cajeput Oil, Celandine, Cocculus Indicus, Colocynth, Ergot, Mustard, Nutmeg, Nux-vomica, Oil of Bay, Rhus, Senega, St. Ignatius' Bean, Strychnia, Winter's Bark.

PARALYSIS OF THE TONGUE: Cinnamon, Cloves, Ginger, Horse-radish, Pepper, Mezereon.

PARAPLEGIA: Cocculus Indicus.

PARTURITION: Belladonna, cimicifuga, Cotton-root, Dill, Ergot, Indian Hemp, Marsh Mallow, Uva-ursi.

PEDICULI: Anise oil, Cocculus Indicus, Stavesacre seeds.

PERICARDITIS: Aconite, Bryony, Digitalis, Spigelia Anthelmia.

PERIOSTITIS: Stavesacre.

PERITONITIS: Aconite, Cimicifuga, Cocculus Indicus, Oil of Turpentine, Opium, Quinine.

PHAGEDENIC ULCERS: Opium.

PHOSPHORUS, POISONING BY: Turpentine.

PHTHISIS (See COUGH, BRONCHITIS, &c.): Bibiru Bark, Calumba, Cherry-Laurel, Coniine-vapor, Dulcamara, Grape-cure, Iceland Moss, Logwood, Myrrh, Narceia.

PHTHISIS: Pomegranate, Sanguinaria, Sumbul, Tannic Acid, Tar.

PHYSOSTIGMA - POISONING: Atropia.

PILES: (See HÆMORRHOIDS).

PLETHORA: Aconite.

PLEURISY: Aconite, Bryony, Burgundy Pitch, Digitalis, Opium.

PLEURODYNIA: Cimicifuga, Mustard.

PNEUMONIA: Aconite, Belladonna, Bryony, Celandine, Digitalis, Ipecacuanha, Quinine, Sanguinaria, Senega, Turpentine, Veratrine, Veratrum Viride.

POISONING, ANTIDOTES TO: By Alkaloids, Tannic Acid; by Irritants, Olive Oil; by Metallic Substances, Tannic Acid, White of Eggs; by Narcotics, Mustard, Lemon-juice; by Opium, Coffee, Mustard; by Phosphorus, Turpentine; by Strychnine, Valerian, Curara.

POLYPI: Sanguinaria.

PREGNANCY, DISORDERS OF: Aloes, Calumba, Castor-oil, Cherry-Laurel, Cocculus Indicus, Sumbul.

PROLAPSUS ANI: Nux-vomica, Hydrastis, Pomegranate bark, Podophyllum, Strychnia, Tannic Acid.

PROLAPSUS UTERI: Pomegranate-bark.

PROSTATE, DISEASES OF THE: Buchu.

PRURIGO: Laurel-water, Tar.

PRUSSIC ACID POISONING: Atropia.

PSEUDO - NEURALGIC AFFECTIONS: Chamomile.

PUERPERAL CONVULSIONS: Belladonna, Opium.

PUERPERAL FEVERS; Aconite, Calumba, Quinine, Stramonium, Turpentine.

PUERPERAL HYPOCHONDRIASIS: Cimicifuga.

PUERPERAL MANIA: Cimicifuga, Stramonium.

PUERPERAL PERITONITIS: Opium, Aconite.

PULMONARY AFFECTIONS: Almond, Arnica, Burgundy Pitch, Coltsfoot, Elecampane, Mustard, Sarsaparilla, Squills, Tar.

PURPURA: Ergot.

PYÆMIA: Aconite, Digitalis. Morphia, Quinine.

PYREXIA: Digitalis.

PYREXIAL DISEASES: Aconite, Veratrine, Veratrum Viride.

PYROSIS: Calumba, Catechu, Cherry-Laurel, Nux-vomica, Opium, Strychnia.

QUINSY: Belladonna, Tannic Acid.

RABIES: Stramonium.

RECTUM, DISEASES OF THE: Castor-oil, Pepper, Stramonium, Tannic Acid.

RECTUM ULCER OF THE: Belladonna.

REFLEX EXCITABILITY: Camphor, Chamomile, Valerian.

REGURGITATION: Strychnia.

RENAL CALCULI: Calumba.

RETENTION OF URINE: Aconite, Buchu, Strychnia.

RHEUMATIC FEVER: Aconite, Digitalis, Rhus, Spigelia Anthelmia, Trimethylamine.

RHEUMATIC GOUT: Arnica.

RHEUMATIC PAINS: Arbor Vitæ, Burgundy Pitch.

RHEUMATISM: Aconite, Cimicifuga, Arnica, Belladonna, Bryonia, Buckthorn, Buchu, Cajeput-oil, Canada Balsam, Chima-

phila, Codein, Colchicum, Cotton-wool, Cubebs, Dulcamara, Horse-radish, Hyoscyamus, Lemon-juice, Lupuline, Mezereon, Nutmeg, Oil of Turpentine, Pyrethrum, Rhus, Sarsaparilla, Sassafras, Senega, Serpentaria, Stavesacre, Trimethylamine, Turpentine, Veratrine, Veratrum Viride.
RIBS, FRACTURED: Arnica.
RICKETS: Quinine.
RINGWORM: Cocculus Indicus, Pepper, Tar.

SALIVATION: Belladonna.
SCABIES: Cocculus Indicus, Dulcamara, Stavesacre, Tar.
SCALDS (See BURNS).
SCARLATINA: Aconite, Belladonna, Capsicum, Hellebore, Juniper, Quinine, Rhus.
SCIATICA: Belladonna, Canada Balsam, Codein, Digitalis, Turpentine, Morphia.
SCIRRHUS: Aconite, Belladonna.
SCHIRRHOUS BREAST: Conium.
SCIRRHUS OF THE LIVER: Taraxacum.
SCROFULA: Capsicum, Chimaphila, Coltsfoot, Digitalis, Gentian, Mezereon, Sarsaparilla, Wild Cherry.
SCROFULOUS PHOTOPHOBIA: Coniine.
SCROFULOUS ULCERS: Hyoscyamus.
SCURVY: Canella Bark, Horse-radish, Lemon-juice, Lime-juice, Winter's Bark.
SEA SICKNESS: Stavesacre.
SEMINAL EMISSIONS: Cimicifuga, Digitalis, Lupuline.
SICKNESS OF CHRONIC ALCOHOLISM: Ipecacuanha.
SICKNESS OF PREGNANCY: Calumba, Cherry-Laurel, Ipecacuanha, Stavesacre, Strychnia.
SINGULTUS: (See HICCOUGH).
SORE-THROAT: Aconite, Cimicifuga, Belladonna, Capsicum, Catechu, Kino, Liquorice, Oil of Turpentine, Serpentaria, Tannic Acid.
SPASMS: Aconite, Belladonna, Cardamoms, Chamomile, Chamomile Oil, Cinnamon, Coniine, Hemlock, Ipecacuanha, Lobelia, Lupuline, Morphia, Mustard, Opium, Strychnia, Sumbul, Tobacco.
SPASMODIC COUGH: Cherry-Laurel, Hyoscyamus, Sanguinaria.
SPERMATORRHŒA: Cimicifuga, Digitalis.
SPHINCTERS, PARALYSIS OF THE: Cocculus Indicus.
SPINAL CORD, INFLAMMATION OF THE: Belladonna.
SPRAINS: Arnica, Oil of Bay, Oil of Turpentine.
STRABISMUS: Belladonna.
STRANGULATED HERNIA: Lobelia, Tobacco enemata.
STRANGURY: Cotton-root.
STRICTURE OF THE URETHRA: Belladonna, Buchu, Castor-oil.
STRUMOUS AFFECTIONS: Bibiru Bark, Valerian.
STRYCHNIA POISONING: Curara, Oil of Chamomile, Tobacco, Valerian.
SUNSTROKE: Hydrastis, Hyoscyamus.
SUPPRESSION OF URINE: Digitalis.
SUPPURATION: Sarsaparilla.
SYNCOPE: Aconite, Cinnamon.
SYPHILIPHOBIA: Hyoscyamus.
SYPHILIS: Belladonna, Dulcamara, Hemidesmus, Mezereon, Sanguinaria, Sarsaparilla, Sassafras.
TAPEWORM: Filix-mas, Jalap, Kamala, Koussa, Pomegranate-bark, Pulsatilla, Turpentine.
TEETH, CARIOUS: Collodion.
TEETHING (See DENTITION).
TENESMUS: Marsh Mallow.
TETANUS: Aconite, Apomorphia, Curara, Indian Hemp, Opium, Physostigma, Tobacco.
THROAT AFFECTIONS: Marsh Mallow, Myrrh, Pomegranate-bark, Pyrethrum.
TIC-DOULOUREUX: Cherry-Laurel, Stramonium.
TINEA: Pansy-leaves, Pepper, Sanguinaria.
TONGUE, PARALYSIS OF THE: Cinnamon, Cloves, Ginger, Horse-radish, Mezereon, Pepper, Pyrethrum.
TONSILLITIS: Aconite, Belladonna, Guaiacum, Tannin.
TOOTHACHE: Balsams of Peru and Tolu, Cloves, Ginger, Mezereon, Pyrethrum, Stavesacre.
TYMPANITIS: Asafœtida, Capsicum, Cocculus Indicus, Ginger, Turpentine.
TYPHOID FEVER: Arnica, Coffee, Digitalis, Quinine, Sumbul, Turpentine.
TYPHUS FEVER: Arnica, Belladonna, Coffee, Digitalis, Hyoscyamus, Podophyllum, Quinine, Rhatany, Serpentaria.

ULCERS: Arbor-vitæ, Balsams of Peru and Tolu, Belladonna, Benzoin, Bistort, Calumba, Capsicum, Catechu, Chimaphila, Collodion, Copaiva, Hamamelis, Hydrastis, Hyoscyamus, Myrrh, Opium, Pepper, Rhatany, Rhubarb, Sanguinaria, Savin, Storax, Tannic Acid, Tar, Turmeric Acid, Turpentine.
ULCERATED SORE-THROAT: Belladonna.
UTERINE AFFECTIONS: Aloes, Bay-tree, Benzoin, Cinnamon, Cocculus Indicus, Ergot, Ergotine, Pennyroyal, Pulsatilla, Rhatany, Rosemary, Rue, Savin, Tannic Acid, Uva-ursi.
UVULA, RELAXED: Catechu, Ginger, Pepper, Pyrethrum, Tannin.

VOMITING: Bryony, Cajeput-oil, Calumba, Cherry-Laurel, Cinnamon, Cloves, Cocculus indicus, Ipecacuanha, Nutmeg, Opium, Pepper, Serpentaria.

WARTS: Rue, Savin, Thuja.

WORMS: Aconite, Aloes, Cajeput-oil, Chamomile, Filix-mas, Gentian, Hellebore, Jalap, Kamala, Kousso, Olive-oil, Quassia Wood, Santonica, Scammony, Senna, Spigelia, Stavesacre, St. Ignatius' Bean, Turpentine, Valerian, Wormwood.

WOUNDS: Arnica, Balsams of Peru and Tolu, Benzoin, Collodion, Copaiva, Matico, Opium, Savin.

WRY-NECK: Cimicifuga.

www.ingramcontent.com/pod-product-compliance
Lightning Source LLC
Chambersburg PA
CBHW030012240426
43672CB00007B/920